EUROPEAN ARACHNOLOGY 2000

This volume is the Proceedings of the 19th European Colloquium of Arachnology, held at the University of Aarhus, Denmark, on 17—22 July 2000

Dedicated to Edwin Nørgaard

European Arachnology 2000

Editors: Søren Toft & Nikolaj Scharff
Technical editor: Per G. Henriksen

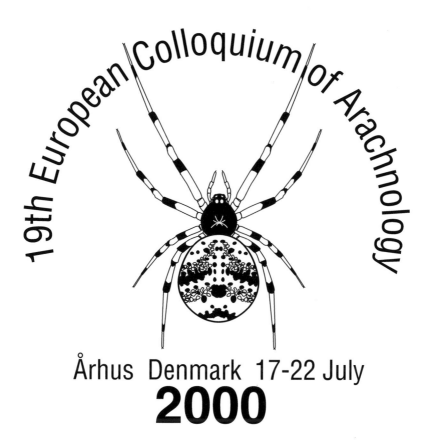

19th European Colloquium of Arachnology

Århus Denmark 17-22 July
2000

AARHUS UNIVERSITY PRESS

European Arachnology 2000
is printed by Scanprint as, Denmark
© The Authors and Aarhus University Press 2002
Cover designed by Birgitte Rubæk
Printed in Denmark 2002

ISBN 87 7934 001 6

Aarhus University Press
Langelandsgade 177
DK-8200 Aarhus N
Phone (+45) 8942 5370
Fax (+45) 8942 5380

73 Lime Walk
Headington, Oxford OX3 7AD
Fax (+44) 1865 750 079

Box 511
Oakville, Connecticut 06779
Fax (+1) 860 945 9468

www.unipress.dk

FOREWORD

The 19th European Colloquium of Arachnology took place at the University of Aarhus, Denmark, from 17-22 July 2000, hosted by the Department of Zoology. The European Society for Arachnology, situated in Paris, is the formal organisational background for these meetings. A total of 113 delegates and 17 accompanying persons attended the colloquium, and 85 presentations (51 papers and 34 posters) were given. Of all these presentations approximately half (43) are represented in these proceedings. The programme included a colloquium excursion on Wednesday, 19 July to the castle ruin at Kalø, and thento the sandy hills of Mols Bjerge. On the post-colloquium excursion we first stopped at the Viking castle, Fyrkat, at Hobro; then further to Bulbjerg, a steep limestone cliff on the Skagerak coast of Northern Jutland. From there we walked along the beach to Vester Thorup, arriving in time to see the fisherboats coming in and being hauled up onto the beach in the old-fashioned way. The programme for accompanying persons included visits to the tourist attractions of Eastern Jutland. We are grateful to Ulla Thyssen for guiding these tours.

The social part of the programme also included an evening reception at the City Hall of Århus. In this regard we are indebted to Århus City Council and especially councillor Tove Tolstrup who was our host that evening. At the colloquium dinner the folk music group 'Fair Wind' (Birgitte Rasmussen, Jens Josephsen and Søren Achim Nielsen) first led the singing of Pierre Bonnet's 'Le Chant des Arachnologistes' and then later entertained and instructed traditional Danish folk dances.

Of course with our meeting taking place at the turn of a millenium, we were presented with an obvious opportunity to consider the 'state' of European arachnology, for example, the situation regarding the recruitment of young scientists to the study of arachnids? Gladly, the answer is positive. With an increase in the number of young university students attending the 19th European meeting, this continued a trend apparent in previous meetings. We were especially delighted to see how many young scientists and students from Eastern European countries attended this meeting. Interestingly, students from all over Europe increasingly chose to complete their Masters or Ph.D. degree with a research project on the arachnids, and the age profile of the colloquium participants was the best indicator that research in arachnids is thriving and will continue to do so in the near future.

Which subjects are presently attracting the interest of researchers and students? Many aspects of arachnid biology are covered by current research; both pure and applied projects are undertaken and represented in this volume. It is inevitable though that in smaller research areas like arachnology, some disciplines - even long established ones - are poorly represented, with only scattered contributions of diverging scopes. Fortunately, however, the 'traditional' disciplines of systematics and taxonomy continue to make up a substantial part of arachnological activities, in spite of increasing funding problems. At our meeting the new Internet key to Central European spiders by Nentwig, Hänggi, Kropf & Blick (http://www.araneae.unibe.ch/) was presented for the first time. This will make it much easier for new arachnologists and students with a budding interest in spiders to get started. It will be particularly useful to students interested in conservation-oriented faunistics which, according to the present proceedings, may be the single most attractive subject at the moment for arachnologists from all countries of the Continent. Many papers are devoted to the effort of using spiders for evaluating the conservation value of natural or man-modified habitats. We look forward to spider monitoring being used on a wider scale in practical environmental planning. The necessary biological knowledge seems to be available now for most of Central and Northern Europe.

Applied arachnology takes many other forms. The usefulness of spiders as biocontrol agents in agricultural systems is widely recognised and also represented in these proceedings. A third area of great applied potential is the study of the physical and chemical properties of silk. This is a field in which, perhaps more than other areas of applied arachnology, the practical use depends heavily on progress in pure research. Spider webs have always fascinated arachnologists and will probably continue to do so into the distant future. Whereas research on orb-webs continues to attract students, analysis of other web types is still in its infancy. Reproductive behaviour, including sexual selection and sperm competition, is another 'hot' area of arachnological research which continues to furnish biological stories with great appeal also to the general public. Several articles in this volume prove that good science can also produce exciting natural history.

Except for the scorpions, the usual complaint can be repeated here about the absence of studies on the minor arachnid groups. As a European meeting it is no wonder that most of the exotic groups were not represented, but the almost complete absence of opilionids is astounding. This is a group that ecologists and behavioural biologists should soon "discover"!

The utility of the European colloquia as a regular forum for meeting old friends and new colleagues, and for young students to become acquainted with the arachnological community is evident, but what about the printed Proceedings? In most cases they are published as separate books, often by local institutional printing facilities or publishing companies. This means that they are not easily accessible for purchase, the articles are not indexed in the literature databases normally used, and they are often impossible to get through university libraries. Fortunately they are listed in ISA's (and formerly CIDA's) annual list of arachnological papers (still called 'Liste des Travaux Arachnologiques'). Though citations may be a poor indicator of scientific quality, they do reveal the extent to which the proceedings of our meetings are subsequently used. A survey of the reference lists of the articles in the present volume gave 27 citations from proceedings of earlier European colloquia, and 33 citations from proceedings of international arachnological conferences. Close to three-quarters of all papers have a reference to the proceedings of an arachnological meeting. These figures are very encouraging. They indicate that at least the researchers who attend these meetings benefit considerably, though the dissemination of the papers outside this narrow circle is more doubtful.

Around the time of the colloquium a live spider exhibition was held at the Natural History Museum of Århus, running throughout the months of July and August. During these two months, 15000 people visited the museum, many of them obviously attracted by the spiders. A daily feeding time at 11 a.m. often resulted in an overwhelming crowd of all ages in front of the spider cages. Questions from the audience clearly showed that people came not only for the 'horror' of the big tarantulas, but also out of a true fascination with the biological details of the spiders. We are greatly indebted to Kurt Nikolajsen and Peter Klaas for lending us a large collection of their mygalomorphs. A great deal of publicity also surrounded the colloquium, which was covered by most Danish newspapers, radio and television stations. To further stimulate an interest in spiders, a public evening was arranged at the Natural History Museum. A lecture hall full to capacity again demonstrated the enormous public interest in our animals. We are indebted to Paul Selden and Fritz Vollrath for giving popular lectures on spider diversity and biology. We also thank museum director Thomas Secher Jensen for hosting this event, including the reception afterwards.

Financial support for the colloquium came from the European Society for Arachnology, the Department of Zoology and the Faculty of Natural Sciences, Aarhus University, the Danish Natural Science Research Council, and G.E.C. Gads Foundation. We are particularly grateful for the grants from the Faculty and from G.E.C. Gads Foundation, which were specifically awarded to support the participation of Eastern European researchers and students.

This is also the place to thank several technicians as well as present and former students who helped with the practical accomplishment of the colloquium: Else Bomholt Rasmussen, Per G. Henriksen, Birgitte Dahl, Cecilie Holm, Anja Petersen, Lene Møller Krag, Nina Dideriksen, Pernille Thorbek, Peter Kruse, Thomas Nørgaard.

Apart from the authors, several people have contributed to these proceedings. All papers have been reviewed by two anonymous referees. Many colloquium participants have taken their share of the work, and several people who did not participate (including non-arachnologists) kindly reviewed papers within their field of expertise. We thank everyone for this invaluable help. Special thanks to Birgitte Rubæk, Zoological Museum of Copenhagen, who designed the colloquium logo that covers the front page of this volume.

The full programme of the Colloquium and abstracts of the presentations, including those that are not published in these proceedings, as well as other information about the conference, can be found on the colloquium web site: http://support.bio.au.dk/spider

We look forward to meeting everybody again at the next meeting to be held in Hungary in 2002!

      SØREN TOFT   NIKOLAJ SCHARFF

COLLOQUIUM PHOTO

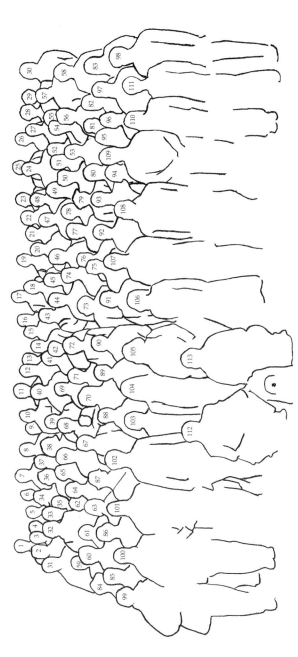

1 Thomas Norgaard, 2 Kjetil Aakra, 3 Ambros Hänggi, Lene Kragh Møller, 5 Elena Mikhailova, 6 Boyan Petrov, 7 Konrad Thaler, 8 Otto Kraus, 9 Rauno Alatalo, 10 Janne Kotiaho, 11 Nikolaj Scharff, 12 Per de Place Bjørn, 13 Friedrich Wallenstein, 14 Suresh Benjamin, 15 Sirichai Divasiri, 16 Kirill Mikhailov, 17 Peter Koomen, 18 Lars Jonsson, 19 Christian Kropf, 20 Vygandas Relys, 21 Ole Bøggild, 22 Peter Horak, 23 Ole Gudik-Sørensen, 24 Wilson Lourenço, 25 Tamás Szűts, 26 Gernot Bergthaler, 27 Boris Striffler, 28 Sebastian Frische, 29 Stanislav Pekar, 30 John Kenney, 31 Peter Gajdoš, 32 Pavel Kasal, 33 Seppo Koponen, 34 Barbara Knoflach-Thaler, 35 Maciej Bartos, 36 Pekka Lehtinen, 37 Margarete Kraus, 38 Lene Sigsgaard, 39 Leon Baert, 40 Domir De Bakker, 41 Jean-Pierre Maelfait, 42 Karin Schütt, 43 James Bell, 44 Finn Erik Klausen, 45 Michael Saaristo, 46 Wolfgang Nentwig, 47 Theo Blick, 48 Charlotte Rouaud, 49 Sabine Merkens, 50 Henning Clausen, 51 Marco Moretti, 52 Marie Herberstein, 53 Paul Selden, 54 Fréderic Ysnel, 55 Christoph Muster, 56 Torbjörn Kronestedt, 57 Christian Komposch, 58 Jakob Walter, 59 Jozefína Jedličková, 60 Artur Baranowski, 61 Carmen Fernandez-Muntraveta, 62 Franceska Di Franco, 63 Christa Deeleman-Reinhold, 64 Søren Achim Nielsen, 65 Izabela Hajdamowicz, 66 Yuri Marusik, 67 Dmitri Logunov, 68 Gabriele Uhl (and daughter), 69 Shirley Gurdebeke, 70 Tina Berendonck, 71 Reidun Pommeresche, 72 Trine Bilde, 73 Roland Stockman, 74 Alain Canard, 75 Ferenc Samu, 76 Mark Judson, 77 Søren Langemark, 78 Kaj Nissen, 79 Joachim Haupt, 80 Jörg Wunderlich, 81 Jason Dunlop, 82 Samuel Zschokke, 83 Christin Zschokke, 84 Else Bomholt Rasmussen, 85 Jørgen Beck, 86 Peter van Helsdingen, 87 Gunnar Alroth, 88 Marzena Stańska, 89 Søren Toft, 90 Mie Thers, 91 Gabor Lövei, 92 Anja Petersen, 93 Richard Cutts, 94 Astrid Heiling, 95 Elke Jantscher, 96 Michael Ziegler, 97 Peter Klaas, 98 Dieter Scholz, 99 Nina Didriksen, 100 Cecilie Holm, 101 Peter Kruse, 102 Oscar Rohte, 103 Maria Wolak, 104 Sidsel Larsen, 105 Pernille Thorbek, 106 Christine Rollard, 107 Ferenc Toth, 108 Balazs Kiss, 109 András Szirányi, 110 Yelena Gorbunova, 111 Olga Bartosh, 112 Fritz Vollrath, 113 Ernst-August Seyfarth

Table of Contents

*Invited presentations

European Arachnology 2000 (S. Toft & N. Scharff eds.), pp. 13-16.
© Aarhus University Press, Aarhus, 2002. ISBN 87 7934 001 6
(Proceedings of the 19th European Colloquium of Arachnology, Århus 17-22 July 2000)

These proceedings are dedicated to the Danish arachnologist

EDWIN NØRGAARD
b. 26 July 1910

in honour of his outstanding contributions to the study of spider ecology and behaviour

As the motif for the logo of the 19th European Colloquium we selected the spider *Achaearanea riparia*. It was chosen in recognition of the magnificent study of the behaviour of this species, published by Edwin Nørgaard in 1956 (under its former name *Theridion saxatile*). This was the last of a series of groundbreaking scientific papers which gave Nørgaard a pioneer position, not only in the history of arachnology, but general ecology as well.

Like several other great arachnologists at that time, Edwin Nørgaard was an amateur scientist. He was educated as a schoolteacher and did his research in the summer holidays when school was closed. Early in his first post in the northern Jutland town of Løgstør, he found populations of *Eresus niger* on the nearby heathlands, which started a life-long passion. His first publication on *Eresus* is from 1936. This and following papers (see publication list below) still form the basis of research on this species.

After moving to Århus, Nørgaard became associated with the Natural History Museum, and in the early 1940s he started doing research at the Mols Laboratory. It was here that he performed the studies leading to his now classical papers in spider ecology and behaviour.

Probably the most widely known of his works is the one on two wolf spiders, *Pirata piraticus* and *Pardosa pullata*, living alongside each other in a Sphagnum bog. In this study, published in 1951, Nørgaard demonstrated the close correspondence between the species' temperature preferences and the thermal properties of their separate microhabitats, in spite of the fact that they were separated in space by only a few millimeters. At the time it was performed, this work was unprecedented with respect to the detailed understanding it gave of the separate habitat niches of two closely related species and the behavioural mechanisms by which they are realised. It was also unprecedented in the scientific approach, which combined field observations with detailed laboratory experimentation in a very 'modern' fashion. Here Nørgaard turned natural history into an experimental science of ecology. In one of the earliest of the modern ecology text books *Animal Ecology. Aims and Methods*, A. Macfadyen states: 'Among recent work on terrestrial animals that of ... Nørgaard is outstanding' (cited from the 1963 edition, p. 63), referring to five papers of which two were published in Danish!

The *Theridion saxatile* study is still one of the most elegant investigations in spider ecology and behaviour. Nørgaard had read in Emil Nielsen's book *The Biology of Spiders* that females of this species when guarding eggs in the nest sometimes hang their egg sacs out in the sun to warm them, and Nørgaard wanted to experimentally test this hypothesis. Again, he did this by a unique combination of field observations and lab experimentation. In the field Nørgaard placed his thermocouples in the various microhabitats of the spider's nest: inside the nest, below the nest where the female would hang the sac, on the ground, etc. He found that when the sun was shining the temperature within the nest would rise to far above 40 °C, whereas the temperature just below the sac remained at c. 30 °C. This observation alone turned Nielsen's hypothesis upside-down: the nest heats up like a greenhouse, and the spider most likely hangs out the sac to avoid overheating. However, Nørgaard did not stop here. In the lab he showed that heating to > 40 °C would kill the eggs; thus, hanging the eggs out for cooling was really a good idea in terms of offspring survival. Nørgaard was also interested in the specific cues that trigger this highly adaptive behaviour and performed several additional experiments to elucidate the matter.

I would fully encourage anyone interested in spider biology to get hold of the original papers, now half a century old. You will be impressed by how readable and exciting they still are, and at the same time realize how much is lost if your literature search is restricted to the electronic databases that only date 20 years back in time.

In the late 1950s Nørgaard transferred to a Teachers College in Århus, and this left him little time to continue his scientific work. Fortunately, this did not stop his arachnological activities, they just took a different course. From that time up until the present, Nørgaard wrote several faunistic notes on Danish spiders. He also authored many popular articles about spiders and contributed to several books. As a matter of fact, almost everything about spiders that is available in the Danish language has been written by Nørgaard! At the same time he wrote an ecology textbook for Teacher College students, and for 30 years he was the editor of the journal *Flora og Fauna*, which is now a gold mine of information about faunistics, floristics and natural history in Denmark.

Nørgaard's most important contribution to popular science is the series of wonderful booklets on various aspects of spider biology, that has been published by the Natural History Museum in Århus in their *Natur og Museum* series. Titles are (in translation): 'Care of young in spiders'; 'Spider webs'; 'Spiders in house and garden'; 'Dangerous spiders'; 'Courtship behaviour of spiders'. These booklets have been a rich source of information for students and other people interested in spiders. The latest one is from 1998. Nørgaard's publication list spans 62 years!

We were pleased that Edwin Nørgaard attended the opening session of the Århus Colloquium. The following week he celebrated his 90th birthday. Could we find a more appropriate time to honour a great Danish arachnologist? The Colloquium was happy to send him a collective birthday greeting in the form of a participants list signed by all colloquium attendants.

SØREN TOFT

LIST OF EDWIN NØRGAARD'S PUBLICATIONS

Original contributions

1936   Iagttagelser af biologien hos *Eresus niger* (Petagna) Simon. *Flora og Fauna* 42, 7-13.

1940   En sjælden dansk hjulspinder, *Cercidia prominens* Westr. *Flora og Fauna* 46, 9-16.

1941   On the biology of *Eresus niger* Pet. *Entomologiske Meddelelser* 22, 150-179.

1942   Bidrag til hedeskrattens biologi (*Bryodema tuberculata* F.). *Flora og Fauna* 48, 1-17.

1943   Investigations on the feeding habits of *Linyphia*. *Entomologiske Meddelelser* 23, 82-100.

1945   Økologiske undersøgelser over nogle danske jagtedderkopper. *Flora og Fauna* 51, 1-37 (English summary: Ecological investigations on some Danish lycosid spiders).

1948   Bidrag til de danske edderkoppers biologi, 1. *Lithyphantes albomaculatus* (De Geer). *Flora og Fauna* 54, 1-14 (English summary: Contributions to the biology of Danish spiders. I. *Lithyphantes albomaculatus* (De Geer)).

1951   Notes on the biology of *Filistata insidiatrix* (Forsk.). *Entomologiske Meddelelser* 26, 170-184.

1951   On the biology of two lycosid spiders (*Pirata piraticus* and *Lycosa pullata*) from a Danish sphagnum bog. *Oikos* 3, 1-21.

1952   The habitats of the Danish species of *Pirata*. *Entomologiske Meddelelser* 26, 415-423.

1954   Kokondannelsen hos snyltehvepsen *Meteorus scutellator* Nees. *Flora og Fauna* 60, 109-113.

1955   *Scytodes thoracica* Latr. En ny edderkop for Danmark. *Flora og Fauna* 61, 19-21. (English summary).

1956   Environment and behaviour of *Theridion saxatile*. *Oikos* 7, 159-192.

1960   *Physocyclus simoni* Ber. Ny edderkoppeart for Danmark. *Flora og Fauna* 66, 101-102. (English summary).

1988   *Eresus niger* (Pet.) i Danmark. *Flora og Fauna* 94, 3-8 (English summary: *Eresus niger* (Pet.) in Denmark).

Popular articles, books and book chapters

1950   Edderkopper (Araneæ). In: *Vort lands dyreliv* (F.W. Bræstrup et al. eds.), bd. 2, pp. 397-418. København.

1951   (with H.A. Rasmussen & H. Moth) *Lejrskole i Fuglsø*, pp. 1-93. De danske Gymnastik-foreninger, Odense.

1952   Edderkoppernes fangnet. *Natur og Museum* 1 (4), 3-11.

1956   Peter edderkop som industriarbejder. *Vor Viden* 172, 488-493.

1957   Edderkopper med bola of kastenet. *Vor Viden* 215, 201-207.

1958   Kvadratedderkoppen. In: *Glimt af naturen* (H.H. Seedorff ed.), pp. 19-21. København.

1961   Studier over en enkelt art. In: *Jeg ser på insekter* (A. Nørrevang & T.J. Meyer eds.), pp. 248-258. Politikens Forlag, København.

1969   Strandens og klittens edderkopper. In: *Danmarks natur* (Nørrevang et al. eds.), bd. 4, pp. 331-337. Politikens Forlag, København.

1970   Spindlere, Arachnida. In: *Biologi for videregående uddannelser* (R. Munk & A. Munk eds.), bd. 2, pp. 147-157.

1970   Livsvilkårene på heden. In: *Danmarks natur* (Nørrevang et al. eds.), bd. 7, pp. 107-117. Politikens Forlag, København.

1971   Spindlere. In: *Danmarks dyreverden* (H. Hvass ed.), bd. 1, pp. 211-296. Rosenkilde & Bagger, København.

1975   *Elementær økologi*. Biologforbundets forlag, Gedved. 80 pp.

1979   *Edderkopper*. AV-media. 64 pp.

1982   *Elementær økologi*, 2nd ed. Biologforbundets forlag, Hammel. 104 pp.

1984   Flora og Fauna's historie 1884-1984. *Flora og Fauna* 90, 75-90.

1985   Edderkoppers yngelpleje. *Natur og Museum* 24 (1), 1-32.

1987   Spindelvæv. *Natur og Museum* 26 (2), 1-32.

1987   Silkebroer - lige til at spise. *Ingeniøren* 13 (43), 24-25.

1990   *Eresus niger*, Hedens sorte edderkop. *Kaskelot* 87, 2-21.

1991   Edderkopper i hus og have. *Natur og Museum* 30 (3), 1-32.

1993   Farlige edderkopper. *Natur og Museum* 32 (2), 1-32.

1996   Edderkopper. In: *Danmarks Nationalleksikon* (J. Lund ed.), bd. 5, pp. 366-369. Gyldendal, København.

1996   Langemosen i Mols Bjerge. *Kaskelot* 111, 54-55.

1998   Edderkop på frierfødder. *Natur og Museum* 37 (4), 1-36.

PHYSIOLOGY

ECOLOGY

LIFE HISTORY

BEHAVIOUR

European Arachnology 2000 (S. Toft & N. Scharff eds.), pp. 19-32.
© Aarhus University Press, Aarhus, 2002. ISBN 87 7934 001 6
(Proceedings of the 19th European Colloquium of Arachnology, Århus 17-22 July 2000)

Tactile body raising: neuronal correlates of a 'simple' behavior in spiders

ERNST-AUGUST SEYFARTH

Zoologisches Institut der J.W. Goethe-Universität, Biologie-Campus, Siesmayerstrasse 70,
D-60054 Frankfurt am Main, Germany (Seyfarth@zoology.uni-frankfurt.de)

Abstract
This review summarizes our recent results on the sense organs, the central nervous elements, and the neuronal mechanisms responsible for a relatively simple, tactile behavior of spiders. In *Cupiennius salei* (Keyserling 1877) (Ctenidae), a large tropical hunting spider, stimulation of tactile hairs on the ventral aspects of the body and the legs evokes reflex activity in several leg muscles. Coordinated contraction of these muscles raises the body - as in doing 'push-ups'. Using this reliable reaction we examined the neuronal circuitry underlying 'body raising behavior'. Electrophysiological recordings from particular leg muscles and from single, identifiable neurons in the leg ganglia reveal interneurons whose (electrical) activation causes the muscle reflexes. Depending on the exact stimulus situation (tactile and/or displacement stimuli), we have found local and plurisegmental responses and sequential activation of local and plurisegmental interneurons. The results provide a first glimpse of the architecture and functional hierarchy of single, sensory-motor elements in the fused central nervous system of spiders.

Key words: tactile hairs, motoneurons, local interneurons, plurisegmental interneurons, reflexes, leg muscles

INTRODUCTION

Comparative neurobiology seeks to understand the causal relationships between the nervous systems of various animal species and their behavior. Ultimately, it is expected that the comparative approach will yield results that can be generalized and applied across phyla. At the same time, the analysis of a specialized behavior pattern in a particular species can further our understanding of the functional possibilities and the limits of neuronal systems (Bullock 1984; Huber 1988). Neuroethological research with arthropods has been especially successful in understanding the roles played by single neurons and by relatively simple neuronal networks in controlling behavior (see,

e.g., the work summarized by Burrows 1996). The success is due to the advantageous situation found in many arthropods (such as orthopteran insects and decapod crustaceans): large parts of their nervous systems are clearly partitioned into segmental ganglia; single, prominent neurons can be identified from individual to individual; and much of their behavior is relatively simple if not stereotyped (Hoyle 1977; Breidbach & Kutsch 1995; Katz & Harris-Warrick 1999).

Among the arthropods, spiders are interesting from a neurobiological point of view for at least two reasons: (i) Unlike the insect and crustacean species often studied by neuro-ethologists, spiders do not have a clearly segmented and distributed nervous system; rather their central nervous system (CNS) consists of a fused ganglion complex that is concentrated in the prosoma (Fig. 5). (ii) Spiders use their hemolymph as a hydraulic fluid and extend their distal leg joints by local changes in hemolymph pressure. These two peculiarities of the basic arachnid body plan are interesting in themselves and in comparison with other arthropods. They do, however, also cause problems for experimental work with modern neuroanatomical and electrophysiological methods at the cellular level. The highly cephalized CNS makes orientation difficult for the neuranatomist, and the high hemolyph pressure existing in spiders (near the level of human blood pressure) precludes major dissection to isolate the CNS and hence restricts the possibilities for intracellular recordings from neurons and for dye-injections.

Faced with these peculiarities and difficulties we decided to follow a drastically reduced experimental approach. The present review will summarize experiments analyzing relatively simple leg reflexes involved in 'tactile body raising'. This is a stereotyped behavior of spiders that can easily be elicited in the laboratory. Our guiding principle and main question during this work has been: Can we identify the various neuronal components that control tactile body raising?

TACTILE HAIRS AND TACTILE BODY RAISING

Our experimental animal is *Cupiennius salei* (Keyserling 1877), a large Ctenid spider that we breed in the laboratory (Höger & Seyfarth 1995). *C. salei* is predominantly dark-active in its natural habitat in Central America; hence it has to rely mostly on mechanical and other non-visual clues to locate prey and to find its way about in the dark environment (Seyfarth 1980).

Like most spiders, *C. salei* has a dense coat of cuticular hairs covering the entire body surface. On the ventral aspects of the proximal leg parts, the hair coat is as dense as 400 hairs/mm^2

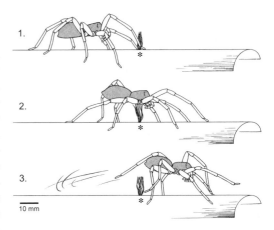

Fig.1. Movement sequence (1. - 3.) of *C. salei* as it approaches and walks across 10-mm high wire obstacle on walkway. The spider raises its body as soon as it touches the obstacle (asterisk) with tactile hairs located ventrally on the sternum and on proximal leg parts (coxa, trochanter and proximal femur). [Drawings adapted from a series of photographs taken with electronic flash]

of body surface (Eckweiler & Seyfarth 1988). The shaft of these hairs is moveable and can be as long as 2.1 mm. The vast majority of cuticular hairs in spiders are 'tactile hairs' (Foelix 1985; Seyfarth 1985). Each is innervated by 3 bipolar neurons that are mechanosensitive and activated by direct touch deflecting the hair shaft in its socket. The exact sensory response (i.e., the electrical discharges of the 3 sensory neurons) depends on the direction and the intensity of the stimulus. The sensory discharges are conducted to the CNS via afferent axons, such as those running in the leg nerves.

When tactile hairs are touched, the spider will retract from the stimulus - either by quickly pulling away a leg or by turning around and walking off. However, touching the hairs located on the proximal ventral leg parts, the sternum, and the ventral opisthosoma will lead to a different, very distinct response. In this case, the animals abruptly raise their body by extending their legs. This happens through muscle contractions and through local increases in hemolymph pressure. We call this reaction 'tactile body raising' (Eckweiler &

Seyfarth 1988). In their natural habitat - *C. salei* is a swift hunter and lives on plants (Barth & Seyfarth 1979; Barth et al. 1988) - the behavior appears to protect the ventral body side against injuries (such as those inflicted by sharp thorns or kicking prey insects) that could cause fatal bleeding of the animals. In the laboratory, tactile body raising can be elicited very reliably and in a stereotyped fashion even at repeated stimulation. In addition to *C. salei*, we have observed stereotyped body raising upon stimulation of ventral hairs in large salticids (*Phidippus regius*), in theraphosids (*Brachypelma* sp.), and in 4 other *Cupiennius* species (Eckweiler & Seyfarth 1988). As further discussed below, deflection of just one hair of the many thousands present suffices to induce a coordinated extension of all 8 legs raising the body.

In order to study the behavior at three successive levels, we examine the tactile reaction in three different experimental situations: (i) in spiders freely walking across an obstacle; here we measure movement patterns and identify sense organs involved with the reaction; (ii) in animals that are tethered and walk on a spherical treadmill; here we record the electrical activity of muscles causing leg extension; and finally (iii) in completely restrained spiders to record neuronal events in the CNS with intracellular electrophysiological techniques.

THE LOCAL TACTILE RESPONSE IS FOLLOWED BY A PLURISEGMENTAL REATION

To determine the exact movement pattern during body raising, spiders were video-filmed while they walked along a narrow walkway and over a flexible, 10-mm high obstacle formed by a row of fine copper wires. The animals raise their body as soon as some of their ventral hairs brush against the wires; they then walk across the obstacle with extended legs as shown in Fig. 1. Ablation of hairs in various body regions confirms that body raising is induced primarily by the deflection of long hairs situated on the sternum and on the ventral surface of the proximal leg parts, i.e., of the coxa,

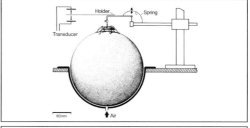
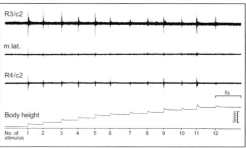

Fig. 2. Experiments on spherical treadmill. **Upper panel:** Arrangement for tethering spiders atop air-suspended styrofoam sphere. The spider (at its prosoma) is attached to a light holder that is pivoted and steadied by a weak spring; changes in body height are recorded via a capacitive transducer. **Lower panel:** Transient muscle activity during body raising. Upper 3 traces are simultaneous myogram recordings from c2-muscles in legs R3 and R4 and from lateral muscles (m. lat.). The latter contract to increase the hemolymph pressure in the prosoma. Bottom trace: transducer signal displaying increase in body height upon 12 successive tactile stimuli applied manually to several hairs at ventral trochanter of leg R3. Each stimulus evokes activity in all 3 muscles. [Adapted from Eckweiler and Seyfarth 1988]

trochanter, and proximal femur. Complete removal of all hairs in these regions leads to collision with the obstacle because the animals do not raise their prosoma until the hairy opisthosoma touches the wires (Eckweiler & Seyfarth 1988).

Our detailed analysis of the behavior from video-footage (filmed at 50 frames/s) shows that the spiders raise their body within 120-160 ms after first touching the obstacle. Consequently, the perception and central processing of tactile information as well as the coordination and execution of movements in the 8 legs must all be accomplished within this brief period.

For experiments under more controlled conditions spiders were tethered dorsally above an air-suspended styrofoam sphere as shown in Fig. 2 (top). In this situation, spiders can freely adjust their body height and walk (by rotating the sphere) but are 'fixed in space'. Changes in body height are recorded with a capacitive position transducer. Deflection of individual hairs (with a small wire loop) reveals that stimulation of a single, isolated hair suffices to elicit the concerted action of all 8 legs resulting in body raising (Eckweiler & Seyfarth 1988). Simultaneous deflection of several hairs (3 to 5 in a group) leads to successive, reliable reactions even when the stimulus is repeated at short intervals. An example of such repetitive action is shown in Fig. 2 (bottom).

In addition to exact tactile stimulation, the treadmill device also allows electrophysiological recordings of reflex activity in muscles bringing about body raising. We implant fine, flexible copper wires as electrodes in leg muscles and observe the appearance of muscle potentials reflecting the activity of several 'motor units' in each muscle ('electromyogram'; Figs. 2, 3). The signals are stored on magnetic tape and can be played back for later analysis. Such measurements show that the strongest and most reliable reflex response upon ventral tactile stimulation occurs in the coxal levator muscle c2 (muscle nomenclature according to Palmgren 1981). Contraction of this muscle pulls the coxa against the prosoma, presumably locking the pleural-coxal joint in place while the leg joints further distally extend via the hydraulic mechanism mentioned above. Fig. 2 (bottom) shows myograms recorded from adjacent legs (R3 and R4) during several successive body raising reactions on the sphere. Electrical activity sets in almost simultaneously in all muscles and is transient, that is, it occurs only during changes in body height. Apparently the spider can maintain its new position without continued electrical activity in the leg muscles. We assume that the 'residual tension' generally found after contractions in arthropod muscle and the newly adjusted hemolymph pressure

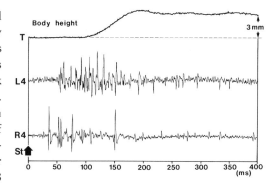

Fig. 3. Myogram recordings from c2-muscles in the two hindlegs (R4, L4), similar to Fig. 2 but displayed here at expanded time scale. Stimulation (St) of ventral tactile hairs at leg R4 elicits body raising (transducer signal, T). A burst of muscle potentials first appears in the stimulated leg (R4), and ca. 30 ms later in the contralateral leg. Each muscle recruits several motor units as indicated by the different amplitude and shape of potentials within the bursts. [Adapted from Kadel et al. 2002]

suffice to hold the body in the newly elevated position (Eckweiler & Seyfarth 1988).

Examination of such myograms at high time-resolution reveals how the muscle activities generated in several legs actually follow a finely tuned sequence. In the experiment shown in Fig. 3, 4 to 5 long hairs were touched beneath the coxa of the right hindleg (R4). Following a delay of ca. 30 ms after stimulus onset, muscle potentials first appear in the same leg being stimulated (R4), and then in the contralateral leg (L4) after an additional delay of 30 ms. Electrical activity in these and in the c2-muscles of all other legs causes muscle contraction and finally body raising - after a further 60-ms delay. The total latency of ca. 120 ms after stimulus onset corresponds to the value from our video-analysis of freely moving spiders crossing the wire obstacle.

The same sequence of events also applies when hairs on the forelegs are stimulated. Muscle c2 of the stimulated leg itself always reacts at least 25 ms prior to the musculature in the remaining, not-stimulated legs, but the latter are activated nearly simultaneously within a

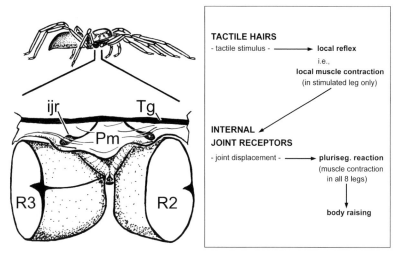

Fig. 4. Left: Topography of internal joint receptors in tergo-coxal joint of legs R2 and R3. Internal joint receptor (ijr) organs are located at the proximal edge of each leg coxa; they are comprised of several multipolar, mechanosensitive neurons which form small sensory ganglia underneath the pleural membrane (pm). These joint receptor organs are stimulated by coxal displacements. [Drawing adapted from Kadel et al. 2002] **Right:** Sequence of events leading from local reflex to body raising. Tactile hairs are involved at the first, local level, while internal joint receptors play a decisive role for initiating the next, plurisegmental stage (see text for detailed discussion).

period of 2 to 4 ms. Hence we identify a short-latency 'local reaction' in the stimulated leg itself, followed by a longer-latency 'plurisegmental reaction' that is virtually simultaneous in all remaining legs. The consistent time relation between stimulus onset and neuromuscular responses indicates that body raising behavior is largely determined by stereotyped reflex pathways.

What determines the sequence from local to plurisegmental reaction? Myogram recordings (such as the ones shown in Fig. 3) in combination with selective sensory ablations demonstrate that the plurisegmental response (that is, actual body raising) is only indirectly induced by the tactile stimulus. The decisive event for plurisegmental activation is displacement of the pleuro-coxal joint - which itself is brought about by short-latency <u>local</u> contraction of the c2-muscle (Kadel 1992; Kadel et al. 2002). If displacement of the joint is precluded experimentally (by immobilizing the joint with beeswax), tactile stimuli will merely result in a local reflex and not activate the remaining legs to raise the body. This is further supported by

experiments with animals that are firmly restrained on their backs so that the experimentor can deflect individual hairs or selectively move single leg joints. There are two notable findings from such experiments: (i) Even under drastically reduced conditions the restrained spiders first react to tactile stimuli with a local reflex (confined to the stimulated leg), followed by the plurisegmental response in all remaining legs. (ii) Passive displacement of the pleuro-coxal joint alone (without prior tactile stimulus) directly induces plurisegmental activity. It turns out that internal joint receptors underneath the articular membrane of this joint perceive such movements (be they passive or active). Fig. 4 shows the topography of internal joint receptors in the tergo-coxal joints of two legs (R2 and R3). After surgical ablation of their sensory axons (through a tiny cut in the tergal membrane), the plurisegmental reaction fails to occur - both on the sphere and in spiders restrained on their backs. The plurisegmental response persists, however, after sham-operations that spare the sensory nerve (Kadel 1992; Kadel et al. 2002).

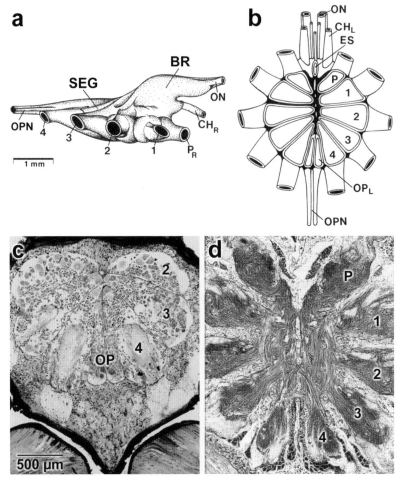

Fig. 5. Central nervous system of *C. salei*. **(a)** Schematized lateral view of subesophageal ganglion complex (SEG) and brain (BR). **(b)** Ventral view shows septal partitions dividing ventral SEG into serially arranged neuromeres. **(c)** Horizontal section through layer of neuronal cell bodies (cortex region) in ventral SEG; leg neuromeres 2 to 4 and smaller opisthosomal neuromeres (OP) are clearly separated by septa (thionine/Nissl stain; juvenile spider). **(d)** Horizontal section made further dorsally through SEG and stained with reduced silver technique; the septal partitions have disappeared; neuronal processes form a complex meshwork (neuropil) and fiber tracts that connect ipsilateral and contralateral regions. CH_R, CH_L: left and right cheliceral neuromere; ES: esophagus; ON: optical nerves; OP_L: left opisthosomal neuromere; OPN: opisthosomal nerves; P: pedipalpal neuromere; P_R: right pedipalpal nerve. [Drawings (a) and (b) modified from Babu & Barth 1984]

The diagram in Fig. 4 (right) summarizes these findings. We observe two successive reflex responses: (i) Touching hairs on the ventral, proximal leg parts evokes a local muscle contraction in the same leg. (ii) The resulting displacement of the coxa stimulates internal joint receptors in the tergo-coxal joint; this induces specific plurisegmental reactions in the other legs and finally body raising. Experimentally, we can directly evoke the plurisegmental response by passively moving the coxa. The two reflex responses appear to be identical in spiders standing on the spherical treadmill and in animals completely restrained on their backs. This latter finding is particularly important because it is the essential basis for our

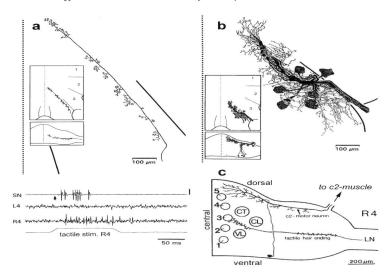

Fig. 6. Afferent and efferent projections in the SEG. **(a)** Primary tactile hair afferent. Top: arborization pattern in dorsal view, reconstructed from serially sectioned Lucifer Yellow preparation. Insets: small-scale dorsal and frontal views. Dotted line: longitudinal midline of SEG. Bottom: local response upon tactile stimulation of hairs in R4. Upper trace: the sensory neuron (SN) fires action potentials beginning 9 ms (arrow) after the hairs are touched (intracellular recording, high-pass filtered; scale bar: 5 mV); 2nd and 3rd trace: myogram recordings from c2-muscle in the two hindlegs; only R4 is activated. Bottom trace: signal driving tactile stimulator. [Adapted from Kadel 1992] **(b)** Motoneurons of c2-muscle in R4, dorsal view reconstructed from axonal backfills with nickel-chloride. The 6 somata are located in the ventral cortex layer (lower inset); the primary neurites ascend dorsally; their dendritic arborizations are confined to dorsal neuropil regions; leaving the CNS, the motor axons reach the c2-muscle via a small, separate nerve. [Preparation and reconstruction: Christiane Bickeböller] **(c)** Schematized frontal section through leg neuromere showing gap between tactile hair endings and c2-motoneuron; CT, CL, VL, and 1 - 5: position of longitudinal fiber tracts connecting ipsilateral neuromeres. [Modified from Milde & Seyfarth 1988]

analysis at the level of single central neurons discussed below.

ANATOMY OF THE SPIDER CNS AND INTRACELLULAR RECORDINGS

Before introducing individual neurons and their activities, I will briefly describe the gross anatomy of the central nervous system (CNS) in *C. salei*. As shown by the lateral view in Fig. 5a, the CNS is comprised of two main parts: (i) ventrally a relatively large subesophageal ganglion complex (SEG) that includes all nerve roots for the pedipalps, walking legs, and the opisthosoma, and (ii) dorsally a supraesophageal ganglion complex or 'brain' (BR) that provides nerves running to the 8 eyes and to the chelicerae. A ventral view of the SEG (Fig. 5b) demonstrates that the original metameric or-

ganization of the fused ganglion complex is still recognizable externally. The segmentally arranged hemiganglia (so-called 'neuromeres') are marked by ventral septa of connective tissue. Histological sections reveal the internal anatomy of the CNS. The majority of neuronal cell bodies are arranged in ventral cortex layers that remain largely separated by the segmental septa (Fig. 5c). Further dorsally in the SEG, the septal partitions between individual segments disappear, and we find a dense meshwork of neuronal processes and tracts of nerve fibers than run between ipsi- and contralateral neuromeres (Fig. 5d). Characteristically, the neurites profusely branch in mid-dorsal and dorsal neuropil regions, where they form synaptic contacts. Further details of the anatomy of the *Cupiennius*-CNS (and of the arachnid CNS in

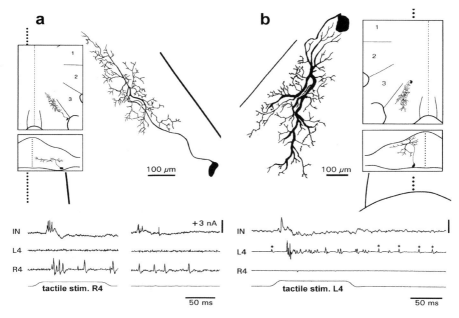

Fig. 7. Local interneurons and their activation following tactile stimuli; electrophysiological responses and anatomical reconstruction after intracellular recording and filling with Lucifer Yellow as in Fig.6a. **(a)** Spiking local interneuron in R4 characterized by arborizations that are confined to the antero-ventral part of the ipsilateral neuromere. Bottom left: tactile stimulus evokes local reaction in interneuron (IN) and in c2-muscle of R4. Bottom right: current injection (+3 nA) elicits action potentials in IN and activates a single motor unit in the ipsilateral c2-muscle. **(b)** Non-spiking local interneuron in L4; arborizations lie further dorsally than in spiking interneurons. Bottom: local reaction upon tactile stimulation; this interneuron (IN) generates only long-lasting, graduated potentials (scale bar: 10 mV); several neuromuscular units are activated in L4; one of them is tonically active due to the long-lasting depolarisation of the pre-motor IN. [Figures modified from Kadel & Seyfarth 2002]

general) are provided by Babu & Barth (1984), Babu (1985), and Seyfarth et al. (1993).

The electrical activity of individual neurons is recorded with glass microelectrodes that are inserted ventrally into the SEG through a tiny cut in the sternum. A small clot of hemolymph usually forms around the inserted electrode shaft securely sealing it in the opening. Using strong light sources, the septal partitions separating the neuromeres of the ventral SEG can be detected through the translucent cuticle (under a dissecting microscope) and are used as landmarks for positioning the electrode (Milde & Seyfarth 1988). For such intracellular recordings to be successful, the spiders must be completely restrained on a sturdy holder. The experimental setup includes custom-built devices for stimulation of tactile hairs and for controlled, passive displacement of individual leg joints. Following electrophysiological recordings of the responses to tactile and/or displacement stimuli, tracer substances (such as the fluorescent dye Lucifer Yellow) are injected into the neuron. Anatomical details of the labeled neurons are then reconstructed from serial histological sections (Milde & Seyfarth 1988; Kadel & Seyfarth 2002).

CENTRAL CORRELATES OF THE LOCAL AND THE PLURISEGMENTAL REFLEX RESPONSES

Intracellular recordings and anatomical reconstruction of labeled neurons reveal three different neuronal correlates of the reflex responses in SEG-neuromeres: (i) primary sensory terminals of the tactile hairs (= afferents), (ii) moto-

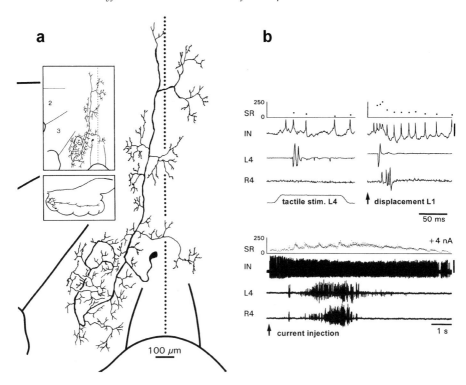

Fig. 8. Plurisegmental interneuron characterized by arborizations that extend into several ipsi- and contra-lateral neuromeres; electrophysiological responses and anatomical reconstruction after intracellular re-cording and filling with Lucifer Yellow as in Figs. 6a, 7. **(a)** The neuron arborizes extensively in L4 (which is also the recording site); a major neurite runs rostrad in parallel to the longitudinal midline (dotted), reaches into the 1st leg neuromere, and branches into contralateral neuromeres. **(b)** Left: local response to tactile stimulus in L4; maximum spike rate in IN is 60 spikes/s (SR, top trace). Right: plurisegmental re-sponse to coxal displacement stimulus (arrow) in L1; spike rate in IN rises up to 240 spikes/s. Bottom: responses to depolarizing current (+4 nA) injected into IN via recording electrode (vertical scale bar: 10 mV); the c2-muscles in both hindlegs (L4, R4) are activated simultaneously at an induced rate of ca. 100 spikes/s. [Modified from Kadel & Seyfarth 2002]

neurons activating muscle c2 (= efferents), and (iii) different types of interneurons that medi-ate between these afferents and efferents.

(i) Primary tactile afferents enter the SEG ventrally via the main leg nerve. They form numerous endings branching off the main neu-rite. Generally, tactile hair projections remain within the ipsilateral leg neuromer. Hence they are called 'local sensory projections'. Fig. 6a shows a typical example in leg neuromere R4. Upon tactile stimulation, the sensory unit (SN in Fig. 6a) is activated first. As expected, after a short delay the local motor response follows in the c2-muscle. There is no plurisegmental reac-

tion because all joints have been immobilized in this experiment.

(ii) Unlike the situation found in verte-brates, muscle fibers of arthropods are typically innervated by several motoneurons ('polyneural innervation'; see Maier et al. 1987, for details of muscle innervation and the func-tional architecture of leg musculature in *C. salei*). In *C. salei*, muscle c2 receives innervation from at least 6 motoneurons. Their shape and location within the SEG are shown in Fig. 6b. Our experiments indicate that not all 6 units but only 3 to 4 are recruited during body rais-ing (Bickeböller et al. 1991; Kadel et al. 2002).

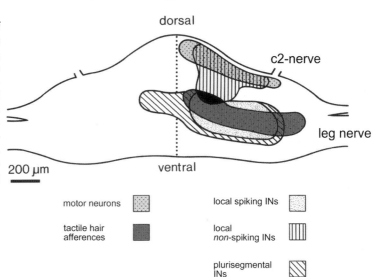

Fig. 9. Summary diagram of the 5 neuronal elements involved in body raising, their projections, and regions of overlap in a leg neuromere (schematized frontal view). Local non-spiking interneurons fill the gap between ventrally located hair afferents/spiking interneurons and the dorsal motor areas (see text for detailed discussion).

The dendritic projection areas of all c2-motoneurons resemble each other and are confined to the dorsal parts of the respective neuromere. Hence direct ('monosynaptic') contacts are not possible with the hair afferents that are located ca. 200 μm more ventrally. The anatomical gap becomes apparent in the frontal section through a neuromere in Fig. 6c. The relatively long latency of ca. 30 ms measured between stimulus onset and motor response is additional evidence for a lack of rapid, monosynaptic contacts in this reflex.

(iii) Various interneurons provide connections between the afferents and the c2-motoneurons. Depending on their electrical behavior and how far their projections reach into adjacent neuromeres, we distinguish local interneurons and plurisegmental interneurons (Milde & Seyfarth 1988; Kadel 1992; Kadel & Seyfarth 2002). Two different kinds of local interneurons are shown in Fig. 7a,b. The neuron in Fig. 7a responds with a burst of action potentials ('spikes') to a (standardized) tactile stimulus; as expected, the local motor reponse follows after a short delay. Interneurons of this type are called 'spiking local interneurons'. Generally, their arborizations are confined to ventro-medial parts of one leg neuromere (see lower inset in Fig. 7a). This example is especially interesting because a depolarizing cur-

rent of +3 nA (injected into the neuron via the microelectrode) causes activation of a single motor unit in the c2-myogram. The finding suggests that this interneuron plays a significant role in the reflex pathway.

The neuron in Fig. 7b does not react with a burst of action potentials, but rather generates longer-lasting, graduated potentials in response to tactile stimuli. This is a 'non-spiking local interneuron'. Injection of depolarizing and hyperpolarizing currents into this neuron directly modulates the strength of the muscle response (data not illustrated in Fig. 7b). Characteristically, the arborizations of such non-spiking local interneurons extend from ventro-medial regions well into the dorsal-most parts of the leg neuromere. Consequently, direct contacts with motoneurons seem feasible here (see also Milde & Seyfarth 1988; Kadel & Seyfarth 2002).

The examples so far have only dealt with local interneurons and local reflexes. We also find various plurisegmental interneurons whose arborizations extend into several SEG-neuromeres; they are candidate neurons for mediating the plurisegmental reactions, that is, actual body raising behavior. Fig. 8 shows an example of a plurisegmental interneuron that is particularly important. Upon tactile stimulation in L4, the neuron responds with several action potentials, and soon after, the local motor response follows in L4. Upon displacement of the pleuro-coxal joint in L1, however, there is a high-frequency spike discharge in the interneuron and a plurisegmental motor reaction (here shown only for L4 and R4). Depolarizing the interneuron with a current of +4 nA (instead of the joint displacement) also causes a massive spike discharge and a strong, plurisegmental motor response in L4 and R4 (Fig. 8b). In fact, by electrically manipulating the spike discharge of this interneuron we evoke motor reactions that greatly resemble the ones occurring during body raising. The arborizations of this type of plurisegmental interneuron are confined to the ventral part of the SEG, so that direct contacts with motoneurons are impossible but feasible via the non-spiking local interneurons discussed above.

Figure 9 summarizes our present data on the architecture of the different neuronal elements in a given leg neuromere. We assume that numerous tactile hair afferents converge onto relatively few interneurons, but that all interneuron types are present in multiple sets so that the whole system is redundant. The projection areas of the hair terminals, c2-motoneurons, and the three types of interneurons and regions of overlap are shown in schematized frontal view in Fig. 9. Based on the dorso-ventral extension of their neurites, non-spiking local interneurons alone are capable of providing local contact between tactile afferents and c2-motoneurons. Latency measurements of the various response times (data not illustrated here) in combination with our anatomical findings suggest the following sequence of neuronal events: In the case of a local reaction, the primary tactile hair afferents directly contact spiking local interneurons. Premotor, non-spiking local interneurons integrate this activity. As soon as a particular depolarization threshold is reached in these pre-motor elements, c2-motoneurons are excited in the same neuromere. In the case of plurisegmental reactions, we assume that afferent input from internal joint receptors (probably together with

some tactile input) is rapidly distributed throughout the ventral SEG via plurisegmental spiking interneurons. Subsequently, and almost simultaneously in each segmental neuromere, the same local pathways are then utilized as described for the local reactions.

OPEN QUESTIONS AND PROSPECTS

The results reviewed here offer a first glimpse of the architecture and functional hierarchy of central neuronal elements responsible for a relatively simple reflexive behavior in spiders. At the same time new perspectives are opened for further research.

Our findings are in general agreement with the situation well known in insects and crustaceans. The sequence of information processing through the various types of interneurons corresponds with that found in the other arthropod groups - including the role of pre-motor, non-spiking local interneurons (see the discussions by Kadel 1992; Laurent 1993; Kadel & Seyfarth 2002). Of particular interest are plurisegmental interneurons of the kind shown in Fig. 8. This neuron responds to tactile stimulation at L4 and to coxal displacement at L1 by generating action potentials with different spike rates (maximum of 60 spikes/s upon tactile stimulation; more than 200 spikes/s at coxal displacement). Electrical manipulation of the neuron also causes a plurisegmental reaction that first appears at an induced activity of ca. 100 spikes/s. The electrical stimulus leads to simultaneous reflex activity in several motor units of both muscles, which closely resembles the situation seen in freely moving animals (while electrical stimulation of local interneurons generally activates merely a single motor unit). Electrical inhibition (by hyperpolarization) of this plurisegemental interneuron, however, does not preclude the motor response. Hence this interneuron behaves like a 'command element' in the CNS (Kupfermann & Weiss 1978; Edwards et al. 1999); it is active during the behavior, and its electrical excitation is sufficient but not essential for evoking the behavior pattern. We assume that the neuron is part of a 'command system' consisting of several (redundant) plurisegmental interneurons, each of which can elicit the behavior alone. So far it is unclear how many elements comprise such a 'command system' in spiders.

Our experimental approach has focussed on tactile responses and hence has concentrated only on one detail in the behavioral repertoire of *C. salei*. Obviously, the animal uses the neuronal elements described here also in other situations and for other behaviors. So far we have not at all considered the role of sensory feedback ('reafference'), which is surely involved in controlling body raising but is totally precluded in fully restrained spiders (see also the discussions by Seyfarth & Bohnenberger 1980; Fabian-Fine et al. 1999, 2000). Moreover, so far we have relatively little information on internal joint receptor afferents, their projection pattern in the SEG, and their electrophysiological behavior. It is also important that we identify distinct functional compartments within individual local and plurisegmental interneurons. Detailed knowledge of the input and output zones along the main neuronal processes would allow more precise predictions about synaptic contacts between the various neuron types. Yet, further analysis of such neuronal circuitry in spiders requires a preparation in which the CNS can be dissected free for unobstructed access with microelectrodes. We have therefore begun to develop such preparations in which the hemolymph circulation is substituted by perfusion with saline.

So far our electrophysiological analysis has concentrated on the SEG-complex because this part of the CNS can be reached rather easily. We know very little about neuronal interactions between neurons in the SEG and so-called 'association centers' in the supraesophageal brain (via ascending and descending pathways). For instance, we observe that the tactile reaction habituates upon long-lasting, repeated stimulation, that is, the strength of the reaction decreases over time (Eckweiler & Seyfarth 1988). We assume that habituation and other 'context-dependent' behavioral adaptations are

controlled by supraesophageal centers. Gronenberg (1990) has described interneurons descending from the brain that may play such a role. There is also recent evidence that neuromodulatory substances - such as the biogenic amines octopamine (OA) and serotonin (5HT) - affect the state of arousal and general excitability in *C. salei*. While it is still unclear where exactly these substances act in the nervous system of *C. salei*, we have identified and mapped numerous OA- and 5HT-immunoreactive neurons and their projections in the CNS (Seyfarth et al. 1990, 1993).

ACKNOWLEDGEMENTS
The work summarized here is based on experiments done by Christiane Bickeböller, Wolfgang Eckweiler, Klaus Hammer, Michael Kadel and Jürgen Milde. I thank them for their dedication and excellent cooperation. Martin Jatho and Axel Stolp helped with computer graphics. Financial support for our research by grants from the Deutsche Forschungsgemeinschaft (DFG) is gratefully acknowledged.

REFERENCES
Babu, K.S. 1985. Patterns of arrangement and connectivity in the central nervous system of arachnids. In: *Neurobiology of arachnids*. (F.G. Barth ed.), pp. 3-19. Springer-Verlag, Berlin, Heidelberg, New York and Tokyo.

Babu, K.S. & Barth, F.G. 1984. Neuroanatomy of the central nervous system of the wandering spider, *Cupiennius salei* (Arachnida, Araneida). *Zoomorphology* 104, 344-359.

Barth, F.G. & Seyfarth, E.-A. 1979. *Cupiennius salei* Keys. (Araneae) in the highlands of central Guatemala. *Journal of Arachnology* 7, 255-263.

Barth, F.G., Seyfarth, E.-A., Bleckmann, H. & Schüch, W. 1988. Spiders of the genus *Cupiennius* Simon 1891 (Araneae, Ctenidae). I. Range distribution, dwelling plants, and climatic characteristics of the habitats. *Oecologia* 77, 187-193.

Bickeböller, C., Kadel, M. & Seyfarth, E.-A. 1991. Coxal muscle c2 in spiders: identification of motoneurons, joint receptors, and their role in body raising behavior. In: *Synapse, transmission, modulation. Proceedings of the 19th Göttingen Neurobiology Conference.* (N. Elsner & H. Penzlin eds.), p. 59. Thieme Verlag, Stuttgart and New York.

Breidbach, O. & Kutsch, W. (eds.) 1995. *The nervous system of invertebrates: an evolutionary and comparative approach.* Birkhäuser Verlag, Basel.

Bullock, T.H. 1984. Comparative neuroscience holds promise for quiet revolutions. *Science* 225, 473-478.

Burrows, M. 1996. *The neurobiology of an insect brain.* Oxford University Press, Oxford, New York and Tokyo.

Eckweiler, W. & Seyfarth, E.-A. 1988. Tactile hairs and the adjustment of body height in wandering spiders: behavior, leg reflexes, and afferent projections in the leg ganglia. *Journal of Comparative Physiology* A162, 611-621.

Edwards, D.H., Heitler, W.J. & Krasne, F.B. 1999. Fifty years of a command neuron: the neurobiology of escape behavior in the crayfish. *Trends in Neurosciences* 22, 153-161.

Fabian-Fine, R., Höger, U., Seyfarth, E.-A. & Meinertzhagen, I.A. 1999. Peripheral synapses at identified mechanosensory neurons in spiders: three-dimensional reconstruction and GABA-immunocytochemistry. *Journal of Neuroscience* 19, 298-310.

Fabian-Fine, R., Meinertzhagen, I.A. & Seyfarth, E.-A. 2000. Organization of efferent peripheral synapses at mechanosensory neurons in spiders. *Journal of Comparative Neurology* 420, 195-210.

Foelix, R.F. 1985. Mechano- and chemoreceptive sensilla. In: *Neurobiology of arachnids.* (F. G. Barth ed.), pp. 118-137. Springer-Verlag, Berlin, Heidelberg, New York and Tokyo.

Gronenberg, W. 1990. The organization of plurisegmental mechanosensitive interneurons in the central nervous system of the wandering spider *Cupiennius salei. Cell and Tissue Research* 260, 49-61.

Höger, U. & Seyfarth, E.-A. 1995. Just in the nick of time: postembryonic development of tactile hairs and of tactile behavior in spiders. *Zoology (ZACS)* 99, 49-57.

Hoyle, G. (ed.) 1977. *Identified neurons and behavior of arthropods.* Plenum Press, New York.

Huber, F. 1988. Invertebrate neuroethology: guiding principles. *Experientia* 44, 428-431.

Kadel, M. 1992. Zentralnervöse Korrelate lokaler und plurisegmentaler Muskelreflexe bei Spinnen: physiologische und morphologische Identifizierung von Einzelneuronen. Doctoral dissertation. Fachbereich Biologie, J.W.Goethe-Universität, Frankfurt am Main.

Kadel, M. & Seyfarth, E.-A. 2002. Body raising in spiders: central correlates of local and plurisegmental reflex activity. (In preparation)

Kadel, M., Bickeböller, C. & Seyfarth, E.-A. 2002. Body raising in spiders: the plurisegmental response is dependent on joint receptors. (In preparation)

Katz, P.S. & Harris-Warrick, R.M. 1999. The evolution of neuronal circuits underlying species-specific behavior. *Current Opinion in Neurobiology* 9, 628-633.

Kupfermann, I. & Weiss, K.R. 1978. The command neuron concept. *Behavioral and Brain Sciences* 1: 3-39.

Laurent, G. 1993. Integration by spiking and nonspiking local neurons in the locust central nervous system. Inportance of cellular and synaptic properties for network function. In: *Biological neural networks in invertebrate neuroethology and robotics.* (R.D. Beer, R.E. Ritzman & T. McKenna eds.), pp. 69-85. Academic Press, London.

Maier, L., Root, T.M. & Seyfarth, E.-A. 1987. Heterogeneity of spider leg muscle: histochemistry and electrophysiology of identified fibers in the claw levator. *Journal of Comparative Physiology* B 157, 285-294.

Milde, J.J. & Seyfarth, E.-A. 1988. Tactile hairs and leg reflexes in wandering spiders: physiological and anatomical correlates of reflex activity in the leg ganglia. *Journal of*

Comparative Physiology A162, 623- 631.

Palmgren, P. 1981. The mechanism of the extrinsic coxal muscles of spiders. *Annales Zoologica Fennici* 18, 203-207.

Seyfarth, E.-A. 1980. Daily patterns of locomotor activity in a wandering spider. *Physiological Entomology* 5, 199-206.

Seyfarth, E.-A. 1985. Spider proprioception: receptors, reflexes, and control of locomotion. In: *Neurobiology of arachnids.* (F.G. Barth ed.), pp. 230-248. Springer-Verlag, Berlin, Heidelberg, New York and Tokyo.

Seyfarth, E.-A. & Bohnenberger, J. 1980. Compensated walking of tarantula spiders and the effect of lyriform slit sense organ ablation. *Proceedings of the International Congress of Arachnology, Vienna* (J. Gruber ed.), pp. 249-255. H. Egermann, Wien.

Seyfarth, E.-A., Hammer, K. & Grünert, U. 1990. Serotonin-like immunoreactive cells in the CNS of spiders. *Verhandlungen der Deutschen Zoologischen Gesellschaft* 83, 640.

Seyfarth, E.-A., Hammer, K., Spörhase-Eichmann, U., Hörner, M. & Vullings, H.G. B. 1993. Octopamine immunoreactive neurons in the fused central nervous system of spiders. *Brain Research* 611, 197-206.

European Arachnology 2000 (S. Toft & N. Scharff eds.), pp. 33-38.
© Aarhus University Press, Aarhus, 2002. ISBN 87 7934 001 6
(Proceedings of the 19th European Colloquium of Arachnology, Århus 17-22 July 2000)

Distance of approach to prey is adjusted to the prey's ability to escape in *Yllenus arenarius* Menge (Araneae, Salticidae)

MACIEJ BARTOS

University of Łódź, Department of Invertebrate Zoology and Hydrology, Banacha 12/16, 90-237 Łódź, Poland. Present address: University of Łódź, Laboratory of Teaching Biology and Studies of Biological Diversity, Banacha 1/3, 90-237 Łódź, Poland (bartos@taxus.biol.uni.lodz.pl)

Abstract

The aim of the study was to investigate, whether *Yllenus arenarius*, a dune dwelling salticid, can adjust its jumping distance when hunting prey of high or low escapability risk. It was found that the spiders possess a conditional hunting strategy, depending on the prey's potential ability to escape. The spiders jumped from significantly longer distance on prey that can escape than on prey that cannot escape, thus decreasing the risk of detection and the escape of prey. There were found no significant differences in relative jumping distances within prey types between juveniles (in first and second year of life) and adults (in third year of life), suggesting, that flexibility in attack behaviour is inherited rather than learned.

Key words: behaviour; predation; jumping distance; conditional strategy; Salticidae; spiders

INTRODUCTION

This article is part of a wider study (Bartos 2000) concerning predatory versatility of a salticid spider, *Yllenus arenarius* Menge, 1868. The aim of the following research was to find out, whether the spider can adjust its jumping distance when attacking prey of different ability to escape.

Yllenus arenarius is a salticid inhabiting open, sandy dunes of mainly Central and Eastern Europe (Żabka 1997). It seems to be a good model for the study, because it is a predator stalking the prey in a habitat with very few places to hide. Thus, it has to depend solely on its cryptic coloration and prey approaching tactics.

Y. arenarius hunts a wide variety of invertebrates (Bartos 2000), which among other things differ in mobility. Some of them can easily escape (e.g. Diptera, Homoptera, Orthoptera) while others practically cannot escape (e.g. Thysanoptera and larvae of Lepidoptera). Irrespective of prey type, spiders approach the prey and jump on it (Bartos 2000). Close approach is advantageous because of more precise identification of prey and more precise jumping and grasping. However, close approach is also connected with a high risk of being noticed by the prey, compared to jumping from longer distances. Jumping distance is, therefore, a trade-off between several factors (Bear & Hasson 1997). It is most profitable to attack the prey that can easily escape from a longer distance, and the prey that cannot escape from a shorter distance. Since the spiders encounter different types of prey randomly, they should apply both of the behavioural tactics flexibly and quickly.

Flexibility in behavioural tactics, in order to optimise the outcome is known as a conditional strategy (Alcock 1993), and the phenomenon has already been reported for salticids (Jackson 1978, 1992; Edwards & Jackson 1993, 1994; Bear & Hasson 1997). Jackson (1977ab, 1978) found, that a male's mating behaviour in *Phidippus johnsoni* depends on female maturity and location. Behavioural flexibility was also found in the genus *Portia*, where spiders were found to tune the mode of hunting to specific conditions, such as prey type and its location (Jackson & Blest 1982; Jackson & Hallas 1986a, b; Jackson 1992).

Most cases of predatory versatility have been reported for the subfamily Spartaeinae. However, other studies suggest that conditional strategies are likely to be found also in other salticids (Edwards & Jackson 1993, 1994; Bear & Hasson 1997).

MATERIALS AND METHODS

A two-year-long diet analysis was carried out prior to the experiment. On the basis of that research five groups of insects were chosen for the experiments. These were: Homoptera, Diptera, Orthoptera, Thysanoptera and Lepidoptera larvae (Bartos 2000). Insects from the three former orders are capable of efficient escape, and were therefore regarded as prey with high escape risk. Thysanoptera and Lepidoptera larvae are unable to move quickly and were regarded as prey with low escape risk. In both groups of different escape potential, the prey types which were most common in the spider's diet, were chosen for the experiments.

By sweep-netting dune grass (*Corynephorus canescens*) prey items were collected either on the day of experiment or the day before. They were brought to the laboratory and kept individually. In order to reduce mortality of the prey, insects were stored in a refrigerator (temp. 5 °C) and taken out 15 min. before the experiment started.

Spiders were collected on the day of experiment or the day before in order to reduce the influences of rearing conditions on the spider's behaviour (Carducci & Jakob 2000; Bartos unpubl.). They were kept in glass chambers (height: 10 cm, Ø: 10 cm) with a 2 cm layer of dune sand on the bottom. Spiders from three age groups were used in the experiments: juveniles in first year of life (juv-I), juveniles in second year of life (juv-II) and adults in third year of life (ad). Spiders were assigned to the age groups on the basis of their size and maturity according to a previously developed method (Bartos 2000). To include all three age groups in the study the data had to be standarised. Therefore the jumping distance (JD) was divided by the abdomen length (AL) and the relative jumping distance (JD/AL) was further included in the tests. Abdomen length was used to standarise the jumping distance to correct not only for body size but also for the condition of different spiders of the same age. Spiders in bad condition have shorter abdomens than spiders in good condition. The worse the condition the shorter jumping distance, therefore the use of abdomen length as a standarising factor reduces the influence of spider condition on the relative jumping distance (JD/AL). Abdomen length was measured with a stereomicroscope on living spiders. A linear relationship was found between jumping distance and abdomen length ($r = 0.70$; $df = 222$; $P = 0.001$), which allowed employing the relative jumping distance for the analysis. Spiders as well as prey items were chosen randomly for experiments and used only once in the whole set of experiments.

All the experiments were carried out within a white cardboard arena (height: 15 cm, Ø: 20 cm) with a 1cm sand layer on the bottom. Attack behaviour was recorded with a camera placed above the arena and connected to a computer. A scale was also recorded, which allowed measuring of the jumping distance.

The significance of the differences in jumping distance was tested with one-way ANOVA and the Tukey test with unequal sample sizes. To test differences between the three age groups one-way ANOVA was applied. The significance of skewness (G_1) was also calculated. Data are presented as: mean $\pm$ SD.

RESULTS

Prey differences

Significant differences in relative jumping distance on the five types of prey were found (one-way ANOVA: $F_{0.05;4.219}$ = 85.56; P = 0.001). Homoptera, Diptera and Orthoptera were attacked from significantly longer distances than Thysanoptera and larvae of Lepidoptera (Fig. 1, Table 1). The preferences in jumping distance are also seen for relative jumping distance distribution (Fig. 1). In cases of hunting prey with a high ability to escape, all distributions are positive; they are significant, however, only for Diptera (G_1 = 2.70) and Homoptera (G_1 = 2.62) (Table 2). Significant difference (P = 0.02) was also found between relative jumping distance on Thysanoptera and Lepidoptera larvae (Table 1).

Age differences

There were no significant differences in relative jumping distance between spiders in different age groups attacking Homoptera (one-way ANOVA: $F_{0.05;2.64}$ = 1.57; P = 0.2), Diptera (one-way ANOVA: $F_{0.05;2.74}$ = 1.53; P = 0.2) or Lepidoptera larvae (one-way ANOVA: $F_{0.05;2;14}$ = 0.33; P = 0.7). Regarding other prey types, the number of data in age groups was not big enough to be tested. Thysanoptera were hunted only by juveniles in the first year of life, and Orthoptera were eaten only by adults.

DISCUSSION

The data suggest that *Y. arenarius* is able to discriminate between different types of prey and applies different behavioural tactics to hunt them. Spiders jumped on Diptera, Homoptera and Heteroptera from significantly longer dis-

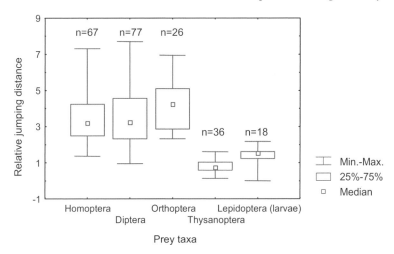

Fig. 1. Relative jumping distance (JD/AL) on 5 prey taxa. Homoptera, Diptera and Orthoptera are prey with high risk of escape, whereas Thysanoptera and Lepidoptera larvae are prey with low risk of escape.

Table 1. Results of the Tukey test with unequal sample sizes showing significant differences (marked with asterisk) in pair-wise comparison between relative jumping distances (JD/AL) on different prey types.

Prey type	Diptera	Orthoptera	Thysanoptera	Lepidoptera larvae
Homoptera	P = 0.9	P = 0.4	P = 0.0001*	P = 0.0001*
Diptera	-	P = 0.5	P = 0.0001*	P = 0.0001*
Orthoptera		-	P = 0.0001*	P = 0.0001*
Thysanoptera			-	P = 0.02*

Table 2. Descriptive statistics of relative jumping distance (jumping distance / abdomen length) of *Yllenus arenarius* on five prey taxa.

Taxon of prey	N	Mean	Median	SD	Min.	Max.	Skewness	G_1
Homoptera	67	3.46	3.19	1.32	1.37	7.31	0.76	2.62
Diptera	77	3.60	3.23	1.53	0.95	7.69	0.73	2.70
Orthoptera	26	4.25	4.22	1.40	2.32	6.94	0.28	0.61
Thysanoptera	36	0.83	0.74	0.39	0.13	1.61	0.43	1.10
Lepidoptera larvae	18	1.43	1.54	0.50	0	2.17	-1.46	-2.70

tance than on Thysanoptera and larvae of Lepidoptera, which probably decreases the risk of prey escape and increases hunting success.

The distribution of jumping distance (Fig. 1) on Diptera and Homoptera, where asymmetry is significantly positive (Table 2), suggests that the optimal jumping distance is shorter than the average one. Approaching closer than the mode value is probably connected with a very high risk of prey escape and therefore observed in very few cases. To approach further than the mode value is much more common and probably less disadvantageous (lower precision in grasping the prey is less disadvantageous than losing it).

It is interesting that the spiders in fact jump when hunting Thysanoptera and larvae of Lepidoptera, which obviously cannot escape. Tactile contact between the spider and its prey may cause rapid movement of the prey. Jumping, even from very close distance, allows precise grasping and venom injection. None of the observed jumps made the prey escape before venom injection, whereas in two cases, when spiders walked directly to a caterpillar and tried to stab it, the prey started to move rapidly with its head, which prevented the predator from attacking.

The difference in distance of attack on Thysanoptera and larvae of Lepidoptera suggests that there are other factors influencing the distance of approach to prey than the prey's escape potential. The most striking differences between the two prey types are their size and mode of movement. Thysanoptera are the smallest (1.2 ± 0.30 mm) and larvae of Lepidoptera the largest of all five types of prey ($8.5 \pm$ 4.16 mm). Since the spiders approached the prey from their front and stabbed their thorax and not the head (Bartos 2000), they probably could not approach larvae of Lepidoptera too closely because of the prey's comparatively large head which is difficult to jump over. In addition, side-to-side movements of the caterpillar's head were often found to make the spiders withdraw. In attacking Thysanoptera this was not the case because of their flat bodies and different type of locomotion. Caterpillars, having larger bodies and moving slowly in a predictable way, were much easier to grasp from a distance. Thysanoptera changed direction of movement much more often and rapidly, which made an attack from a close distance more successful.

The evidence that *Y. arenarius* is a versatile predator switching quickly and flexibly between different behavioural tactics is consistent with similar results given by Edwards & Jackson (1993, 1994) for several species of *Phidippus*, and also by Bear & Hasson (1997) for *Plexippus paykulli*. The present study gives evidence of conditional strategies in non-Spartaeinae salticids. This suggests, that conditional strategies may be common not only in Sparteinae but also in other, non-specialised salticids.

The question, whether the hunting tactics in Salticids are learned over the spider's life or they are genetically determined was considered by Edwards & Jackson (1994). Their experiments with first-instar and adult *Phidippus regius* (Edwards & Jackson 1993, 1994) showed, that inexperienced juveniles were versatile predators, just like adults. In the presented study the lack of differences in relative jumping distance between the three age groups supports this idea.

Differences in the number of data for particular prey types result from three sources. The first one is, that the number of different prey items in the pool prepared for the experiments was not equal, but reflected the proportion of the prey in nature. Secondly, some prey types (Orthoptera, Thysanoptera, larvae of Lepidoptera) were ignored by the spiders during the experiments more often than others (Diptera, Homoptera). Thirdly, in the experiments only Diptera, Homoptera and Lepidoptera larvae were eaten by spiders of all three age groups. This prevented the inclusion of Thysanoptera and Orthoptera in tests of differences in jumping distance between age groups. The ignoring of Thysanoptera, a very small prey, by juv-II and adult spiders, and a large prey (Orthoptera) by all age groups except adults, suggests that during the life cycle there is a change not only in the prey size most commonly eaten but there is also a qualitative change in diet.

Another interesting problem is how Salticidae recognise their prey (see review in Jackson & Pollard 1996) and especially the prey's ability to escape. The prey's behaviour in the experiments suggests that prey mobility is not the only cue the predator exploits. In both groups of different ability to escape there were prey which moved a lot after being placed on the sand (Diptera, Thysanoptera), and prey which either remained where placed while cleaning legs and antennae (Diptera, Homoptera, Orthoptera) or moved slowly only after a period of time (Lepidoptera larvae).

REFERENCES

Bartos, M. 2000. Cykl życiowy i strategia polowania pająka *Yllenus arenarius* Menge, 1868 (Araneae, Salticidae). Ph.D. Thesis. University of Łódź, Łódź.

Alcock, J. 1993. *Animal behavior: an evolutionary approach*. Sinauer Associates, Sunderland.

Bear, A. & Hasson, O. 1997. The predatory response of a stalking spider, *Plexippus paykulli*, to camouflage and prey type. *Animal Behaviour* 54, 993-998.

Carducci, J.P. & Jakob, E.M. 2000. Rearing environment affects behaviour of jumping spiders. *Animal Behaviour* 59, 39-46.

Edwards, G.B. & Jackson, R.R. 1993. Use of prey-specific predatory behaviour by North American jumping spiders (Araneae, Salticidae) of the genus *Phidippus*. *Journal of Zoology, London* 229, 709-716.

Edwards, G.B. & Jackson, R.R. 1994. The role of experience in the development of predatory behaviour in *Phidippus regius*, a jumping spider (Araneae, Salticidae) from Florida. *New Zealand Journal of Zoology* 21, 269-277.

Jackson, R.R. 1977a. An analysis of alternative mating tactics of the jumping spider *Phidippus johnsoni* (Araneae, Salticidae). *Journal of Arachnology* 5, 185-230.

Jackson, R.R. 1977b. Courtship versatility in the jumping spider, *Phidippus johnsoni* (Araneae: Salticidae). *Animal Behaviour* 25, 953-957.

Jackson, R.R. 1978. An analysis of alternative mating tactics of the jumping spider *Phidippus johnsoni* (Araneae, Salticidae). *Journal of Arachnology* 5, 185-230.

Jackson, R.R. 1992. Conditional strategies and interpopulation variation in the behaviour of jumping spiders. *New Zealand Journal of Zoology* 9, 99-111.

Jackson, R.R. & Blest, A.D. 1982. The biology of *Portia fimbriata*, a web-building jumping spider (Araneae, Salticidae) from Queensland: utilization of webs and predatory versatility. *Journal of Zoology, London* 196, 255-293.

Jackson, R.R. & Hallas, S.E. 1986a. Comparative biology of *Portia africana, P. albimana, P. fimbriata, P. labiata, P. schultzi*, araneophagic, web-building jumping spiders (Araneae: Salticidae): utilisation of webs, predatory versatility, and intraspecific interactions. *New Zealand Journal of Zoology* 13, 423-489.

Jackson, R.R. & Hallas, S.E. 1986b. Predatory versatility and intraspecific interactions of spartaeinae jumping spiders (Araneae: Salticidae): *Brettus adonis, B. cingulatus, Cyrba algerina,* and *Phaeacius sp.* indet. *New Zealand Journal of Zoology* 13, 491-520.

Jackson R.R. & Pollard S.D. 1996. Predatory behavior of jumping spiders. *Annual Review of Entomology* 41, 287-308.

Żabka, M. 1997. *Salticidae. Pająki skaczące (Arachnida: Araneae).* Muzeum i Instytut Zoologii PAN, Warszawa.

European Arachnology 2000 (S. Toft & N. Scharff eds.), pp. 39-44.
© Aarhus University Press, Aarhus, 2002. ISBN 87 7934 001 6
(Proceedings of the 19th European Colloquium of Arachnology, Århus 17-22 July 2000)

Cocoon care in the social spider *Stegodyphus dumicola* (Eresidae)

S. M. KÜRPICK

Zaunäcker 13, D-74740 Adelsheim, Germany (sc-kuerpick@t-online.de)

Abstract
The evolution of cocoon care in *Stegodyphus dumicola* seems to be influenced by a high migration rate between the colonies. Intruding conspecifics are tolerated by resident spiders and discriminating behaviour is observed only under special conditions. To prevent social parasitism, females care only for their own cocoons and not for those of other females from the same or foreign colonies. I describe biparental cocoon care and two different reproductive strategies that males have developed due to the social organisation in the colonies of *S. dumicola*. This is the first description of biparental care in spiders. Possible causes of biparental care are discussed.

Key words: Social spiders, cocoon recognition, biparental cocoon care, male reproductive strategies

INTRODUCTION

Ecological studies have revealed both the costs and benefits of sociality and have shown that social behaviour is maintained only under special ecological conditions. Spiders are usually solitary, exhibiting aggressive behaviour towards other animals, including conspecifics. Communal and cooperative living patterns have only been observed in a few species from several families.

The genus *Stegodyphus*, which is common in arid regions of Africa, Asia and South America, has three species groups (the *miranda*, *dufouri* and *africanus* groups), each including both social and solitary species, making it a very interesting spider genus for the study of social behaviour (Kraus & Kraus 1988, 1992). The development of permanent social life patterns in *Stegodyphus* seems to be the result of extending the early social stage of brood-caring subsocial species. This could finally lead from communities of juveniles to permanently social colonies. In this context parental care seems to be one of the main steps in the evolution of sociality in *Stegodyphus* (Kraus 1988).

Discrimination of conspecific unrelated individuals is a general phenomenon of true societies and was explained by the concepts of 'inclusive fitness' and 'kin selection' by Hamilton (1964). Several references (Kullmann 1974; Kraus 1988; Seibt & Wickler 1988) to the previous suggestions that spider societies are open systems without kin recognition, contradict the concept of societies in which individuals invest in valuable cooperative efforts. A condition for effective kin selection is that natural selection will favour social or altruistic behaviour if individuals are able to develop in ways that affect their parental care or helping behaviour. No observations of cooperative behaviour should be expected without kin recognition because of the increased risk of social parasitism (Hamilton 1964). But cooperative brood care has been shown several times in permanent social species of *Stegodyphus* (Kullmann 1974; Kraus 1988).

I will show (1) that colonies of *S. dumicola* are open systems without discriminating behaviour towards conspecifics, and (2) that females of *S. dumicola* care only for their own cocoons.

METHODS
Natural History
Stegodyphus dumicola (Fig. 1) is a common social spider in the thornbush savanna of Southern Africa (Kraus 1988). Colonies consist of 2 to 400 spiders but solitary individuals are often found. Spiders sitting underneath the silky nest show activities like mating, prey capture and guarding of cocoons both during the day and night. Web maintenance activities are performed cooperatively and mainly at night. Feeding takes place below the nest and the spiders mostly share the prey. The first mature males were found in January, the first cocoons were produced in February. The reproductive cycle ends in May/June when all adults are dead, sucked out by the young.

Observations
Field work was done in 1996 and 1997 in Namibia on the farm Otjiseva in the Khomas Highland 40 km north of Windhoek, and in Khorixas, Damaraland (about 480 km from Otjiseva). The colonies were from different sites and different populations (24 colonies from Otjiseva, 21 colonies from 23 km north from Otjiseva and 40 colonies from Khorixas). Of these, 35 colonies were observed in the natural situation near Khorixas. Some colonies were transported from their place of origin and observed at Otjeseva. During transportation the spiders remained within their nest. Spiders from about 85 colonies were collected and marked individually with fluorescent hair gel (Pop color, Jofrika Kosmetik GmbH) and their length and weight were determined. Cocoons were marked with silk colour (Seidicolor, Germany). The colonies were observed 1996/97 from January to May. They were checked once

Fig. 1. Female (left) and male (top) of *Stegodyphus dumicola* in the colony.

every hour between sunrise and sunset with respect to: immigration/emigration of males/females, aggressive behaviour, copulations, production of cocoons, cocoon guarding (contact, defensive behaviour, handling), and hatching of spiderlings.

Experiments
Cocoon guarding. To test the benefits of guarding cocoons, field experiments were carried out in Otjiseva. Females in colonies and solitary females (mothers) were removed from their cocoons, which were then inspected during the following 5 to 7 days. As controls, females were removed, marked and returned to their cocoons.

Cocoon care. Cocoon exchange experiments were carried out with 15 colonies and 25 solitary females. Cocoons were removed from the colony, marked and put into a foreign colony. Other cocoons were removed, marked and returned to their original colony. A similar experiment was made with cocoons of solitary females: they were removed and either transferred to the nest of another solitary female, whose own cocoon had been removed, or returned to their original nest (control).

To answer the question whether females care for other females' cocoons in their own

Table 1. Behaviour of *Stegodyphus dumicola* towards intruders in the colony.

Intruding individual	Behaviour of colony members
Juveniles (conspecific)	No aggression: 0% (N = 40)
Females (conspecific)	No aggression: 0% (N = 120)
Males (conspecific)	Sometimes aggression from guarding females and males: 26% (N = 75)
Other spider species	Aggressive attacks: 100% (N = 54)
Foreign animals	Aggressive attacks: 100% (N = 43)

colony, females were removed from the colony and their cocoons were observed for more than a month. The control experiment was to put the removed and individually marked females back into their colony.

RESULTS
Kin recognition
Immigration and emigration of males and females was observed in 94% of the colonies (N = 115). The remaining 6% of the colonies died away due to infection with a fungus. There was no difference in migration behaviour between moved colonies and colonies in their natural habitat. There was no discriminating behaviour from resident spiders towards immigrating

Table 2. Hatching success (%) of *Stegodyphus dumicola* cocoons of solitary and colonial females with respect to guarding behaviour

	Solitary		Colonial	
	N	%	N	%
Guarded cocoons	18	12%	15	18%
Unguarded cocoons	20	0%	17	0%

Table 3. Guarding time and hatching success in *Stegodyphus dumicola* in the cocoon exchange experiment.

	Own cocoon N = 27	Exchanged cocoon N = 25	χ^2-test
Solitary females			
Time spent guarding (165 observations)	54%	2%	P < 0.001
Hatching spiderlings	11%	4%	P < 0.05
Colonial females			
Time spent guarding (276 observations)	17%	0%	P < 0.001
Hatching spiderlings	19%	0%	P < 0.001

conspecific spiders, whether juveniles or adult females (Table 1). Only intruding males were attacked several times by female spiders guarding their cocoons. Immigrated males and females were able to reproduce in the new colony.

Intruding individuals of other spider species (*Nephila senegalensis, Stegodyphus bicolor, Gandanameno echinatus*) or insects (ants, termites, wasps, bugs) were vigorously attacked by the members of the colony.

Cocoon guarding
Unguarded cocoons were not able to survive for longer than 3 days. Some cocoons dried out, others were emptied by ants. Spiderlings hatched significantly more often from guarded cocoons than from unguarded cocoons (Table 2; χ^2-test, P < 0.001).

Cocoon care
The experiments showed that no exchanged cocoon was guarded in the colonies. Only one solitary female (out of N = 52) guarded an exchanged cocoon and only in this case did spiderlings hatch (Table 3). Cocoons from which the mother had been removed were not guarded by the remaining females in the colony, and no hatching of spiderlings was observed from motherless cocoons. Most cocoons were dried out.

Previously only females have been described as having guarded cocoons. During field observations in 1996 and 1997, however, I found that males were also guarding cocoons. They attacked and killed ants and other intruders that tried to remove the cocoons, even in the absence of the female. In cases (N = 11) where all females were killed, the males stayed by the cocoons and the spiderlings. This indicates that the males were actually guarding the cocoons, not just guarding the females.

In the situations where males were observed guarding, females and males shared the guarding time (Fig. 2). There was no significant difference between the guarding time of the mother and that of the male. Solitary females and females without a guarding male invested significantly more time in cocoon care than females with a male helping them.

Females whose cocoons were guarded by males were significantly heavier than females without guarding males (females with guarding male: mean ± SD: 157.3 ± 38.0 mg (N = 30); females without guarding male: 116.3 ± 10.3 mg (N = 28); t-test: P < 0.001). Many observations showed that the guarding males copulated several times with the mothers on their cocoons (N = 118).

Males of *S. dumicola* seem to follow two different reproductive strategies: guarding and wandering. Guarding males (20% of N = 150) were significantly larger than wandering males (body length 5.6 ± 0.56 mm vs. 5.0 ± 0.61 mm; N = 30 and 120; t-test: P < 0.001). They remained in the colony, guarding the cocoons of the larger females and showed aggressive behaviour towards other (intruding) males. Smaller males (80% of all males; N = 150) migrated together in groups of 2-9 to other colonies and copulated with the females there. Afterwards they left the colony and wandered, still as a group, to the next colony. The longest journey observed for a male was about 17 m. I never observed any male-male aggression within wandering groups from the same colony during copulations in foreign colonies. Intruding males first simulated guarding behaviour towards cocoons in foreign colonies (N = 12) and copulated with the mothers on the cocoons. Later they tried to remove these cocoons (N = 9).

Females with a guarding male produce a second cocoon more often than females without a guarding male (13 of N = 30 vs. 1 of N = 28; χ^2-test: P < 0.001).

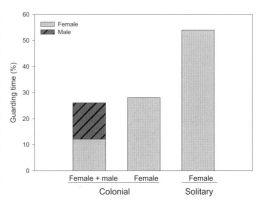

Fig. 2. Cocoon guarding times in *Stegodyphus dumicola*.

DISCUSSION

In many species reproductive success is highly dependent on the quality of parental care (Clutton-Brock 1991). In spiders the adaptive value of cocoon care seems to be the protection of the eggs from predators and parasitoids and the provision of a favourable climate, and also (in some species) helping spiderlings to hatch out of the cocoon. Guarding behaviour in *S. dumicola* is necessary to secure the survival of the eggs, to protect them from desiccation and predators. Previous observations of cooperative parental care in *Stegodyphus* were based mainly on laboratory studies (Kraus 1988, Kullmann 1988). In the field, females of *S. dumicola* cared only for their own cocoons. Cocoon exchange experiments showed that females were able to distinguish between their own cocoons and foreign cocoons from the same or other colonies. This behaviour seems to be adaptive in view of the high migration activity of the spiders and the lack of discrimination towards intruding females. Both factors create a high risk of social parasitism. The ability to distinguish between own cocoon and foreign cocoons was also found in vagrant wolf spiders (Kürpick & Linsenmair in prep.). Observations about cocoon care of social spiders are rare. In species such as *Archaeranea wau* and *Agelena consociata* females care for their own cocoons and all those colonies also show no kin recognition (Buskirk 1981).

Biparental care of cocoons is unusual in spiders. In *S. dumicola* females and males share the guarding time. It is possible that females without a guarding male have less time for foraging and need more energy to ward off predators and parasitoids. Thus females with a guarding male could exploit the advantage to produce a second cocoon. One reproductive benefit for guarding males seems to be the defence of the cocoons against ants and intruding males of *S. dumicola* from foreign colonies. This behaviour is necessary because wandering males showed infanticidal behaviour. Another advantage is, that guarding males also copulated several times with the mothers on their cocoons. Thus, they try to monopolise the fertilisation of the second cocoon.

Smaller males wandered in groups from the same colony to other colonies, copulating with the females there. The intruding males first simulated guarding behaviour, tried to copulate with the females and afterwards removed the female's first cocoon. Infanticidal males were also found in *Stegodyphus lineatus* (Schneider & Lubin 1996). The advantage of wandering in groups of (presumed) relatives to other colonies could be explained by the competition hypothesis: for the smaller males it may be advantageous to find a new colony where they may compete successfully with recident males and avoid competition with large males from their own colony (Moore & Ali 1984). They would increase their inclusive fitness (Hamilton 1964) by inseminating the receptive females in foreign colonies together. Cooperating groups of related males also have better chances to compete with the guarding males in already occupied colonies. Differences in behaviour are of vital importance for variable reproductive advantage and give individuals the chance to choose a promising strategy. In *S. dumicola* alternative strategies according to male size enables optimal reproductive success.

Parental care as in *S. dumicola* is unusual among arachnids and shows an interesting variety of behaviour patterns in the interactions

between the sexes to secure reproductive success. It is urgently necessary to start molecular-biological work on paternity and relationship in colonies of *S. dumicola* to confirm or complete the results presented above.

ACKNOWLEDGEMENTS
I am very grateful to Helmut Stumpf and Dr. Sabine Gebele for commenting on the manuscript.

REFERENCES
Buskirk, R.E. 1981. Sociality in Arachnida. *Social insects* 2, 281-367.

Clutton-Brock, T.H. 1991. *The evolution of parental care*. Princeton University Press, Oxford.

Hamilton, W.D. 1964. The genetical evolution of social behaviour, I + II. *Journal of Theoretical Biology* 7, 1-52.

Kraus, M. 1988. Cocoon-spinning behaviour in the social spider *Stegodyphus dumicola* (Arachnida, Araneae): Cooperating females as 'helpers'. *Verhandlungen des naturwissenschaftlichen Vereins in Hamburg* (NF) 30, 305-309.

Kraus, O. & Kraus, M. 1988. The genus *Stegodyphus* (Arachnida, Araneae). Sibling species, species groups, and parallel origin of social living. *Verhandlungen des naturwissenschaftlichen Vereins in Hamburg* (NF) 30, 151-254.

Kraus, O. & Kraus, M. 1992. Eresid spiders in the neotropics: *Stegodyphus manaus* n.sp. (Arachnida, Araneae, Eresidae). *Verhandlungen des naturwissenschaftlichen Vereins in Hamburg* (NF) 33, 15-19.

Kullmann, E.J. & Zimmermann, W. 1975. Regurgitationsfütterung als Bestandteil der Brutfürsorge bei Haubennetz- und Röhrenspinnen (Araneae, Theridiidae und Eresidae). In: *Proceedings of the 6th International Arachnological Congress*, pp. 120-124. Amsterdam.

Moore, J. & Ali, R. 1984. Are dispersal and inbreeding avoidance related? *Animal Behaviour* 32, 94-112.

Schneider, J.M. & Lubin, Y. 1996 . Infanticidal male eresid spiders. *Nature* 381, 655-656.

Seibt, U. & Wickler, W. 1988. Bionomics and social structure of 'Family Spiders' of the genus Stegodyphus. *Verhandlungen des naturwissenschaftlichen Vereins in Hamburg* (NF) 30, 255-303.

European Arachnology 2000 (S. Toft & N. Scharff eds.), pp. 45-49.
© Aarhus University Press, Aarhus, 2002. ISBN 87 7934 001 6
(Proceedings of the 19th European Colloquium of Arachnology, Århus 17-22 July 2000)

Fungal and rickettsial infections of some East Asian trapdoor spiders

JOACHIM HAUPT

Technische Universität Berlin, FR 1-1, Franklinstr. 28/29, D-10587 Berlin, Germany
(hptjeiic@sp.zrz.tu-berlin.de)

Abstract

While published data on fungal and rickettsial infections of spiders give the impression of a wide range of hosts infected by particular agents, field collections of trapdoor spiders from East Asia reveal a clear distinction: representatives of Mesothelae were infected by rickettsiae, while specimens of the genus *Latouchia* (Mygalomorphae, Ctenizidae) were found to be infected by the hyphomycete fungus *Nomuraea atypicola*.

Key words: Mesothelae, Ctenizidae, infection, Rickettsiales, *Nomuraea*

INTRODUCTION

Mesothelae have not been spared by predators, parasites and diseases. Besides ravaging animals like skinks and centipedes which regularly come across those soil dwelling spiders, there are reports on parasitic laelapid mites like *Liunghia bristowei* Finnegan, 1933 living on *Liphistius malayanus* Abraham, 1923 and there are parasitic flies and wasps (Sawaguti and Ozi 1937; Bristowe 1976; Schwendinger 1990) which attack Mesothelae along with other spiders. Mesothelae are also subject to parasitation by mermithid nematodes (pers. obs.).

Infections found in some species of Mesothelae are compared to those in other trapdoor spiders of the genus *Latouchia* (Mygalomorphae: Ctenizidae).

MATERIAL AND METHODS

Specimens of Mesothelae and Ctenizidae (genus *Latouchia*) collected in Kyushu, in Okinawa and in Malaysia were brought to the laboratory and kept there for ethological and morphological studies. The determination of mesothele spiders is according to Haupt (1983). Specimens of *Latouchia* were only determined to the rank of genus which is sufficient for the present work. The fungus *Nomuraea atypicola* (Yasuda, 1915) Samson, 1974 (syn. *Isaria atypicola, Cordyceps atypicola*) (Deuteromycotina: Hyphomycetes) was identified according to Kobayasi (1982). Rickettsiales could only be identified as an order, according to the electron microscopic appearance of the cells.

As in several species the collected material was quite numerous, it represented a good occasion to study the prevalence of different pathogens. The present study was carried out on living spiders, except for *Nomuraea* infections in which the perithecial stroma grows out from the dead spider and opens the trapdoor from inside the burrow, thus indicating the infection.

Trapdoor spiders live inside their burrows in moist soil, therefore the observer easily misses the date when a specimen dies. In the

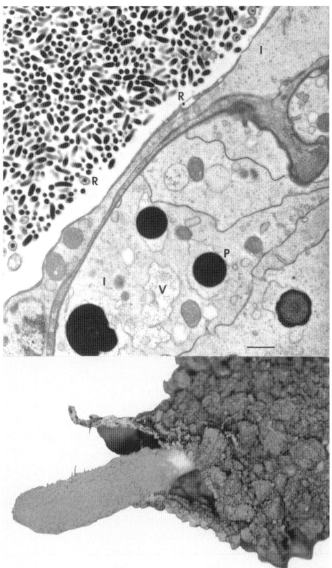

Fig. 1. Electron microscopic section through the opisthosoma of *Liphistius malayanus cameroni* with intestinal cells (I) and rickettsiae in intermediate cell (R). P: vacuoles containing protein, V: vacuoles. Scale 10 μm.

Fig. 2. Perithecial stroma of *Nomuraea atypicola* emerging from the burrow of *Latouchia* sp. (Ctenizidae) by opening the trapdoor from inside. Scale 2 mm.

moist environment dead spiders may become infected secondarily by fungi from the soil: such cases were not studied in the present paper.

For light microscopic study, the opisthosoma of infected specimens was submersed in Bouin's fluid and embedded in paraplast. Sections were stained with Haematein, Azan or Haematoxylin/Chromotrop after Dobell. Procedures for electron microscopy were as described previously (Haupt 1996), semi-thin sec-

tions (1 μm) were stained with Kristallviolett (Merck). Specimens of *Nomuraea atypicola* were deposited in the Botanical Museum, Berlin.

RESULTS
Histological and electron microscopic results
In Mesothelae, rickettsial infection is very obvious in the intermediate cells of the hepatopancreas (Fig. 1), while neighbouring cells of the midgut remain uninfected. The intermediate tissue is known to function as storage organ

Table 1. Number of specimens of Mesothelae and *Latouchia* (Ctenizidae) collected (N) compared to those infected by different pathogens.

	N	Infected by		%
		Rickettsiales	*Nomuraea*	
Heptathela kimurai kimurai (Kishida, 1920)	52	1	-	1.9
Heptathela kimurai yanbaruensis Haupt, 1983	164	3	-	1.8
Ryuthela nishihirai nishihirai (Haupt, 1979)	199	3	-	1.5
Liphistius malayanus cameroni Haupt, 1983	134	6	-	4.5
Latouchia sp.	22	-	4	18.2

analogous to the fat body of insects. Mainly lipids and glycogen are found in these cells (Ludwig & Alberti 1988), but the histological study also reveals numerous vesicles containing protein.

The first rikettsial stages to be detected are narrow and electron dense. Apparently, they grow inside the host cells to form larger, less electron dense cells. This stage may or does undergo binary fission thus multiplying the parasitic population in the host cell. Finally, all intermediate cells in the hepatopancreas are filled with nothing but rickettsial cells: all organelles and even the nuclei have disappeared. Apparently, the rikettsiae use up the storage products for their own purpose, and moulting becomes impossible for the spider.

These later stages of infection can be recognized easily by the whitish opaque opisthosoma of host spiders, which is typical for rikettsial infections. Finally, the intermediate cells rupture and as the spider dies and the whole opisthosoma dissolves, the pathogens spread into the soil. In this final phase the opisthosoma is so inflated that dissection becomes impossible. Even a careful touch results in the disruption of the cuticle and a milky fluid appears from inside.

So far, no experimental infection of Mesothelae along with food has been successful.

The other group of trapdoor spiders, living in the same habitats, belongs to the genus *Latouchia* (Ctenizidae). Among them, one may regularly observe infections by the fungus *Nomuraea atypicola*. Its hyphae grow through the whole body of the spider and finally, within a few hours, form a long stalk with conidium production. Generally, this stalk, the perithecial stroma, grows along the burrow, opens the trapdoor from inside, and reaches a height of 20 to 30 millimeters. In the upper part it bears numerous purple ascospores (Fig. 2), which now, outside the spider's burrow, are subject to aerial dispersal.

Instead, during 20 years of breeding Mesothelae, out of 549 specimens taken from the field not a single one was found to be infected by *Nomuraea atypicola*.

Infection rates

There are regular infections by Rikettsiales in Mesothelae (Table 1) found in Kyushu, Ryukyu, or Malaysia. The infection rate must be considered low, as far as most localities are concerned. Nevertheless, when comparing infection rates of the same species from different localities, there were striking differences: Rickettsial infection rates of *Ryuthela nishihirai nishihirai* were almost 6% (n = 52) in Suyeyoshi, Naha, Okinawa, but almost negligible in Ryutan, Naha, Okinawa. Both places, only a few kilometers from each other, are or were residues of rather natural habitats with similar soil conditions, but the locality in Suyeyoshi is situated in a valley of a brooklet, while Ryutan is close to a hill top, more exposed to wind and this place appears much drier.

Trapdoor spiders of the genus *Latouchia* (Ctenizidae) were infected by the hyphomycete *Nomuraea atypicola*, and to a much larger degree. The 22 specimens of *Latouchia* proved to be infected at a rate of about 18%.

DISCUSSION

Many fungal infections have been described in connection with spiders (Evans & Samson 1987), but unfortunately, in many cases either the spider or the fungus remained undetermined. The hyphomycete fungus *Nomuraea* certainly occurs in a wide range of soil dwelling mygalomorphs, and it has also been found in araneomorph spiders (for literature, see Coyle et al. 1990). In North America a broad range of spider species could be infected experimentally by an isolate of *Nomuraea* from ascospores originating from an infected Brazilian trapdoor spider. By applying a conidial suspension containing a detergent, 20 out of 27 spider species were successfully infected (Greenstone et al. 1987). In Okinawa, on the other hand, infectious diseases in trapdoor spiders seem to be clearly linked to different families: while *Latouchia* (Ctenizidae) specimens were found to be infected by *Nomuraea atypicola*, rickettsial infections were limited to Mesothelae. So far, the experimental infection of Ryukyuan mesothele spiders by *Nomuraea* (without using detergents) turned out to be impossible.

Among Rikettsiales, the genus pathogenic in arthropods has been named *Rikettsiella*. Very similar pathogens as in Mesothelae have been reported from intermediate (interstitial) cells of a Japanese funnel web spider (*Paracoelotes luctuosus*) (Osaki 1973) and from a linyphiid spider (Suhm 1995). These pathogens are comparable to infections described from the hepatopancreatic caeca in a buthid scorpion (Morel 1976).

In acarids (Reinhardt et al. 1972), insects, diplopods (Schlueter & Seifert 1985) and woodlice (Vago et al. 1970) the occurence of similar pathogens is quite frequent, and they have been reported as intracellular parasites from different tissues, such as labial glands (*Lepisma saccharina*, pers. obs.), Malpighian tubules (Schlueter & Seifert 1985) and adipous tissue (Götz 1972; Vago et al. 1970). In the case of woodlice, there is strong evidence that specimens from moist and periodically wet habitats are most likely to be infected (Federici 1984), which is in line with our observations on *Ryuthela nishihirai*.

The wide distribution does not automatically allow any conlusion on specific identity. At present it seems doubtful whether pathogenic Rickettsiales represent only one species. Transmission of Rickettsiales from a scorpion to other arachnids, even to members of other scorpion families, turned out to be impossible (Morel 1978). Our studies on infections of trapdoor spiders point to the same direction. The redefinition of the group based on serological data has been strongly suggested previously (Louis et al. 1977).

In general, the infection rate with Rickettsiales is low. A 5% infection rate in scorpions (Morel 1976) is well in line with our results. Such low prevalence of infection may also be the reason that early stages of infections are hardly found in the intermediate cells of free-living Mesothelae; only the late stage of infection is obvious.

There are interesting reports on Rickettsiales and unidentified virus particles occurring in acarid ovaries (Lewis 1979), in insect sperm, spider coenospermia, and mite spermatozoa (Afzelius et al. 1989). Apparently, still other paths of infection are used beside food, and *Wolbachia pipiens* is known to commonly infect ovary cells, thus being transmitted in a vertical manner (Bourtzis & Braig 1999).

REFERENCES

Afzelius, B.A., Alberti, G., Dallai, R., Godula, J. & Witalinski, W. 1989. Virus- and rickettsia-infected sperm cells in Arthropods. *Journal of Invertebrate Pathology* 53, 365-377.

Bourtzis, K. & Braig, H.R. 1999. The many forms of *Wolbachia*. In: *Rickettsiae and rikettsial diseases at the turn of the third millenium* (D. Raoult & P. Brouqui eds.), pp. 199-219. Elsevier, Paris.

Bristowe, W.S. 1976. A contribution to the knowledge of liphistiid spiders. *Journal of Zoology* 178, 1-6.

Coyle, F.A., Goloboff, P.A. & Samson, R.A. 1990. *Actinopus* trapdoor spiders (Araneae, Actinopodidae) killed by the fungus, *Nomuraea atypicola* (Deuteromycotina). *Acta Zoologica Fennica* 190, 89-93.

Evans, H.C. & Samson, R.A. 1987. Fungal pathogens of spiders. *Mycologist* 21 (4), 152-154.

Federici, B.A. 1984. Diseases of terrestrial Isopods. *Symposia of the Zoological Society London* 53, 233-245.

Finnegan, S. 1933. A new species of mite parasitic on the spider *Liphistius malayanus* Abraham, from Malaya. *Proceedings of the Zoological Society London* 413-417.

Götz, P. 1972. 'Rickettsiella chironomi': an unusual bacterial pathogen which reproduces by multiple cell division. *Journal of Invertebrate Pathology* 20, 22-30.

Greenstone, M.H., Ignoffo, C.M. & Samson, R.A. 1987. Susceptibility of spider species to the fungus *Nomuraea atypicola*. *Journal of Arachnology* 15, 266-268.

Haupt, J. 1983. Vergleichende Morphologie der Genitalorgane und Phylogenie der liphistiomorphen Webspinnen (Araneae: Mesothelae). I. Revision der bisher bekannten Arten. *Zeitschrift für zoologische Systematik & Evolutionsforschung* 21, 275-293.

Haupt, J. 1996. Fine structure of the trichobothria and their regeneration during moulting in the whip scorpion *Typopeltis crucifer* Pocock, 1894. *Acta Zoologica* 77, 123-136.

Kobayasi, Y. 1982. Keys to the taxa of the genera *Cordyceps* and *Torubiella*. *Transactions of the Mycological Society of Japan* 23, 329-364.

Lewis, D. 1979. The detection of *Rickettsia*-like organisms within the ovaries of female *Ixodes ricinus* ticks. *Zeitschrift für Parasitenkunde* 59, 295-298.

Louis, C., Croizier, G. & Meynadier, G. 1977. Trame cristalline des inclusions protéique chez une Rickettsiella. *Biologie Cellulaire* 29, 77-80.

Ludwig, M. & Alberti, G. 1988. Digestion in spiders: histology and fine structure of the midgut gland of *Coelotes terrestris* (Agelenidae). *Journal of Submicroscopical Cytology and Pathology* 20, 709-718.

Morel, G. 1976. Studies on *Porochlamydia buthi* g.n., sp.n., an intracellular pathogen of the scorpion *Buthus occitanus*. *Journal of Invertebrate Pathology* 28, 167-175.

Morel, G. 1978. Les maladies microbiennes des Arachnides (Acariens exceptés). *Symposium of the Zoological Society London* 42, 477-481.

Osaki, H. 1973. Electron microscopic observations of Chlamydia-like microorganism in hepatopancreas cells of the spider, *Coras luctuosus*. *Acta Arachnologica* 25, 23 - 36.

Reinhardt, C., Aeschlimann, A., & Hecker, H. 1972. Distribution of *Rickettsia*-like microorganisms in various organs of an *Ornithodorus moubata* laboratory strain (Ixodoidea, Argasidae) as revealed by electron microscopy. *Zeitschrift für Parasitenkunde* 39, 201-209.

Sawaguti, Y. & Ozi, Y. 1937. Kimuragumo ni tsuite (Kishida, 1920). *Acta Arachnologica* 2, 115-123.

Schlueter, U. & Seifert, G. 1985. Rickettsiales-like microorganisms associated with the malpighian tubules of the millipede, *Polyxenus lagurus* (Diplopoda, Penicillata). *Journal of Invertebrate Pathology* 46, 211-214.

Schwendinger, P.J. 1990. On the spider genus *Liphistius* (Araneae: Mesothelae) in Thailand and Burma. *Zoologica Scripta* 19, 331-351.

Suhm, M. 1995. Rickettsien in der Mitteldarmdrüse von *Oedothorax apicatus* (Araneae, Linyphiidae). *Verhandlungen der Deutschen Zoologischen Gesellschaft* 88, 232.

Vago, C., Meynadier, G., Juchault, P., Legrand, J.-J., Amargier, A. & Duthoit, J. 1970. Une maladie rickettsienne chez les Crustacés Isopodes. *Comptes rendues hebdomadaires des Séances de l' Académie des Sciences*, Paris (D) 271, 2061-2063.

European Arachnology 2000 (S. Toft & N. Scharff eds.), pp. 51-56.
© Aarhus University Press, Aarhus, 2002. ISBN 87 7934 001 6
(Proceedings of the 19th European Colloquium of Arachnology, Århus 17-22 July 2000)

The influence of the 1997-1998 El Niño upon the Galápagos lycosid populations, and a possible role in speciation

LÉON BAERT[1] & JEAN-PIERRE MAELFAIT[2]

[1]*Koninklijk Belgisch Instituut voor Natuurwetenschappen, Departement Entomologie, Vautierstraat 29, B-1000 Brussel, Belgium* (leon.baert@natuurwetenschappen.be)
[2]*Laboratorium voor Ecologie, Ledeganckstraat 35, B-9000 Gent, Belgium*

Abstract

El Niño is a worldwide climatological event occurring every 2 to 8.5 years. This event is associated with high sea-surface temperatures across the tropical Pacific and weak or reversing easterly trade winds. The combination of both leads to abnormally strong convective storms in the eastern Pacific and heavy rainfall in western Latin America (especially Ecuador and Peru) and in the Galápagos. The authors had the opportunity to witness the recent 1997-1998 El Niño event during their stay in the Galápagos in March and April 1998. The effect of extremely wet conditions upon the *Hogna albemarlensis* populations of the islands Santa Cruz, Santiago and Volcán Cero Azul (Island Isabela) were observed.

Hogna albemarlensis, a coastal species, normally lives in saline habitats near lagoons and in permanent wetlands below 600m of altitude. It exhibited an extremely aggressive expansion of its distribution all over the islands and volcanoes, occurring everywhere in very high densities and even outnumbering the highland species *Hogna galapagoensis* which presumably evolved from the founder species *H. albemarlensis*). A situation was created in which both the coastal and the highland species, met each other for a certain period of time. Every El Niño event is followed by a number of extremely dry years (called La Niña) resulting in the drying out of all temporary wetlands produced during the El Niño period and once again restricting the lycosid populations to their former areas. El Niños have certainly played an important role in the speciation of the lycosid species on islands where the coastal species and highland species occur. A hypothesis is proposed.

Key words: Galápagos, Lycosidae, speciation, El Niño

INTRODUCTION

Seven different lycosid species belonging to the genus *Hogna* can be distinguished morphologically in the Galápagos archipelago. Their distributions over the islands have been studied since 1982 and are well known (Baert & Maelfait 1997).

The authors witnessed the 1997-1998 El Niño during their stay on the Galápagos from March-April 1998, the second strongest El Niño of the 20th century. One of the species, *Hogna*

albemarlensis (Banks 1902), normally confined to the coast, showed an extremely aggressive expansion of its distribution throughout the three islands visited (Santa Cruz, Santiago and Volcán Cerro Azul of Isla Isabela).

There have been five strong El Niño events on the Galápagos since 1965: 1975-76, 1982-83, 1986-87, 1992-93 and 1997-98 (Snell & Rea 1999). An El Niño event is a worldwide climatological and oceanographic phenomenon resulting in extreme drought in the western part

Table 1. Catches of two *Hogna* spp. from Isla Santa Cruz. Numbers caught: 1 or 2: 1 or 2 specimens, +: 3-10 specimens, ++: 10-100 specimens, +++: > 100 specimens. Collectors: P: S. Peck, R: L. Roque, A: S. Abedrabbo. Blank: not sampled that year, * sampled in November and December, ** sampled in November. Grey columns: Years of El Niños. The localities are ordered according to a South-North transect.

Annual rainfall (mm)		640	2769	63.6	277.6	1254	78.5	85.5	503.3	856.3	747	187	1655	1752		
Year		1982	1983	1985	1986	1987	1988	1989	1991	1992	1993	1996	1997	1998	1998	2000
Collectors				P			P			P, A			R *		**	
	Alt. (m)															
Hogna albemarlensis																
Littoral zone: Coastal lagoons	0-1	++			+			+++				+++	+++		+	++
Southern Dry arid zone	2-100	++		0	0			0	1				0	++		+
Southern Transition forest	190	+													+	+
Southern Culture zone	230-500	1			++								++	++		1
Southern *Miconia* shrub	500				0		+	1					+	1		1
Southern *Scalesia* forest	570	0		0	2		+	0					+			++
Los Gemelos pampa	570	0			0		++						++	++	+	++
Media Luna pampa	600	0		0	0		0	0	0				0	++		1
Northern Transition forest zone	300-560				0		1			+	++		0			
Northern Dry arid zone	0-300				0		0		0	0			0	1		
Hogna galapagoensis																
Southern *Miconia* zone	500-620			++			1	+					+	0		0
Southern *Scalesia* forest	570	1		++	++		+++	+++					+++			
Los Gemelos pampa	570	++			1		++						++	+++	+	
Pampa zone	600-875	++			++		+++	+++	++				+++	++		+++
Northern *Scalesia* forest	650							+++								

of the Pacific Ocean and extreme heavy rainfall along the western coast of the American continent, especially in Peru, Ecuador and on the Galápagos islands.

In this paper we evaluate the possible importance that past periods of strong consecutive El Niño events might have had upon the speciation of the lycosid species of the Galápagos.

MATERIAL AND METHODS

An overview of all the sites where *Hogna albemarlensis* and *Hogna galapagoensis* have been caught or observed by us and other collectors since 1982 on the central island Santa Cruz is given in Table 1. The differentiation in the occurring vegetation zones is after Baert et al. (1991). Most sampling was done during the months of March , April and May. Those made in December and/or November are indicated with an asterisk. The annual rainfall (in mm) given for each year is the rainfall measured at the Charles Darwin Research Station (CDRS), a meteorological station situated on the southern coast of the island. The annual rainfall at Bellavista (at 200 m of altitude) is on average 500 mm higher than on the coast (Snell & Rea 1999). The years characterized as El Niño years are indicated with grey background. A number of sites are sampled by means of pitfall traps each time we visit the islands.

Table 2 and 3 give an overview of the sites were *Hogna* species were caught on Isla Santiago and on the Volcán Cerro Azul (Isla Isabela) respectively. Sampling was with pitfall traps or by hand catches.

RESULTS

Isla Santa Cruz

Distribution of the Hogna species in climatically normal years

In normal, climatically dry years, *Hogna albemarlensis* is found in all coastal salt marsh habitats around lagoons. It can also be encountered around the unique permanent water pool called 'El Chato', situated along the southern flanc of the island in the Transition forest zone at an altitude of 190 m, and also around temporary pools situated at higher altitudes in the small pampa area of Los Gemelos in the Scalesia forest zone (Table 1). Here rainfall is more frequent so that the temporary pools do not

Table 2. Catches of two *Hogna* spp. from Volcán Cerro Azul (Isla Isabela). Ha: *Hogna albemarlensis*, Hg: *Hogna galapagoensis*, H1: *Hogna* species 1. Numbers caught cfr. Table 1. n: not present.

Cerro Azul (Isabela)	Alt. (m)	1986	1991	1998
Caleta Iguana	5		Ha(++)	
Caleta Iguana, open	5		Ha(+)	Ha(++)
Western slope	80		Ha(+)	
Western slope	150		Ha(+)	
Western slope	200		Ha(+++)	
Western slope	300		Ha(++)	
Western slope	400		Ha(++)	
Western slope	450	Ha(++)		
Western slope	530			Ha(++)
Western slope, pampa	620		Ha(+)	
Western slope, pampa	680		H1(++)	H1(++),Ha(+)
Western slope, pampa	700	Ha(++)		
Western slope, pampa	760		H1(+);Ha(+)	
Cerro Gavilan	850		H1(+);Ha(+)	
Western slope, pampa	1000		H1(+);Ha(+);Hg(1)	
Western slope, pampa	1100	H1(1);Ha(+),Hg(+)	H1(1);Hg(+)	Ha(++)
Western slope, ferns	1200	Hg(1)	n	
Western slope, xerophylic	1300	Hg(+)	n	
Top, xerophylic	1530	n	n	Ha(++)

stay dry for too long. It is never encountered above the altitude of 600 m (the temperature inversion zone situated not far below this altitude on Santa Cruz makes upward ballooning of lycosid pulli from below impossible, while the broad forest zones between the arid and the pampa zones are surely barriers) nor along the northern side of the island (except around coastal lagoons) where rainfall is extremely low (the upper borders of the vegetation zones reach higher along the northern side of the island. Arid zone: S ca. 100 m, N ca. 300 m; Transition zone: S ca. 250 m, N ca. 500 m).

H. galapagoensis is a typical highland pampa species encountered at altitudes above 600 m and becoming more abundant with increasing altitude. It is very rarely found in the lower *Miconia* zone above 500 m. Both species occur in mixed populations at Los Gemelos (570 m altitude).

Situation observed during the 1997-1998 El Niño (March-April 1998)
As a result of frequent rainfall there was an abundance of permanent pools all over the island for the duration of the El Niño. The presence of *H. galapagoensis* could only be confirmed above 800m altitude and at the site of Los Gemelos (see further). *H. albemarlensis* was found in high numbers up to the pampa zone near Media Luna (600 m altitude) where *H. galapagoensis* seemed to have disappeared (see Table 1). The same occurred at Los Gemelos where normally both species live in a mixed population. *H. galapagoensis* was found in high numbers in November 1997, in March-April 1998 there was a much higher density of *H. albemarlensis* than of *H. galapagoensis* and in November of the same year no more *H. galapagoensis* were found. This situation was unchanged in March-April 2000, while *H. galapagoensis* reappeared at Media Luna. Table 1 shows further that both species were only found along the northern side of the island (between 300 and 500 m altitude) during an El Niño year or in the year just after.

Isla Isabela – Volcán Cerro Azul
Based on observations in 1986 and 1991 (Table 2), *H. albemarlensis* occurs here from the coast up to an elevation of 1100 m (temperature inversion zone). Another *Hogna* species (H1) lives more or less between the elevations of 650 m and 1100 m. A few specimens of *H. galapagoensis* (the species found at the highest elevation of Santa Cruz) have been found between 1000 and 1300 m altitude.

During the 1997-1998 El Niño (March—

Table 3. Catches of two *Hogna* spp. from Isla Santiago, Isla Rabida and Isla Bartolomé. Ha: *Hogna alber-marlensis*, Hg: *Hogna galapagoensis*, H6: *Hogna* species 6. Numbers caught: cfr table 1. P: collector S. Peck. n: not present..

	Alt. (m)	1986	1991(P)	1992(P)	1998
Isla Santiago					
Playa Espumila	2	Ha (++)			Ha(++)
Southwestern slope	300		Ha(+)		
Southwestern slope	600		Ha(++)		
Aguacate camp	650		Ha(++)	Ha(+++)	Ha(++)
3 km NE of Aguacate	740		Ha(+++)		
Los Jabboncillos	820				Ha(++)
Top of island	900	H6(+);Ha(+);Hg(+)	Ha(++)		
La Central	700	Ha (++)			Ha(++)
Isla Rabida		n	n	n	Ha(++)
Isla Bartolomé			n	n	Ha(++)

April 1998) *H. albemarlensis* had invaded the whole of the followed trajectory (only a few sites (see Table 2) were really sampled) from coastline up to the volcano rim (1530 m alt.) and occurred everywhere in high densities.

DISCUSSION
Facts
H. albemarlensis is clearly an eurytopic species. Although it lives preferentially in coastal salt marsh habitats, it is able to thrive and build high populations in other wet situations in the vicinity of fresh water pools (occurring temporarily during an El Niño year), as long as these pools subsist. As these pools dry out, their population density diminishes strongly and after a short time they disappear completely. This species displays a quick expansion of its distribution during very wet years. This might explain its wide distribution over most of the islands of the archipelago (Baert & Maelfait 1997).

During the 18 years (between 1982 and 2000) the populations of the coastal *Hogna* species (*albemarlensis*) and the highland *Hogna* species (*galapagoensis*) were monitored on the island of Santa Cruz, two 'very strong' El Niño events occurred, the first was in 1982-83 (December 1982 – July 1983), the second in 1997-98 (November 1997 – May 1998). Quin et al. (1987) stated that 'very strong' El Niño events, as the one in 1982-83, occur on average nearly every 50 years, as the previous event was in 1925-26 and before that in 1891. Enfield (1987) stated that the 1891 event most closely

resembled the 1982-83 event and therefore concluded that 'very strong' El Niños occur on average every 100 years.

It is only within the 1997-98 El Niño event (March 1998) that *H.albemarlensis* was recorded in high numbers in the pampa zone above the *Miconia* border at Media Luna (Table 1), while *H. galapagoensis* seemed to have disappeared from this locality. The situation was still the same two years later (March 2000). Both species can sometimes be found in mixed populations at Los Gemelos (alt. 570 m) (Table 1). During the 1997-98 event *H. galapagoensis* was wiped out of this small pampa area by *H. albemarlensis*. There are two facts that indicate that the 1997-98 El Niño was more prolific for *H. albemarlensis*. The difference between both 'very strong' El Niño events is that the 1982-83 event was extremely wet, but only during a short period of time (from December 1982 till July 1983), while the 1997-98 event (November 1997-May 1998) was preceded by an extremely wet 'dry season' (March – June 1997) having the effect of two consecutive El Niños.

The records made on Volcán Cerro Azul (Table 2) show that *H. albemarlensis* can reach much higher altitudes (up to the summit of the volcano situated at 1530 m, while the top of Santa Cruz is at 875 m altitude) and that this species might have reached the summit pampa zone of Santa Cruz. Unfortunately the summit of Santa Cruz could not be reached in 1998 due to the heavy rainfall. Possibly, it could not reach the summit of the island, as most of the pampa zone nowadays is converted into a dense *Cinchona* forest, which might have been a limiting barrier for the upward progress of the species.

That the 1997-98 event was favourable to *H. albemarlensis* is corroborated by the high densities this species attained everywhere it was found, as on Isla Santiago (Table 3) and on smaller drier islands as Isla Rabida and Bartolomé where it was never encountered before.

How did *H. albemarlensis* and *H. galapagoensis* speciate?

Hypothesis of speciation

The founder species must have reached the island of Santa Cruz (it may of course have first reached one of the islands situated in the East of Santa Cruz) by way of rafts of flotsam. It is not known when and how many times colon- ization may have occurred. It colonized the salt marshes along the coast and during wet El Niño years could expand to higher altitudes where permanent fresh water pools were present in rather open habitats. During a period of consecutive 'very strong' El Niños (an interglacial period, see further), it could even bridge the *Scalesia* forest zone and *Miconia* shrub zone, and establish itself in the upper permanent moister pampa zone. A very long period of drought (a glacial period, see further) resulted in the separation of the two populations, one coastal salt marsh population and one highland pampa population. Due to that separation, both populations could evolve during that long period of drought into two different species. By the time a next period of 'very strong' El Niños (interglacial period) occurred, two different species met at the border of their distribution area.

This hypothesis of allopatric speciation based on a climatologically and altitudal barrier is only plausible if the dry period between both periods of very strong El Niños is very long, i.e. long enough to permit two separated populations to evolve into two different species which only meet after substantial speciation had occurred (Maelfait & Baert 1986). There are paleoclimatological evidences that can sustain this hypothesis.

Evidence corroborating this hypothesis

Cox (1983) has estimated that the earliest Galápagos islands probably emerged 3 to 5 million years ago. Around 3.5 to 3.2 million years ago (Keigwin 1978) the Isthmus of Panama closed resulting in a complete reorganization of the oceanic palaeocirculation (Romine 1985) producing the oceanographic and atmospheric conditions necessary for intense El Niño events (Colgan 1990). Since the closure of the Panamanian seaway and the onset of the glacial cycles, the eastern Pacific has faced two different climatic states: one during glacial periods with cool waters and lowered sea-levels and the other during interglacial periods with higher sea-levels, warmer waters and El Niño events (Colgan 1990). Over the last 700,000 years, there have been seven interglacial periods, and at these times sea-level and climate were similar to those occurring at present, and should have experienced El Niños. At the moment the earth is in an interglacial period and sedimentological evidence from Peruvian coastal plains show that in the last 7,500 years a minimum of 15 'very strong' El Niño events have occurred (Wells 1987). From the early 1500s to 1982-83, 8 'very strong' El Niño events have affected the eastern Pacific (Quinn et al. 1987). The 20th century has experienced 3 'very strong' El Niños, of which two occurred within an interval of 18 years and therefore it seems highly plausible that the Galápagos archipelago could be entering one of the wettest periods in its recent history (Snell & Rea 1999).

There is evidence that during glaciation times, i.e. periods of lowered sea-level, El Niño events were absent (Colgan, 1990). The stratigraphy of El Junco Lake on San Cristóbal Island suggests that rainfall was greatly reduced between 34,000 to 10,000 years ago (Colinvaux 1984) . In conclusion, we may say that Galápagos has faced large periods of drought without El Niño events (equal to glacial period) alternating with very wet periods with El Niño events (equal to interglacial period), which supports the possibility of our hypothesis.

ACKNOWLEDGEMENTS

Excellent cooperation and field logistic support were provided by the Charles Darwin Research Station (CDRS, Isla Santa Cruz, Galápagos), the directors F. Koestner, G. Reck, D. Evans, C. Blanton and R. Bensted-Smith, and their staff; the Galápagos National Park Service (SPNG Superintendents M. Cifuentes, Ir.H.Ochoa, F. Cepeda, A. Izurieta and E. Cruz), Department of Forestry, Ministry of Agriculture of Ecuador; TAME airline kindly issued reduced price travel tickets. Our investigations and field work were financially supported by (1) the Belgian DWTC (initially Ministry of Education), (2) the national Fund for Scientific Research (FWO) and (3) the Léopold III Foundation. Help in field sampling was provided by S. Abedrabbo, K. Desender, G. Estevez, M-A. Galarza, L. Roque, S. Sandoval and P. Verdyck.

REFERENCES

Baert, L. & Maelfait, J.-P. 1997. Taxonomy, distribution and ecology of the lycosid spiders occurring on the Santa Cruz island, Galápagos Archipelago, Ecuador. In: *Proceedings of the 16th European Colloquium of Arachnology, Siedlce Poland 1996* (M. Żabka ed.), pp. 1-11. Wyższa Szkoła Rolniczo – Pedagogiczna, Siedlce, Poland.

Baert, L., Desender, K. & Maelfait, J.-P. 1991. Spider communities of Isla Santa Cruz (Galápagos, Ecuador). *Journal of Biogeography* 18, 333-340.

Banks, N. 1902. Papers from the Hopkins Stanford Galapagos expedition, 1898-1899. VII. Entomological results (6). *Proceedings of the Washington Academy of Sciences* 4, 49-86.

Colgan, M.W. 1990. El Niño and the history of Eastern Pacific reef building. In: *Global eco-logical consequences of the 1982-83 El Niño-southern oscillation* (P.W. Glynn ed.), pp. 183-232. Elsevier Oceanography Series, Amsterdam.

Colinvaux, P.A. 1984. The Galápagos climate: present and past. In: *Key environments, Galapagos*. (R. Perry ed.), pp. 55-70. Pergamon Press, Oxford.

Cox, A. 1983. Ages of the Galapagos Islands. In: *Patterns of evolution in Galapagos organisms* (R.I. Bowman, M. Berson & A.E. Leviton eds.), pp. 11-24. Pacific Division, AAAS, San Francisco, California.

Enfield, D.B. 1987. Progress in understanding El Niño. *Endeavour* 11, 197-204.

Keigwin, L.D. 1978. Pliocene closing of the Isthmus of Panama based on biostratigraphic evidence from nearby Pacific Ocean and Caribbean Sea cores. *Geology* 6, 630-634.

Maelfait, J.-P. & Baert, L. 1986. Observations sur les Lycosides des îles Galapagos. *Mémoires de la Société Royale Belge d'Entomologie* 33, 139-142.

Quinn, W.H., Neal, V.T. & Antunez de Mayolo, S.E. 1987. El Niño occurrences over the past four and a half centuries. *Journal of Geophysic Research* 92, 14.449-14.461.

Romine, K., 1985. Radiolarian biogeography and paleoceanography of the North pacific at 8 ma. *Memoranda of the Geological Society of America* 163, 237-272.

Snell, H. & Rea, S. 1999. The 1997-98 El Niño in Galápagos: can 34 years of data estimate 120 years of pattern? *Noticias de Galápagos* 60, 11-20.

Wells, L.E. 1987. An alluvial record of El Niño events from northern coastal Peru. *Journal of Geophysic Research* 92, 14.463-14.470.

European Arachnology 2000 (S. Toft & N. Scharff eds.), pp. 57-64.
© Aarhus University Press, Aarhus, 2002. ISBN 87 7934 001 6
(Proceedings of the 19th European Colloquium of Arachnology, Århus 17-22 July 2000)

Early season natural biological control of insect pests in rice by spiders - and some factors in the management of the cropping system that may affect this control

LENE SIGSGAARD

International Rice Research Institute (IRRI), EPPD, MCPO 3127, 1271 Makati City, Philippines
Present address: *Royal Veterinary and Agricultural University, Dept. of Ecology, the Zoology Section, Thorvaldsensvej 40, DK-1871 Frederiksberg C, Denmark* (les@kvl.dk)

Abstract
There are relatively few insect pest problems in unsprayed, irrigated rice. Spiders are numerous early season predators and are thought to play an important role in the suppression of insect pests such as plant- and leafhoppers. Pest resurgence after insecticide spraying has been clearly linked to the negative impact of insecticides on spiders and other natural enemies. In particular, recent studies show that spiders depend on detritivores for food during fallow periods. The relatively low prey quality of pest species suggests that alternative prey serve as important food supplements. Future changes in the irrigated rice cropping system, such as direct seeding, chemical rather than manual weed control, mechanization and larger field sizes, will have significant impacts on spiders and other beneficials, thus disrupting natural biological control in rice.

Key words: *Nilaparvata lugens, Nephotettix virescens, Pardosa pseudoannulata, Atypena formosana,* spider, rice, natural biological control

INTRODUCTION

Small fields characterize traditional rice farming in Asia. Rice is the dominant staple food in the developing world. More than 90% of the world's rice is produced and consumed in Asia. Water is a prominent factor in the irrigated rice system making it different from other agricultural productions. The biodiversity of irrigated rice is higher than in many natural ecosystems (Schoenly et al. 1998).

Several insects feed on rice. Until the 1960s stemborers were considered the most important rice pests, in particular *Chilo suppressalis, Scirpophaga incertulas* and *S. innotata* (Lepidoptera: Pyralidae). With the widespread introduction during the green revolution in the sixties and seventies of fertilizers, of improved varieties and of pesticides to rice crops, leaf- and planthoppers became important pests, most notably the leafhoppers *Nilaparvata lugens* Stål, the brown planthopper (BPH), and *Sogatella furcifera* (Horvath) (Homoptera: Delphacidae), and the green planthopper (GLH) *Nephotettix virescens* (Distant) (Homoptera: Cicadellidae). Other herbivores in rice that can be insect pests include the rice gall midge (*Orselia oryzae* (Wood-Mason) (Diptera: Cecidomyiidae)), the rice leaffolder complex of which three have attained pest status: *Cnaphalocrocis medinalis* (Guenee), *Marasmia patnalis* Bradley and *M. exigua* (Butler) (Lepidoptera: Pyralidae), rice bugs, in particular the genus *Leptocorisa*

(Hemiptera: Coreidae), pentatomid bugs, rice hispa (*Dicladispa armigera* (Oliver) (Coleoptera: Hispidae), rice thrips (Thysanoptera: Thripidae), rice caseworm (*Nymphula depunctalis* (Guenee) (Lepidoptera: Pyralidae) and whorl maggot flies (several species of the genus *Hydrellia* (Diptera: Ephydridae)) (see for example Pathak & Kahn 1994).

However, unsprayed, irrigated rice fields have relatively few insect pest problems. This is largely attributed to natural biological control, which keep planthoppers, most notably BPH, and other potential pests in check (Kenmore et al. 1984; Way & Heong 1994).

SPIDERS IN RICE

Three guilds of spiders are found in rice: orb-weaving spiders, hunting spiders and space-web spiders. Orb-weavers include the families Araneidae, Tetragnathidae and Theridosomatidae. The most common orb-weaver genera are *Tetragnatha*, *Araneus* and *Argiope*. Lycosids dominate the guild of hunters, while the guild of space-web spiders contain three families Theridiidae, Linyphiidae and Agelenidae (Barrion and Litsinger 1995).

Natural biological control in irrigated rice at the early crop stages can mainly be attributed to spiders. Orb-weaving spiders are the most abundant spiders assessed across the cropping season, with *Tetragnatha* spp. being the single most common genus in South East Asian countries, except the Philippines where *Pardosa pseudoannulata* is the more common species. Heong et al. (1992) found a relative abundance of *P. pseudoannulata* of 25 to 54% of all spiders at five rice sites in the Philippines across the season. Lowest abundance of *A. formosana* was found at the two sites at higher elevations of 800 and 1500 m (7 and 9%), highest at the lower elevations (23, 35 and 40%). Three species of tetragnathids, *Tetragnatha virescens* Okuma, *T. maxillosa* Thorell, and *T. javana* (Thorell) together comprised 10 to 39% of the spiders in these sites. However, orb-weavers usually become abundant when insect damage has already occurred (Barrion and Litsinger 1984).

In the first 35 days after transplanting the dominant predators in irrigated rice are the lycosid *Pardosa pseudoannulata* (Bösenberg & Strand) and the linyphiid *Atypena formosana* (Oi) (Sigsgaard et al. 1999, the Philippines; Sahu et al. 1996, Northern Bihar, India). Both spiders are considered important predators of BPH and *Nephotettix virescens* (Distant) (Hemiptera: Cicadellidae), the green leafhopper (GLH). *P. pseudoannulata* is perhaps the single most important predator of BPH, and can effectively regulate the pest population of leafhoppers and planthoppers (Kiritani et al. 1972; Gavarra & Raros 1975; Kiritani & Kakiya 1975; Kenmore 1980; Kenmore et al. 1984; Shepard et al. 1987; Ooi & Shepard 1994). Both spiders occur throughout the year. *P. pseudoannulata* is most common among the tillers at the base of the plants. It preys on a wide array of insect pests, including leafhoppers and planthoppers, whorl maggot flies, leaffolders, caseworm and stem borers (Barrion & Litsinger 1984; Shepard et al. 1987, Rubia et al. 1990). Field densities of both spiders co-vary with hopper densities (Reddy & Heong 1991). References to the importance of the smaller and less conspicuous *A. formosana* have been few until recently (Shepard et al. 1987; Inthavong et al. 1996; Barrion 1999; Sigsgaard & Villareal 1999). *A. formosana* adults and immatures prefer to live among the rice stem or at the base of rice hills. They have been observed to hunt for nymphs of planthoppers and leafhoppers, Collembola, and small dipterans, such as whorl maggot flies (Barrion and Litsinger 1984, Shepard et al. 1987, Sigsgaard and Villareal 1999a).

Later in the cropping season predatory bugs become the most numerous predators. The most abundant of these are *Microvelia douglasi atrolineata* Bergoth (Veliidae), *Mesovelia vittigera* (Horvath) (Mesoveliidae), and *Cyrtorhinus lividipennis* Reuter (Miridae) (Heong et al. 1991).

PEST MANAGEMENT IN RICE

Until the green revolution BPH was considered a minor pest, but during the seventies it became a major pest in rice. This demonstrated the effects of 'turning off' the biological control of this pest, which is normally controlled at low levels by the many spiders and other natural enemies (Matteson 2000). Kenmore et al. (1984) showed that BPH populations increased drastically when spider and veliid predators were removed. When insecticide use was intensified, insecticide resistant strains of insect pests emerged. Rice varieties resistant to some pests including BPH were developed, but planting over large areas created pests, most notably BPH, which could overcome the plant resistance (Heinrichs & Mochida, 1984). Subsequently new pest management strategies, Integrated Pest Management (IPM) were developed that emphasized host plant resistance, biological control and minimal use of insecticides (Waage 1999).

An increasing amount of research evidence from tropical irrigated rice areas shows that there is little or no crop loss in insecticide untreated fields (Kenmore 1991; Litsinger 1991; Way & Heong 1994). This includes defoliators and stem borers that were recorded as important pests even before the green revolution. This inconsistency may be explained by: a) earlier estimates of yield loss were based more on damage than on actual yield, b) moderate resistance against insect pests in many modern varieties, c) the ability of some modern varieties to compensate for damage, because they produce more tillers (Rubia et al. 1989), and d) better control of insect pests by natural enemies with less use of insecticides (Way & Heong 1994).

Findings that moderately BPH-resistant and BPH-susceptible rices grown by a large number of farmers have had low and stable BPH populations for several years suggest that the pest control strategy in rice should be revised to put higher priority on natural biological control (Heong & Schoenly 1998).

Apart from the fact that insecticide use is rarely necessary, it also poses a risk to farmer health and the environment (Heong et al. 1995). Continued insecticide use stresses the need to bridge the gap between research and farmers. FAO (Food and Agriculture Organization of the United Nations) has supported Farmers' Field Schools in many countries and provided farmers with a practical understanding of integrated pest and nutrient management (Matteson 2000). The expectation is that the farmers who receive training will pass their new knowledge on to other farmers. Another approach was developed by Heong et al. (1998). Here farmers were motivated to 'test' a simple rule of thumb (no spray necessary in the first 40 days after sowing) by the use of communication media, including the radio. The practice of no early spray is now adopted by many farmers in southern Vietnam, and recommended by the National Agricultural Research and Extension Agencies in Malaysia, the Philippines, and Thailand (K.L. Heong, pers. comm.).

SPIDERS AND THEIR ROLE IN THE IRRIGATED RICE AGROECOSYSTEM
Detritivores and organic material
The population build-up of natural enemies is dependent on the availability of suitable host/prey. The abundant detritivores early in the season may be one key to the success of the current rice agroecosystem (Settle et al. 1996). Being polyphagous predators, spiders can prey on alternative prey such as Collembola during fallow periods, hereby maintaining high population levels. (I here use the term alternative prey to describe all suitable prey other than the target species). The levels of these alternative prey in turn depend on decaying organic material available in the field. Field and laboratory data from research at the International Rice Research Institute in the Philippines (IRRI) and elsewhere indicate that spiders survive and build up their populations on alternative prey, such as Collembola and dipterans, before the crop is established and in the first weeks after crop establishment (Guo et al. 1995; Settle et al. 1996).

Settle et al. (1996) were able to increase the number of detritus feeders, such as collembola, and of plankton feeders by adding organic material to the rice field in the treated plots. Most interestingly the number of spiders increased in the same plots. Plankton feeders in that study included mosquito larvae and chironomid midge larvae, of which many species also feed on detritus (Settle et al. 1996). In a study at IRRI, the addition of rice straw bundles in the rice field after harvest increased the number of *A. formosana* and *P. pseudoannulata* as well as plant- and leafhoppers (Shepard et al. 1989). Though the study by Shepard et al. (1989) did not report effects on Collembola density, high Collembola density can be observed in recently cut straw, so probably the beneficial effect was also due to an increase in Collembola. In upland, rice weed residues placed within the rice fields can significantly increase spider densities (Afun et al. 1999). Apart from providing refuges for predators and increasing the density of alternative prey, organic material will also influence plant nutrition, which in turn can influence herbivores feeding on the crop. One can speculate that this in turn could indirectly affect predators.

Dietary value of insect pests and alternative prey

Spiders may not be as polyphagous as earlier thought (Toft 1999). The dietary value of alternative prey would determine its role in maintaining a high population of spiders early in the cropping season. The dietary value of alternative prey in terms of immature survival and development and adult fecundity can be high, as found in a recent study at IRRI (Sigsgaard et al., 2001a). In contrast BPH and GLH are of low quality to *A. formosana*. Similar results were obtained for *P. pseudoannulata* with fecundity as a fitness parameter, but BPH was of intermediate to high quality for this predator (Sigsgaard et al. 2001b). Earlier, Toft (see for example Toft 1995, 1996, 1999), found aphids to be a generally poor quality prey for linyphiid and lycosid spiders. These findings extend this to other Homopterans, like the BPH and GLH. Results suggest that spiders would perform less well in an agroecosystem with little alternative prey.

Intraguild predation has been documented by Heong et al. (1990), in cage experiments with *P. pseudoannulata* preying upon BPH and the mirid bug *C. lividipennis,* and by Fagan et al. (1998), with *P. pseudoannulata* preying upon hoppers and mesoveliid bugs. Predator prey-switching, intraguild predation and cannibalism are thought to help predator survival when prey is scarce (Way & Heong 1994).

Bunds and surrounding habitats

Between the irrigated rice fields there are usually bunds, which may be narrow and low and reconstructed often with low and poor vegetation, or which may be wider and higher and with more permanent vegetation. Some bunds are used for growing vegetables or fruits. The bunds surrounding the rice fields provide refugia for predators during fallow periods as well as during farm operations. Bunds may be particularly important as a source of colonization by ground dispersing predators, such as large *P. pseudoannulata* spiderlings and adults, and may be less important for linyphiids as *A. formosana*, which colonizes the rice field by ballooning. Preliminary results from a study of the directional movement of predators between the rice field and the bund show that *P. pseudoannulata* is an early colonizer of newly established rice, with the highest relative abundance of *P. pseudoannulata* in the bund, stressing the importance of this habitat (Sigsgaard et al. 1999). The same study showed that three or four weeks after transplanting of rice the directional movement changed and the early planted field may have become a source of *P. pseudoannulata* to later planted fields. Even within the soil cracks of the fallow rice field some spiders like *P. pseudoannulata* are commonly found (Arida and Heong 1994). The management of bunds can also affect spiders. Grazing of bunds reduced the density of web-building spiders as well as of two hunting spider families, Lycosidae and Oxyopidae, probably due to loss of webbing sites for the web-building spiders and hunting grounds for the hunting spiders (Barrion 1999). Rice fields are usually intermingled with other crops and habitats such as coconut or banana, and houses, gardens, fallow fields and forests, creating a varied landscape mosaic. Rice is often grown in rotation with vegetables such as onions, or with legumes.

Surrounding habitats may also serve as a source of spiders for the rice field. Barrion (1999) found, that the most abundant species in some non-rice habitats (irrigation canal, set-aside rice field, edge of bund, a common roadside habitat (the grass *Saccharum spontaneum* L.), coconut, banana, and coconut-banana mixed) were *Theridion* sp. (family Theridiidae), *P. pseudoannulata* and *A. formosana*. Two key spider species in rice, *A. formosana* and *P. pseudoannulata* are thus utilizing non-rice habitats. Of these habitats the bunds held the highest densities of *A. formosana* followed by the un-cropped rice field, while *P. pseudoannulata* was almost equally abundant in all habitats except relatively low densities in the roadside habitat and banana plantation.

THE RICE CROPPING SYSTEM IS FACING NEW CHANGES

Resource conserving strategies such as the use of compost and the integration of fish and duck production with irrigated rice, practices that contribute to the control of weeds and insect pests (Zhang 1992), are now being actively promoted in some countries such as Malaysia (Ibrahim 1999). These practices decreased substantially with the introduction of pesticides and other partially or fully incompatible technologies.

Today, the irrigated rice cropping system is facing changes, which may have equally strong effects on the characteristics of the system as the green revolution had, and may in turn affect the natural biological control of insect pests in rice. With the rapid growth of cities there is less available water and labour for rice farming. It is foreseen that production will change towards more direct seeding and less transplanting, and towards other potentially water and labour saving methods, such as mechanization, larger fields and more synchronous

cropping (IRRI 2000). In peninsular Malaysia direct seeding has now become the predominant method of crop establishment (Normiyah & Chang 1997). Larger fields and more synchronous planting may delay colonization by predators, also reducing the benefit gained from the abundant early season alternative prey. A delay in colonization by predators in large monocultures of rice has been shown by Settle et al. (1996) in Indonesia. Continuous flooding has been found to be associated with higher spider numbers (Lam et al. 1997), suggesting a possible lower density of spiders with less available water. The growth of areas under directly-seeded rice, as well as the increasing cost of hand weeding, causes the use of herbicides to rise. Genetically modified rice may also affect the cropping system in ways not yet fully anticipated. For example, we do not know the consequences of creating a field with no lepidopterans, as may be the case in rice genetically modified to contain the *Bacillus thuringiensis* toxin.

Understanding of the biology of insect pests, their natural enemies and the factors in the management of the cropping system, which may affect this control, can be an important tool in maintaining the desirable traits of the current irrigated rice ecosystem, as the rice cropping system changes.

ACKNOWLEDGEMENTS
Thanks to K.L. Heong, K.G. Schoenly, A. Barrion, S. Toft and M. Cohen for valuable discussions. Thanks to K.L. Heong, S. Toft and two anonymous referees for commenting on an earlier draft of this paper.

REFERENCES
Afun, J.V.K., Johnson, D.E., & Russell-Smith, A. 1999. The effects of weed residue management on pests, pest damage, predators and crop yield in upland rice in Cote d'Ivoire. *Biological Agriculture & Horticulture* 17 (1), 47-58.

Arida, G.S. & Heong, K.L. 1994. Sampling spiders during the rice fallow period. *International Rice Research Notes* 19 (1), 20.

Barrion, A. & Litsinger, J. 1984. The spider fauna of Philippine rice agroecosystems. II. Wetland. *Philippine Entomologist* 6, 11-37.

Barrion, A. & Litsinger, J. 1995. *Riceland spiders of South and Southeast Asia*. CAB International/International Rice Research Institute. University Press, Cambridge.

Barrion, A. 1999. Ecology of spiders in selected non-rice habitats and irrigated rice fields in two southern Tagalog provinces in the Philippines. Ph.D. Thesis. University of the Philippines at Los Banos.

Fagan, W.F., Hakim, A.L., Ariawan, H. & Yuliyantiningsih, S. 1998. Interactions between biological control efforts and insecticide applications in tropical rice agroecosystems: the potential role of intraguild predation. *Biological Control* 13 (2), 121-126.

Gavarra M. & Raros R.S., 1975. Studies on the biology of the predatory wolf spider, *Lycosa pseudoannulata* Boesenberg & Strand. *Philippine Entomologist*, 2 (6), 427-444.

Guo, Y.J., Wang, N.Y., Jiang, J.W., Chen, J.W. & Tang, J. 1995. Ecological significance of neutral insects as a nutrient bridge for predators in irrigated rice arthropod communities. *Chinese Journal of Biological Control* 11, 5-9.

Heinrichs, E.A. & Mochida, O. 1984. From secondary to major pest status: the case of the insecticide induced rice brown planthopper, *Nilaparvata lugens*, resurgence. *Protection Ecology* 7, 201-218.

Heong, K., Bleih, S. & Rubia, E. 1990. Prey preference of the wolf spider, *Pardosa pseudoannulata* (Boesenberg et Strand). *Researches on Population Ecology* 32, 179-186.

Heong, K.L., Aquino, G.B. & Barrion, A.T. 1991. Arthropod Community Structures of Rice Ecosystems in the Philippines. *Bulletin of Entomological Research* 81 (4), 407-416.

Heong, K., Aquino, G. & Barrion, A. 1992. Population dynamics of plant- and leafhoppers and their natural enemies in rice ecosystems in the Philippines. *Crop Protection* 4, 371-379.

Heong, K.L., Escalada, M.M. & Lazaro, A.A. 1995. Misuse of pesticides among rice farmers in Leyte, Philippines. In: *Impact of pesticides and farmer health and the rice environment* (P.L. Pingali & P.A. Roger eds.), pp. 97-108. IRRI, Los Banos, Laguna.

Heong, K.L., Escalada, M.M., Huanh, N.H. & Mai, V. 1998. Use of communication media in changing rice farmers' pest management in the Mekong Delta, Vietnam. *Crop Protection* 17 (5), 413-425.

Heong, K.L. & Schoenly, K.G. 1998. Impact of insecticides on herbivore-natural enemy communities in tropical rice ecosystems. In: *Ecotoxicology: pesticides and beneficial organisms*, (P.T. Haskell & P.McEwen eds.), pp. 381-403. Kluwer, Dordrecht, Netherlands.

Ibrahim, R. 1999. Malaysia. *Far East Agriculture* May/June 1999, 16-17.

Inthavong, S., Inthavong, K., Sengsaulivong, V., Schiller, J.M., Rapusas, H.R., Barrion, A. T. & Heong, K.L. 1996. Arthropod biodiversity in Lao irrigated rice ecosystem. In: *Proceedings of the Rice Integrated Pest Management (IPM) Conference: Integrating science and people in Rice Pest Management*, 18-21 November 1996, Kuala Lumpur, Malaysia (A. A. Hamid et al., eds.), pp. 69-85. Malaysian Agricultural Research and Development Institute, Kuala Lumpur, Malaysia.

IRRI, 2000. *IRRI Annual Report 1999-2000. The Rewards of Rice Research*. International Rice Research Institute, Los Banos, Philippines.

Kenmore, P.E. 1980. Ecology and outbreaks of a tropical insect pest of the green revolution, the rice brown planthopper, *Nilaparvata lugens* (Stål), PhD dissertation. University of California, Berkeley, USA.

Kenmore, P.E., Carino, F., Perez, C., Dyck, V. & Gutierrez, A. 1984. Population regulation of the rice brown planthopper (*Nilaparvata lugens* Stal) within rice fields in the Philippines. *Journal of Plant Protection in the Tropics* 1, 1-37.

Kenmore, P.E., 1991. *Indonesia's integrated pest management – a model for Asia*. Food and Agriculture Organisation, Manila, Philippines.

Kiritani, K. & Kakiya N. (1975) An analysis of

the predator-prey system in the paddy field. *Research in Population Ecology* 17 (1), 29-38.

Kiritani, K., Kawahara, S., Sasaba, T. & Nakasuji, F. (1972) Quantitative evaluation of predation by spiders on the green leafhopper *Nephotettix cincticeps* Uhler, by sight-count method. *Research in Population Ecology* 13, 187-200.

Lam, P.V., Huong, T.H. & Lan, T.T. 1997. A study on spider fauna of rice fields. *Nong Nghiep Cong Ngiep Thuc Pham* 3, 107-109.

Litsinger, J.A. 1991. Crop loss assessment in rice. In: *Rice insects: management strategies* (E. A. Heinrichs & T.A. Miller eds.), pp. 1-65. Springer Verlag, New York.

Matteson, P.C. 2000. Insect pest management in tropical Asian irrigated rice. *Annual Review of Entomology* 45, 549–574.

Normiyah, R. & Chang, P.M. 1997. Pest management practices of rice farmers in the Muda and Kemubu irrigation schemes in peninsular Malaysia. In: *Pest management of rice farmers in Asia*, (K.L. Heong & M.M. Escalada eds.), pp. 115-127. IRRI, Los Banos, Philippines.

Ooi, P.A.C. & Shepard, B.M. 1994. Predators and parasitoids of rice insect pests. In: *Biology and management of rice insects*, (E.A. Heinrichs ed.), pp. 585-612. Wiley Eastern Limited [For] IRRI, New Delhi.

Pathak, M.D. & Kahn, Z.R. 1994. *Insect pests of rice.* International Rice Research Institute, Manila, Philippines.

Reddy, P.S. & Heong, K.L. 1991. Co-variation between insects in a ricefield and important spider species. *International Rice Research Notes* 16 (5), 24.

Rubia, E.G., Shepard, B.M., Yamba, E.B., Ingram, K.T., Arida, G.S., Penning de Vries, F. 1989. Stem borer damage and grain yield of flooded rice. *Journal of Plant Protection in the Tropics* 6, 205-211.

Rubia, E., Almazan, L., & Heong, K. 1990. Predation of yellow stem borer (YSB) by wolf spider. *International Rice Research Newsletter* 15, 22.

Sahu, S., Singh, R. & Kumar, P. 1996. Host preference and feeding potential of spiders predaceous in insect pests of rice. *Journal of Entomological Research* 20 (2), 145-150.

Schoenly, K.G., Justo, H.D., Barrion, A.T., Harris, M. & Bottrell, D.G. 1998. Analysis of invertebrate biodiversity in a Philippine farmer's irrigated rice field. *Environmental Entomology* 27 (5): 1125-1136.

Settle, W.H., Ariawan, H., Astuti, E., Cahyana, W., Hakim, A.L., Hindayana, D., Lestari, A. S. & Pajarningsih 1996. Managing tropical rice pests through conservation of generalist natural enemies and alternative prey. *Ecology* 77 (7), 1975-1988.

Shepard, B.M., Barrion, A.T., & Litsinger, J.A. 1987. *Friends of the rice farmer. Helpful insects, spiders and pathogens.* International Rice Research Institute, Manila, Philippines.

Shepard, B.M., Rapusas, H.R., & Estano, D.B. 1989. Using rice straw bundles to conserve beneficial arthropod communities in the ricefield. *International Rice Research Notes* 14, 30-31.

Sigsgaard, L. & Villareal, S. 1999. Predation rates of *Atypena formosana* (Araneae: Linyphiidae) on brown planthopper, and green leafhopper. *International Rice Research Notes* 24 (3), 18.

Sigsgaard, L., Villareal, S., Gapud, V., & Rajotte, E. 1999. Directional movement of predators between the irrigated rice field and its surroundings. In *Biological control in the Tropics. Towards efficient biodiversity and bioresource management for effective biological control. Proceedings of the Symposium on Biological Control in the Tropics* (L.W. Hong, S.S. Sastroutomo, I.G. Caunter, J. Ali, L.K. Yeang, S. Vijaysegaran & Y.H. Sen eds.), pp. 43-47. The National Council for Biological Control (NCBC)/CAB International, South East Asian Regional Centre, Malaysia.

Sigsgaard, L., Toft, S. & Villareal, S. 2001a. Diet-dependent survival, development and fecundity of the spider *Atypena formosana* (Oi) (Araneae: Linyphiidae) – Implications for biological control in rice. *Biocontrol Science and Technology* 11, 233-244.

Sigsgaard, L., Toft, S. & Villareal, S. 2001b. Diet-dependent fecundity of the spiders *Atypena formosana* and *Pardosa pseudoannulata*, predators in irrigated rice. *Agricultural and Forest Entomology* 3, 285-295.

Toft, S. 1995. Value of the aphid *Rhopalosiphum padi* as food for cereal spiders. *Journal of Applied Ecology* 32, 552-560.

Toft, S. 1996. Indicators of prey quality for arthropod predators. In: *Arthropod natural enemies in arable land. II. Survival, reproduction and enhancement* (C.J.H. Booij & L.J.M.F. den Nijs eds.), pp. 107-116. Aarhus University Press, Aarhus.

Toft, S. 1999. Prey choice and spider fitness. *Journal of Arachnology* 27, 301-307.

Waage, J. 1999. Beyond the realm of conventional biological control: harnessing bioresources and developing biologically-based technologies for sustainable pest management. In: *Biological control in the Tropics. Towards efficient biodiversity and bioresource management for effective biological control. Proceedings of the Symposium on Biological Control in the Tropics* (L.W. Hong, S.S. Sastroutomo, I.G. Caunter, J. Ali, L.K. Yeang, S. Vijaysegaran & Y.H. Sen eds.), pp. 5-17. The National Council for Biological Control (NCBC)/CAB International, South East Asian Regional Centre, Malaysia.

Way, M.J. & Heong, K.L. 1994. The role of biodiversity in the dynamics and management of insect pests of tropical irrigated rice: A review. *Bulletin of Entomological Research* 84, 567-587.

Zhang, Z.Q. 1992. The Use of Beneficial Birds for Biological Pest Control in China. *Biocontrol News and Information* 13 (1): 11-16.

European Arachnology 2000 (S. Toft & N. Scharff eds.), pp. 65-70.
© Aarhus University Press, Aarhus, 2002. ISBN 87 7934 001 6
(Proceedings of the 19th European Colloquium of Arachnology, Århus 17-22 July 2000)

Responses of a detoxification enzyme to diet quality in the wolf spider, *Pardosa prativaga*

SØREN ACHIM NIELSEN[1] & SØREN TOFT[2]

[1]*Department of Life Sciences and Chemistry, Roskilde University, P.O. Box 260, DK-4000 Roskilde, Denmark* (san@ruc.dk)

[2]*Department of Zoology, University of Aarhus, Bldg. 135, DK-8000 Århus C, Denmark* (soeren.toft@biology.au.dk)

Abstract

A previous study revealed a significantly increased respiration rate in wolf spiders, *Pardosa prativaga*, fed diets of low-quality prey through 2-4 weeks compared to spiders fed high-quality prey. We tested here the hypothesis that a higher metabolic rate was due to activity of detoxification enzymes, induced by presumed toxins in the low-quality prey. Two aspects of the activity of a detoxification enzyme (Glutathione S-transferase (GST) and its peroxidase activity (GSTpx)) were measured on the same individuals as in the previous study. The activity of both enzymes was significantly affected by the diet treatments, but in different ways. GSTpx activity was more or less reduced by all low-quality diets compared to high-quality diets and therefore showed a near-significant relationship with respiration rate and daily weight change over the experimental period. GST activity was reduced only by two aphid species and showed no correlation with the above-mentioned parameters. There was no relationship between the two enzyme activities. Since all significant responses to low-quality prey were inhibitive, the results did not confirm our hypothesis.

Key words: Araneae, biomarker enzymes, food, generalist predator, prey quality

INTRODUCTION

Insects vary in their qualitiy as food for generalist predators, including both high-quality, low-quality as well as toxic prey (Toft 1995; Toft & Wise 1999a,b; Marcussen et al. 1999; Bilde & Toft 2001). Quality statements are based here on feeding experiments determining the fitness consequences of keeping the spiders on diets of single prey species or specified mixed diets. Though in most cases the chemical identity is unknown, deterrents or toxins in the prey are most likely to be responsible for low quality. This may be true both when prey consumption is low or practically zero (in which case the prey is highly deterrent – a predigestive effect) and when consumption is substantial but food utilization is poor (in which case the prey is most likely to contain toxin(s) with postdigestive effects). However, metabolic effects on the predator must be very different in the two situations. Toxin containing prey should induce a variety of physiological responses depending on the chemistry of the toxin, whereas highly deterrent prey that are hardly eaten at all may have no other physiological effects than those of starvation. Most low-quality prey types are probably intermediate between these extremes, i.e. they are both deterrent and toxic to varying degrees, so starvation effects and responses to toxins will often occur simultaneously. One effect may therefore mask the other.

Turnbull (1960) recorded 153 prey species eaten by *Linyphia triangularis* (Clerck) in the field. This exemplifies the high diversity of prey eaten by generalist predators such as spiders. In view of this, spiders should be well equipped with detoxifying enzymes to deal properly with the diversity of defensive compounds that many of these insects may contain. We have earlier reported on activities of Glutathione S-transferase (GST), and two peroxidases (GSTpx and Glutathione peroxidase (GPOX)) in *L. triangularis* and two *Pardosa* spp. (Nielsen et al. 1997, 1999; Nielsen & Toft 1999) and some of the factors influencing their magnitude, incl. feeding level and insecticide treatment. In another study (Toft & Nielsen 1997) we reported on the effects of various single- and mixed-species diets of different quality on weight change and respiratory rate in the wolf spider *Pardosa prativaga* (L. Koch), indicating an increased respiration rate when the spiders were fed low-quality prey. We hypothesized that the increased metabolism might be due to the handling of toxins by detoxification enzymes induced by the low-quality food. The individuals of this previous experiment were stored in a deep-freezer for later measurements of detoxification enzyme activity, the results of which are presented here.

GST is a ubiquitous family of isozymes that are able to handle natural toxins and xenobiotics via conjugation with glutathione (GSH), forming derivatives that are easily excreted (e.g. Timbrell 1991). GSTpx, which is the peroxidase activity of GST (Ahmad 1995), serves to eliminate peroxidised derivatives and superoxide anions formed during metabolic transformation of toxins or food (Ahmad 1992, 1995).

MATERIALS AND METHODS

Since all methods have previously been published we restrict ourselves here to presenting the basic details for understanding the new results, and refer to Toft & Nielsen (1997) for details about respiration measurements, and to Nielsen et al. (1998, 1999) regarding enzymatic measurements and determination of protein content. For the enzyme assays the following substrates were used: 1-chloro-2,4-dinitrobenzene (CDNB) for GST; and t-butyl-hydroxyperoxide (TBH) for GSTpx.

Procedures

Pardosa prativaga were collected as juveniles in spring and kept for 2-4 weeks on 9 prescribed diet treatments representing the full range of qualities from high-quality to toxic, and using starved animals as controls (Table 1). Low-quality prey types gave negative or only slightly positive growth rates (DWC in Table 1), not significantly different from that of the starvation group, while high-quality prey sup-

Table 1. Diet treatments to which *Pardosa prativaga* was subjected prior to measurements of detoxification enzyme activity. Quality statements are based on the mean daily specific weight change (DWC: Δmg/mg/day) during the experimental period (see Toft & Nielsen 1997). N: number of spiders in each treatment group.

Diets	DWC	Quality	Culture medium/host plant	N
Aphid *Rhopalosiphum padi* (L.)	-2.34	Low	Wheat seedlings (mixed cultivars)	6
Aphid *Sitobion avenae* (F.)	-0.72	Low	Wheat seedlings (mixed cultivars)	13
Aphid *Metopolophium dirhodum* (Walker)	3.55	Low	Wheat seedlings (mixed cultivars)	15
A mixture of aphids *R. padi, S, avenae, M. dirhodum*	1.18	Low	Wheat seedlings (mixed cultivars)	15
Aphid *Aphis nerii* (B. de F.)	-14.02	Low	*Asclepias curassavica*	8
Collembola *Folsomia candida* (Willem)	-6.50	Low	Baker's yeast	12
Collembola *Isotoma anglicana* Lubbock	22.72	High	From the field (fed Baker's yeast)	22
Fruit fly *Drosophila melanogaster* (Meigen) wild type	30.93	High	Carolina fruit fly medium	17
Mixture of *I. anglicana* and *D. melanogaster*	30.71	High	(See above)	17
Starvation (control)	-7.60	-	-	15

ported high growth rates (Toft & Nielsen 1997). From other studies we have evidence that *F. candida* is toxic but only little deterrent to wolf spiders (Toft & Wise 1999ab; Mayntz & Toft 2000), while the aphids are more deterrent and have no overall toxic effects, possibly because deterrency keeps the spiders' consumption rates low.

All prey types were raised in laboratory cultures, except *Isotoma anglicana* which was collected in the field. Prey was offered ad libitum (except *Sitobion avenae* due to problems with the culture). In the mixed diets each species was offered in approximately equal numbers, and in amounts so that all species were constantly available. The spiders were weighed at the start of the feeding period, and again when respiration measurements were made and the animals were sacrificed for enzymatic measurements. Percent daily weight change was used as an overall measure of the quality of the diets.

Statistical analyses

Variance homogeneity with respect to the ten diet treatments was obtained or approximated by squareroot-transformation for both series of specific enzyme activities (nmol/min/mg protein) (Levene's test; GST: P = 0.069, GSTpx: P = 0.026), and the data were analysed with parametric ANOVA. Because of the variation in food quality, the spiders of the various treatments grew to different final sizes. Size-adjusted activities were therefore obtained by entering an indicator of final spider size as covariate and presenting the results as least squares means. Final body weight and total protein content were strongly correlated (regression analysis, P < 0.001), and the same conclusions were reached whether one or the other served as the size-indicator. The results presented (Fig. 1) used protein content (mg protein/spider) as covariate. Pairwise post-hoc comparisons of treatments were made using the Tukey HSD test.

Regression analysis was used in search for relationships between enzyme activity, respiration

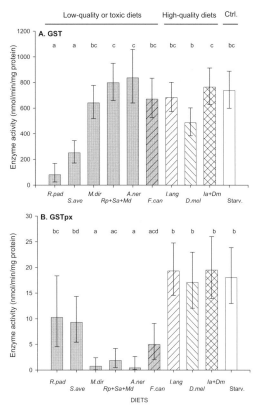

Fig. 1. Least squares means ± 95% confidence limits of detoxification enzyme activities (**A**: GST; **B**: GSTpx) of *Pardosa prativaga* under different diet treatments, adjusted for size (= protein content). Values were obtained by back-transforming means and lower/upper confidence limits of square-root transformed data, obtained by ANCOVA. Grey columns: low-quality prey; white column: high-quality prey and starvation. Left-hatched: Collembola; right-hatched: Diptera; cross-hatched: mixed Collembola/Diptera; unhatched-grey: aphids; unhatched-white: starvation. Letters above bars: bars indicated by the same letter are not significantly different (Tukey HSD).

Treatments: R.pad: *Rhopalosiphum padi*; S.ave: *Sitobion avenae*; M.dir: *Metopolophium dirhodum*; Rp+Sa+Md: mixed diet of *R. padi*, *S. avenae* and *M. dirhodum*; A.ner: *Aphis nerii*; F.can: *Folsomia candida*; I.ang: *Isotoma anglicana*; D.mel: *Drosophila melanogaster*; Ia+Dm: mixed diet of *I. anglicana* and *D. melanogaster*; Starv.: starvation.

rate, and diet quality (% weight change), using treatment-group means as data points (N = 10).

RESULTS

GST

Spiders fed *R. padi* and *S. avenae* had significantly lower GST activity than spiders of all other diet treatments (Fig. 1A). Though the statistical analysis indicates significant variation among some of these latter treatments, the actual differences are small and not easily interpretable. They may be fortuitous.

Mean specific GST activity correlated with no other measured parameter, i.e. diet quality (P = 0.66) or respiration rate (P = 0.64). Also, there was no relationship between GST activity and GSTpx activity (P = 0.58).

GSTpx

The responses of GSTpx were very different from that of GST. Four low-quality treatments (*M. dirhodum*, 3-aphid mix, *A. nerii*, *F. candida*) showed reduced activity of this enzyme. *R. padi* and *S. avenae* treatments did not inhibit this enzyme significantly compared to the high-quality diets, but this may be due to the high variances of the measurements.

However, all low-quality diets had lower activity than all high-quality diets and starvation (to the right in Fig. 1B). As a result of this, specific GSTpx activity correlates positively with diet quality (P = 0.050) and negatively with specific respiration rate (P = 0.054).

DISCUSSION

GST showed no correlation with the overall measure of food quality (% weight change) over the range of diets investigated. The GST activity may depend not only on the presence of certain toxins in the food, but also on the general feeding condition of the spider. Thus, we have previously demonstrated an increased activity with degree of starvation (Nielsen & Toft 1998). This effect is not apparent in the present results. Anyway it is probably safe to conclude that the very low GST activity in the *R. padi* and *S. avenae* diet groups is not due to the starvation effect of these two low-quality prey, because the starvation group had high GST activity. More likely it is due to the spe-

cific chemical components of the two aphid species. Specificity of the effects on GST is underscored by the fact that other low-quality, whether deterrent or toxic, prey did not influence GST activity. GSTpx showed a more consistent reduction of activity with low-quality prey, though this was not significant in all cases. The results are surprising because it was not the same low-quality diets that inhibited the two enzyme activities the most. Also the GST and GSTpx activities did not correlate with each other. Both of these results may be due to their different substrate specificities, in spite of being mediated by the same enzyme. GSTpx activity was inhibited by three aphid treatments (*M. dirhodum,*mixed cereal aphids and *A. nerii*) and by the toxic Collembola *F. candida*, but only slightly (and non-significantly) by *R. padi* and *S. avenae*. Thus, the effects were not related to whether the low-quality prey were primarily deterrent or toxic.

The mixed-aphid diet contained prey species that as single-species diets were inhibitive to GST and others that were inhibitive to GSTpx. However, the mixed-aphid diet itself inhibited GSTpx but not GST. Both enzyme activities responded as the *M. dirhodum* single-species diet. Probably the spiders consumed more of *D. dirhodum* than of the two other species of the mixed diet, because of the three *M. dirhodum* is of the highest food quality for *P. prativaga* (Toft 2000). The spiders eat more *D. dirhodum* than *R. padi* or *S. avenae* before a prey aversion is developed (Toft 1997).

The similar effects of *M. dirhodum*, *A. nerii* and *F. candida* on the one hand, and *R. padi* and *S. avenae* on the other, are not easily reconcilable with other ecophysiological results. *F. candida* is toxic to wolf spiders (Toft & Wise 1999 a, b) and *S. avenae* was toxic to the linyphiid *Erigone atra* (Bl.) for some fitness parameters (Bilde & Toft 2001). In most studies, the aphids have turned out as low-quality but not toxic, e.g. *A. nerii* (Toft & Wise 1999a) and the three cereal aphids (Toft 1995, 2000). Mayntz & Toft (2000) found *R.padi* to be highly deterrent, whereas *F. candida* was less deterrent. When presented together with high-quality insect prey in mixed diets, *R. padi* has even given positive fitness effects both in spiders (Toft 1995; Bilde & Toft 2001) and in partridge chicks (Borg & Toft 2000).

Also, the enzyme responses do not show any consistent pattern in relation to the systematic classification of the aphids. All are members of the Aphididae, but *R. padi* and *A. nerii* both belong to the tribe Aphidini, whereas *Sitobion* and *Metopolophium* belong to the tribe Macrosiphini (Heie 1986,1994).

Earlier we found that GST (but not GSTpx) correlated significantly with feeding level, with activity increasing with degree of starvation. Spiders on the low-quality diets grew little or not at all, reflecting their low consumption rates of these prey (Toft 1995, 1997; Bilde & Toft 1997; Toft & Wise 1999b). In the present study, GST activity of starved spiders did not differ significantly from groups fed high-quality prey ad libitum (i.e. kept at satiation level) and also did not differ from several low-quality treatments (which are likely to cause starvation in the spiders due to low consumption rates). The same can be said here about GSTpx, except that it was different types of low-quality treatments that resembled starvation in the two cases. One interpretation that may explain these results is that the enzyme responses are determined by specific chemical compounds of the different prey types and are unrelated to prey quality as such.

Toft & Nielsen (1997) hypothesized that the increased respiration rate found in the low-quality diet treatments might be due to increased metabolic costs of detoxifying supposed toxins in the low-quality prey. If that were true we would have expected to find increased activities of detoxification enzymes in the low-quality diet treatments due to induction (Terriere 1984), but all significant differences were in the opposite direction: the measured enzyme activities were either inhibited or unaffected by low-quality prey. It is possible, however, that other enzyme systems, not measured here, were induced by the low-quality prey.

While the effects of diets on detoxification enzyme activity may provide indirect confirmation that secondary substances are involved in determining the food quality of insects to spiders, they give no direct clue as to what compounds are involved. A full understanding of the effects of prey on generalist predators probably requires both direct chemical identification of prey toxins and analysis of a wider range of detoxifying enzyme systems.

ACKNOWLEDGEMENTS

We are indebted to Gitte Wikstrøm for doing the enzyme measurements, to Karl-Martin Vagn Jensen for useful discussions, and to David Mayntz and Trine Bilde for critical comments on the manuscript.

REFERENCES

Ahmad, S. 1992. Biochemical defence of pro-oxidant plant allelochemical by herbivorous insects. *Biochemical Systematics and Ecology* 20, 269-296.

Ahmad, S. 1995. Antioxidant mechanisms of enzymes and proteins. In: *Oxidative Stress and Antioxidant Defenses in Biology* (S. Ahmad ed.), pp. 238-272. Chapman & Hall, New York.

Bilde, T. & Toft, S. 1997. Consumption by carabid beetles of three cereal aphid species relative to other prey types. *Entomophaga* 42, 21-32.

Bilde, T. & Toft, S. 2001. Value of three cereal aphid species as food for a generalist predator. *Physiological Entomology* 26, 58-68.

Borg, C. & Toft, S. 2000. Importance of insect prey quality for grey partridge chicks (*Perdix perdix*): a self-selection experiment. *Journal of Applied Ecology* 37, 557-563

Heie, O.E. 1986. *The Aphidoidea (Hemiptera) of Fennoscandia and Denmark. III.* Fauna Entomologica Scandinavica 17. E.J. Brill/ Scandinavian Science Press, Leiden.

Heie, O.E.1994. *The Aphidoidea (Hemiptera) of Fennoscandia and Denmark. V.* Fauna Entomologica Scandinavica 28. E.J. Brill, Leiden.

Marcussen, B.M., Axelsen, J.A. & Toft, S. 1999

The value of two collembola species as food for a cereal spider. *Entomologia Experimentalis et Applicata* 92, 29-36.

Mayntz, D. & Toft, S. 2000. Effects of nutrient balance on tolerance to low quality prey in a wolf spider. *Ekológia (Bratislava)* 19 Suppl. 3, 153-158.

Nielsen, S.A., Clausen, J. & Toft, S. 1997. Detoxification strategies of two types of spiders revealed by cypermethrin application. *Alternatives to Laboratory Animals* 25, 255-261.

Nielsen, S.A. & Toft, S. 1998. Responses of Glutathione S-transferase and Glutathione Peroxidases to feeding rate of a wolf spider *Pardosa prativaga*. *Alternatives to Laboratory Animals* 26, 399-403.

Nielsen, S.A., Toft, S. & Clausen, J. 1999. Cypermethrin effects on detoxification enzymes in active and hibernating wolf spiders (*Pardosa amentata*). *Ecological Applications* 9, 463-468.

Terriere, L.C. 1984. Induction of detoxication enzymes in insects. *Annual Review of Entomology* 29, 71-88.

Timbrell, J.A. 1991. *Principles of Biochemical Toxicology*. 2nd ed. Taylor & Francis, London.

Toft, S. 1995. Value of the aphid *Rhopalosiphum padi* as food for cereal spiders. *Journal of Applied Ecology* 32, 552-560.

Toft, S. 1997. Acquired food aversion of a wolf spider to three cereal aphids: intra- and interspecific effects. *Entomophaga* 42, 63-69.

Toft, S. 2000. Species and age effects in the value of cereal aphids as food for a spider. *Ekológia (Bratislava)* 19 Suppl. 3, 273-278.

Toft, S. & Nielsen, S.A. 1997. Influence of diet quality on the respiratory metabolism of a wolf spider *Pardosa prativaga*. In: *Proceedings of the 16th European Colloquium of Arachnology, Siedlce* (M. Zabka ed.), pp. 301-307. Wyższa Szkoła Rolniczo - Pedagogiczna, Siedlce.

Toft, S. & Wise, D.H. 1999a. Growth, development and survival of a generalist predator fed single- and mixed-species diets of different quality. *Oecologia* 119, 191-197.

Toft, S. & Wise, D.H. 1999b. Behavioral and ecophysiological responses of a generalist predator fed single- and mixed-species diets of different quality. *Oecologia* 119, 198-207.

Turnbull, A. 1960. The prey of the spider *Linyphia triangularis* (Clerck; Araneae, Linyphiidae). *Canadian Journal of Zoology* 38, 859-873.

European Arachnology 2000 (S. Toft & N. Scharff eds.), pp. 71-85.
© Aarhus University Press, Aarhus, 2002. ISBN 87 7934 001 6
(Proceedings of the 19th European Colloquium of Arachnology, Århus 17-22 July 2000)

Reproduction in scorpions, with special reference to parthenogenesis

WILSON R. LOURENÇO

Laboratoire de Zoologie (Arthropodes), Muséum National d'Histoire Naturelle, 61 rue de Buffon, F-75005 Paris, France (arachne@mnhn.fr)

Abstract

A synthesis of all aspects of scorpion reproduction is provided. After a historical introduction, we discuss courtship and mating, embryonic development, birth process, post-embryonic development, life span and several aspects of life history strategies. A particular section is dedicated to the phenomenon of parthenogenesis in scorpions, using selected examples.

Key words: Scorpiones, life history, reproduction, parthenogenesis

HISTORICAL INTRODUCTION

Scorpions are unusual among terrestrial arthropods in several traits of their life-history: ritualized and complex courtship with fertilization by means of a spermatophore; viviparous embryonic development, which can last from several months to almost two years; maternal care, sometimes followed by a degree of social behaviour; and post-embryonic development times that may be extraordinarily long, lasting from 7 to 85 months.

Because of these unusual traits in their life-history, many aspects of the reproductive biology of scorpions were poorly understood by early authors, such as the classical 'promenade à deux' described by Maccary (1810) and Fabre (1907). In the mid-1950s, several researchers discovered, apparently independently, that sperm transfer is accomplished by means of a spermatophore. The first of these were Angermann (1955) and Alexander (1956). Detailed studies of scorpion embryology were carried out by Laurie at the end of the 19th century (1890, 1891, 1896a,b) and were followed by the publications of Pavlovsky (1924, 1925) and

Pflugfelder (1930). After these contributions, little attention was paid to embryology and only a few isolated publications have provided additional information (e.g. Mathew 1956, 1960; Anderson 1973; Yoshikura 1975; Francke 1982; Lourenço et al. 1986a,b; Kovoor et al. 1987).

The first paper to be published on the post-embryonic development of scorpions was by Schultze (1927). Beginning in the mid-1950s, several accounts of various aspects of the reproductive biology, in some cases of the entire post-embryonic development of scorpions have been published. These were mainly by biologists such as Alexander (1956, 1957, 1959), Auber (1959, 1963), Matthiesen (1962, 1969), Maury (1968, 1969), Shulov & Amitai (1958), Shulov, Rosin & Amitai (1960), Varela (1961) and Williams (1969). These citations are certainly not exhaustive; a complete list of references can be found in Polis & Sissom (1990) and Lourenço (1991a).

The mid-1970s saw a renewal of interest in the reproductive biology of scorpions and particularly in their post-embryonic development. Research on this subject was multiplied during

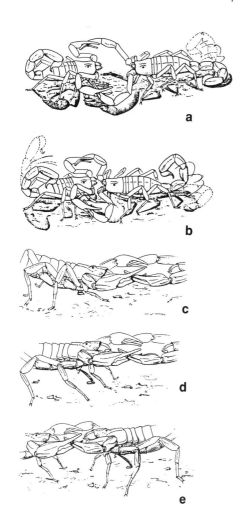

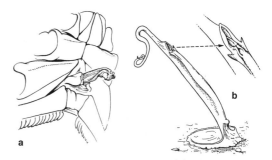

Fig. 2. (a) Detail of male spermatophore extrusion. **(b)** Spermatophore deposited on the substratum (*Tityus fasciolatus* - Buthidae).

methods used and the quality of the observations made. In many cases, the information reported may be speculatory or even fallacious. I will not, however, discuss these aspects here.

Attempts to explain present patterns of phylogeny, biogeography, ecology and, consequently, of the biological diversification of scorpions, cannot be achieved without precise knowledge of the reproductive biology of these animals. For this reason, in the first section of the present paper I will present an account of the different phases involved in the reproduction of scorpions - from mating to the end of post-embryonic development. For a more extensive review of reproduction in scorpions, the reader may refer to Polis and Sissom (1990), where all aspects are presented in full detail.

Fig. 1. Mating behaviour in the buthid scorpion, *Tityus fasciolatus* (male left, female rigth). **(a-b)** 'Promenade à deux'. **(c-d)** Deposition of the spermatophore on the substratum. **(e)** Sperm transfer.

the 1980s and continued throughout the 1990s. Interestingly, most of the authors of this work were primarily taxonomists who, in addition to obtaining biological information, were investigating the ontogenetic variability of the characters used in taxonomy (see the various publications of Armas, Francke, Lourenço and Sissom). Only Polis & Farley (1979, 1980) have attempted to explain reproductive traits in the context of evolutionary ecology.

With regard to known biological data, a great disparity clearly exists concerning the

COURTSHIP AND MATING

Courtship and sperm transfer in scorpions is a very complex process involving several aspects of behaviour. This part of the process of reproduction can be summarised by the classical 'promenade à deux' (Fig. 1) in which the male first approaches the female and then grasps her pedipalp chelae with his own chelae fingers. A form of dance then takes place. This process may last for several minutes until ejection of the spermatophore is prepared and a suitable substratum on which to deposit the spermatophore (Fig. 2) has been found. In the next stage the male leads the female to position her genital aperture over the spermatophore, and the female takes up the sperm. Once sperm trans-

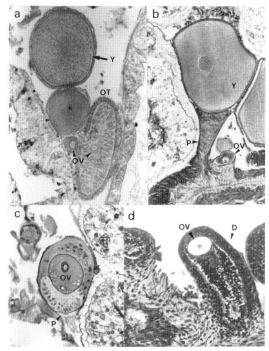

Fig. 3. Modalities of embryonic development in scorpions. **(a)** Early development in a buthid scorpion, *Centruroides barbudensis*. The ovum (OV) starts its development outside the ovariuterine tubule (OT), and later migrates inside it, with subsequent development of yolk-producing cells of germinative epithelium (Y) (see also Fig. 4). **(b-c)** Similar situation in a euscorpiid, *Euscorpius flavicaudis*, and a vaejovid, *Vaejovis vittatus*, but with appearence of a typical 'Peduncle' (P) and a less important production of yolk (see also Fig. 5a-b). **(d)** Embryonic development in ischnurid scorpion *Opisthacanthus cayaporum*, with the formation of a diverticulum (D) and very little production of yolk (see also Fig. 5c).

fer has been completed the partners normally separate. Many details of behaviour may be present in the courtship of particular species (see Polis & Sissom 1990). The classical 'cliché', according to which cannibalism by the female of the male occurs among scorpion species in general, is exaggerated. In fact, this only happens in certain species (39% of 4 families, according to Polis & Sissom 1990), and males are capable of mating more than once. There is considerable evidence that newly mated males can produce new spermatophores and mate again within a short period of time.

EMBRYONIC DEVELOPMENT

Most authors agree that all scorpion species are viviparous (Francke 1982; Polis & Sissom 1990) although a few still believe that they may in part be ovoviviparous. Furthermore, the classical apoikogenic and katoikogenic model proposed by Laurie (1896a) is still retained by most authors (see Polis & Sissom 1990). According to this model there exists a dichotomy in the type of embryonic development in scorpions, i.e. development without diverticula (= apoikogenic, from Greek meaning away from home) and development with diverticula (= katoikogenic or at home).

Lourenço et al. (1986a,b) proposed a new concept of the embryological development of scorpions which modifies the classical apoikogenic and katoikogenic model of Laurie (1896a). According to them, viviparity occurs in all scorpions studied, as previously suggested by Francke (1982). The concept is based on tissue modification of the ovaries and differentiation associated with the formation of the ovarian follicles. Families can be arranged along a gradient of increasing complexity of viviparous development. Trophic exchanges occur between the mother and the embryos from the most simple (at the base of the apoikogenic) to the most complex type (in the katoikogenic). Among scorpion families, the Scorpionidae, Diplocentridae, Ischnuridae, Hadogenidae, Urodacidae, and probably the Hemiscorpiidae, Heteroscorpionidae and Lisposomidae exhibit the most complex gradients of embryonic development with well-developed diverticula (Figs. 3d, 5c). In other families, such as the Buthidae, Bothriuridae, Chactidae, Euscorpiidae, Scorpiopidae, Superstitioniidae, Vaejovidae, Iuridae, Chaerilidae the gradients most certainly range from simple to moderately complex (Figs. 3a-c, 4 and 5a-b). No data are available for the remaining three families, Microcharmidae, Troglotayosicidae and Pseudochactidae.

In Table 1, embryonic development is expressed as number of months necessary for its completion.

A further aspect of the embryonic process in some scorpions is the capacity of the females to produce multiple broods after a single inse-

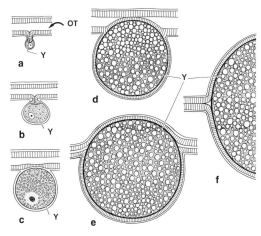

Fig. 4. Schematic embryonic development of a buthid scorpion. **(a-c)** Early development of ovum outside the ovariuterine tubule (OT). **(d-f)** Migration to the tubule and development yolk-producing cells of germinative epithelium (Y).

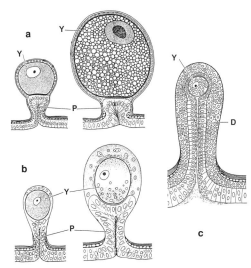

Fig. 5. Schematic embryonic development in **(a)** euscorpiid. **(b)** vaejovid. **(c)** ischnurid scorpions. Yolk-producing cells of germinative epithelium (Y). "Peduncle" (P). Diverticulum (D)

mination. The precise mechanism of this phenomenon was established in the studies conducted by Kovoor et al. (1987) who demonstrated the existence of a novel method of storage of spermatozoa. These are embedded in glandular tissue in the genital tract of the female in species belonging to at least three genera of the family Buthidae, *Centruroides* (Figs. 6a-c), *Tityus* and *Isometrus*. After a single insemination, females of these species are able to give birth to as many as three to five broods when isolated in laboratory conditions. These observations have a bearing on the interpretation of the reproductive strategies of scorpions. Storage of spermatozoa can greatly increase reproductive potential of some species. Moreover, it is significant that two of the three genera concerned (*Centruroides* and *Tityus*) contain species of medical importance which are responsible for thousands of incidents in which human beings are stung and not infrequently killed.

In at least some non-buthid scorpions, such as *Didymocentrus lesueurii*, no effective mechanism of storage of spermatozoa was found. Only a much simpler modality of temporary conservation of spermatozoa in the genital atrium and the proximal region of the ovarian tube was observed (Figs. 6d-f).

BIRTH PROCESS

The process of parturition is very similar in most species of scorpions. The duration of the process, however, varies. Differences are also observed between species with diverticula and those without.

Several hours before the beginning of birth, female scorpions assume a stilting posture. This posture is characterized by the elevation of the anterior portion of the body above the substrate. Flexing of the pedipalps and of the first two pairs of legs underneath the mesosoma, in the proximity of the genital opercula, forms what is called a 'birth basket'. The stilting position is maintained throughout the entire parturition process. The female genital opercula open, and the young emerge one by one (Fig. 7); the birth process is not necessarily carried out at

Table 1. Life history traits in scorpions. (For references see Polis & Sissom (1990), Lourenço (1991a). E.D.: Embryonic development. P.E.D.: Post-embryonic development. Mult. broods: Multiple broods per insemination. **: parthenogenetic species. 15(1-26): average and range values.

Taxon	Region	E. D. (months)	P. E. D. (months)	Brood Size	Life Span (months)	Mult. broods	Body size (mm)
BUTHIDAE							
Caribetityus elii	Caribbean	(3-4)	(10-12)	11 (10-12)	(40-45)	no	(35-40)
Centruroides anchorellus	Cuba	4 (2-9)			(24-36)		(40-60)
C. barbudensis	Caribbean	(3-4)	13	(22-25)		yes	(53-65)
C. gracilis	Americas	(5-6)	11	47(26-91)	(33-50)	yes	(90-110)
C. guanensis	Caribbean	4,5 (2-9)			(24-26)		(35-70)
C. insulanus	Jamaica	5 (4,5-7,5)		50(6-105)			(55-75)
C. margaritatus	Americas			39(26-70)		yes	(80-90)
C. pococki	Caribbean	(3-4)	12	27	(30-35)	yes	(65-75)
C. robertoi	Cuba	4 (3-6)					(40-60)
Isometrus maculatus	Ubiquist	2,5	(7-10)	(12-17)	32	yes	(45-70)
Grosphus flavopiceus	Madagascar	(3-4)	(13-14)	(30-40)	(35-37)	no	(65-75)
Microtityus rickyi	Trinidad	(3-4)	(8-10)	(3-6)	(30-35)	no	(11-15)
M. consuelo	Hispaniola	(3-4)	(8-10)	(4-6)	(30-35)	no	(11-16)
Tityus bahiensis	Brazil	(3-4)	(10-14)	(15-25)	37	yes	(55-75)
T. bastosi	Ecuador	(3-4)	(11-12)	(10-12)	36	no	(35-40)
T. cambridgei	Brazil	(3-4)	(10-12)	(15-25)	(35-40)	yes	(65-85)
T. columbianus	Colombia	3	12	13	36	yes**	(25-30)
T. fasciolatus	Brazil	(3-4)	21	15(1-26)	50	yes	(55-75)
T. fuehrmanni	Colombia	3	(16-17)	(13-16)	45	?	(55-60)
T. insignis	Caribbean	(6-7)	25	(17-20)	56	?	(91-110)
T. mattogrossensis	Brazil	3	13	12	30	no	(30-35)
T. metuendus	Peru	(3-4)	(10-12)	(25-35)	(35-40)	yes	(65-85)
T. morph / serrulatus	Brazil	3	(15-16)	(20-25)	48	yes**	(55-70)
T. strandi	Brazil	4	12	12	46	yes	(50-65)
Rhopalurus garridoi	Cuba	5	12	19(14-27)		no	(60-75)
R. princeps	Haiti	4,5 (2-7)	(13-25)	(15-30)	52	no	(45-65)
DIPLOCENTRIDAE							
Didymocentrus lesueurii	Caribbean	(20-22)	28	52(26-70)	60		(40-50)
D. trinitarius	Caribbean	13,5		46	47		(35-40)
ISCHNURIDAE							
Opisthacanthus asper	Africa	(15-18)	17	(12-22)	65	no	(80-85)
O. capensis	Africa	(15-16)	16	(13-20)	60	no	(45-60)
O. cayaporum	Brazil	(18-22)	24	(15-25)	82	no	(55-65)
O. elatus	S. America	20	26	(15-20)	83	no	(70-90)
SCORPIONIDAE							
Scorpio maurus	Africa	(14-15)		(8-43)		no	(60-70)
Heterometrus longimanus	Asia	12	15	34		no	105-115)
Pandinus gambiensis	Africa		39-83	17	96	no	(140-170)
P. imperator	Africa	7		32		no	(200-220)
URODACIDAE							
Urodacus manicatus	Australia	16	25	(11-18)	80	no	(45-55)
U. yachenkoi	Australia	18	54	8-31)		no	(110-115)
VAEJOVIDAE							
Paruroctonus mesaensis	N. America	12	13	33(9-53)	(60-75)	no	(65-70)
P. utahensis	N. America	11,5		11(2-32)	(72-84)	no	(35-45)
Vaejovis carolinianus	N. America	(12-13)	(22-34)	23(9-36)	72	no	(25-40)
BOTHRIURIDAE							
Bothriurus bonariensis	S. America	12		41(35-48)		no	(50-55)
Urophonius brachycentrus	S. America	6,5		33(21-46)		no	(30-35)
U. granulatus	S. America	(9-10)				no	(35-40)
U. iheringi	S. America	(10-11)		47(31-60)		no	(30-35)

a constant rate. The young drop into the birth basket and, after a short period of time, become active. When born enclosed in a membrane (mainly species without diverticula), they free themselves from the membrane and climb up the female's legs or pedipalps until they reach her back. When parturition is complete and all the young have ascended and settled on the female's back (Fig. 8), her normal activities are renewed. Litter size is variable, ranging from 3-4 to 105-110 young per brood. Table I shows presently available data. The sex-ratio is most often 1:1, but some species have ratios of 3:1 or 4:1 (female: males). The young remain with their mother until their first molt and then disperse. This may be after a further 5 to 30 days, depending on the species. During this period, a rather sophisticated maternal behaviour was observed in all studied species (Lourenço 1991a) (Fig. 9). In species with some degree of social behaviour, e.g. the *Opisthacanthus*, *Pandinus* and *Didymocentrus* species, the young remain with the mother and other adults throughout their lives (Polis & Lourenço 1986).

POST-EMBRYONIC DEVELOPMENT

Post-embryonic development comprises the period after birth until the adult stage has been reached. It can be divided into two phases: projuvenile and juvenile. The pro-juvenile phase consists of a single instar which lasts from the moment of birth until the first molt. During this instar the young remain on their mother's back. The first instar young cannot feed or sting. Their tarsi possess suckers instead of the ungues which appear only after the first molt. The duration of the pro-juvenile instar is variable, ranging in general from 5 to 25 days. The first molt takes place simultaneously in all the young. On average it takes from 6 to 8 hours. The juvenile phase begins after the first molt and comprises a variable number of instars, both among species and also within the same species (Lourenço 1979ab, 1991a; Polis & Sissom 1990). The duration of a given instar is variable among juveniles of the same litter. However, in social species such as *Opisthacanthus cayaporum*

most or all the young of the litter molt during the same night (Lourenço 1985, 1991c). This behaviour suggests a group effect.

Before molting, scorpions become reclusive and inactive until the cuticle has been shed, possibly by blood pressure (Fig. 10). The cuticle ruptures at the sides and front margin of the carapace, while the chelicerae, pedipalps and legs are withdrawn from the exuviae. The body emerges slowly during short periods of vigorous movement which alternate with long periods of relaxation. The process usually takes place in well hidden places or during the night. It lasts from 10 to 14 hours. Immediately after molting the cuticle of scorpions is not fluorescent under UV light, and it does not become so until the new cuticle hardens. The exuviae are, however, fluorescent. The duration of the different instars is variable and depends on the

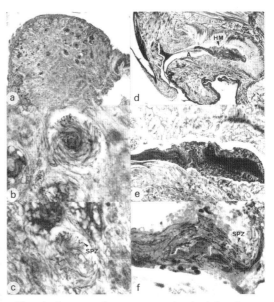

Fig. 6. Storage of spermatozoa in glandular tissue. **(a-c)** Modality of storage in a buthid, *Centruroides barbudensis*. **(a)** Part of the proximal glandular region of the ovarian tube containing piles of spermatozoa. **(b-c)** Details of sperm (SPZ) masses surrounded by glandular cells. **(d-f)** Modality of sperm conservation in a non-buthid scorpion, *Didymocentrus lesueurii* (Diplocentridae). **(d)** Heterogeneous mass (HM) in the genital atrium and the proximal region of the ovarian tube (A). **(e-f)** Bundles of spermatozoa (SPZ) inside the heterogeneous mass.

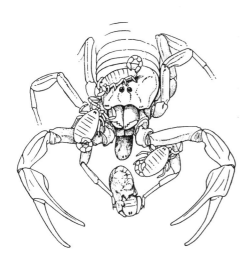

Fig. 7. Process of parturition in a buthid scorpion, *Tityus fasciolatus*. Schematic example of the 'birth basket'.

Fig. 8. Female of *Rhopalurus agamemnon* with first instar brood.

ambient temperature, humidity and food. The total number of instars may vary from 4 to 9.

In several species, males and females can be distinguished only after the last molt when sexual dimorphism becomes visible. This is the case with several species of buthid or chactid genera, such as *Tityus*, *Centruroides*, *Babycurus*, *Brotheas*, *Broteochactas* etc. In other buthid or ischnurid species, such as those belonging to the genera *Grosphus* and *Opisthacanthus*, sexual dimorphism is apparent from birth and the sexes can easily be recognized after the first juvenile instar (instar 2).

LIFE SPAN

The life span of scorpions is variable and may be extraordinarily long, ranging from 4 to 25 years. We still know nothing about the life histories of most small scorpion species, so new data may show more short-lived species.

PARTHENOGENESIS

Parthenogenesis or reproduction from unfertilized eggs was first described in the morph *Tityus serrulatus* (species *Tityus stigmurus*). Contrary to the opinion of Polis and Sissom (1990), this phenomenon is more common than was first thought to be the case.

Of almost 1500 species of scorpions distributed throughout the world, no less than eight are known to be parthenogenetic (Lourenço & Cuellar 1994, 1999). The first of these was reported by Matthiesen (1962) who discovered the phenomenon in the Brazilian species *Tityus serrulatus* Lutz & Mello. Since then, *T. serrulatus* has been transferred to *Tityus stigmurus* (Thorell) (Lourenço & Cloudsley-Thompson 1996), a parthenogenetic species consisting of at least three distinct all-female morphs (Lourenço & Cloudsley-Thompson 1999) of which the original *T. serrulatus* represents one. The other seven species known to be parthenogenetic are *Tityus uruguayensis* Borelli of Uruguay and Brazil, *Tityus columbianus* (Thorell) from Colombia, *Hottentota hottentota* (Fabricius) from West Africa, *Tityus trivittatus* Kraepelin from Argentina, *Liochelis australasiae* (Fabricius) from the South Pacific, *Ananteris coineaui* Lourenço from French Guyana, and *Tityus me-*

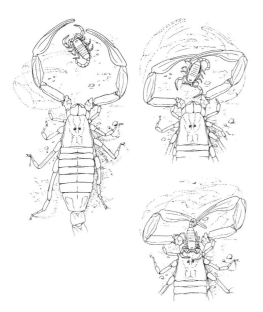

Fig. 9. Aspects of maternal behaviour in *Tityus fasciolatus*. The female helps a displaced first instar young to return to its back.

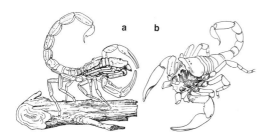

Fig. 10. The molting process in scorpions. **(a)** Schematic example of the molting process in *Tityus fasciolatus*. **(b)** Exuviae of a second instar of *Chactas reticulatus* (Chactidae).

tuendus Pocock from Peru and Brazil (Lourenço & Cuellar 1994, 1999; Lourenço et al. 2000).

The parthenogenetic pattern observed in scorpions corresponds in all cases to the model defined by Vandel (1928) as 'geographic parthenogenesis' and can be tentatively explained in terms of the life history strategies of the populations. A program of studies on the Neotropical parthenogenetic populations has been initiated and will be continued until at least 2006, with the collaboration of researchers in the USA, Brazil and Colombia.

SELECTED EXAMPLES

Tityus columbianus (Thorell, 1876)

Parthenogenesis was demonstrated for the first time in *Tityus columbianus* by Lourenço (1991b). Although only females had been detected in this Colombian species since its description in 1876, Lourenço (1991b) cited the discovery of a single male in the northern part of its range. More recently, a 250 km transect across its entire range, revealed the presence of a sexual population in the northern region, having a sex ratio of about 2:1 in favour of females. The geographic differences in reproductive effort between sexual populations from the area of Iza and parthenogenetic populations from the area of Mosquera, both in Colombia, were compared by Lourenço et al. (1996). The sexual females from Iza were significantly larger and had markedly greater relative litter masses (RLM) than the parthenogenetic ones from Mosquera. In both populations, litter size increased significantly with female body size. Iza receives significantly more precipitation during the critical growing season, and is also significantly warmer than Mosquera. Mosquera, on the other hand, experiences a distinct period of drought in the middle of the summer, which limits productivity to spring and autumn (for details see Lourenço et al. 1996). The heavier body weights and larger litters of the sexual individuals may therefore be due to environmental differences in primary productivity (Lourenço et al. 1996).

Tityus metuendus Pocock, 1897

Tityus metuendus is a rain-forest species distributed mainly in western Amazonia between Brazil and Peru. In the vicinity of Manaus, Brazil, and specifically in the Ducke Reserve, the populations of *T. metuendus* are strictly sexual with a sex ratio of 1:1 (Lourenço 1983, 1997). During recent collections in the Amazonian region of Peru, near Iquitos (the town of Jenaro Herrera), a single pre-adult female of *Tityus metuendus* was collected from a palm tree (*Astrocaryum chambira*; see Kahn, 1997) and brought to the laboratory in Paris. About three months after its last molt this female gave birth to a brood of 21 neonates. Of these, only three (all males) survived to the adult stage. A detailed examination of the size and structure of the pectines of the immature specimens which died and were preserved, revealed that the entire brood consisted of males. Subsequently the same female produced another brood of 32, but all died a few days after the first molt. As with the previous brood, examination of the pectines revealed only males. A third all-male brood was born, bringing further evidence of the possible existence of arrhenotoky (the production of males from unfertilized eggs) in this species. The production of three consecutive all-male broods by the same virgin female may well represent the first known case of arrhenotoky in scorpions, and possibly among all Arachnida other than Acari (Nagelkerke & Sabelis 1991). No data are yet available from scorpions, as exist for other groups such as Hymenoptera (Waage 1986; Cuellar 1987) and mites, either to explain the meiotic mechanism of arrhenotoky or its evolutionary significance (Bull 1983). According to Taylor & Sauer (1980), a major selective advantage of arrhenotoky compared with diploidy is that mothers can precisely determine the sex ratios of their offspring by controlling the fertilization of each egg. This is particularly advantageous in species with finite mating groups in which the probability is high that some clutches may contain no males (Nagelkerke & Sabelis 1991), or in which the sex ratio may be biased in favour of females (Charnov 1982). Precise sex ratios have been documented in the case of several arrhenotokous species of parasitic wasps which lay their eggs either in a single host or in a clumped group of hosts (Waage 1986). In phytoseiid mites, pseudo-arrhenotoky has apparently arisen as a consequence of low mobility and a subdivided population structure. Their dominant prey form patchy infestations which are probably invaded by only a few female mites, leading to very small mating groups (Nagelkerke & Sabelis 1991). Similar mating conditions may exist in the case of *Tityus metu-*

endus, but extensive field work will be needed to explain its life history and behaviour.

The '*Tityus stigmurus*' complex and the *Tityus serrulatus* morph Lutz & Mello, 1922

In the first recorded case of parthenogenesis in scorpions (Matthiesen 1962), pregnant females of *Tityus serrulatus* from Brazil were collected in the field, and their all-female progeny were reared individually, giving virgin birth to a second generation from four to seven months later. Matthiesen's (1962) findings were confirmed some years afterwards by San Martin & Gambardella (1966). Thereafter, *Tityus serrulatus* was considered to be an obligate parthenospecies. Recently, however, a bisexual population has been detected in the state of Minas Gerais in Brazil (Lourenço & Cloudsley-Thompson 1999). In fact, what was initially considered to be the species *Tityus serrulatus* is now known to be one of the morphs within the '*Tityus stigmurus*' complex (Lourenço & Cloudsley-Thompson 1999). The different morphs belonging to this complex can still be found in natural habitats. Some are savanna-dwelling species, while others inhabit palm trees. The *T. serrulatus* morph, however, is virtually restricted to human habitations since the natural savannas have been converted to agriculture and grazing areas. Its original habitat of isolated palm trees within a vast savanna, conforms with the concept of insular parthenogenesis proposed by Cuellar (1977) for the origin of parthenogenetic lizards. According to this author, the chances of colonizing remote or isolated habitats 'are greater if the colonizer can reproduce without a member of the opposite sex, for it dispenses with the need for both sexes to reach the same place simultaneously'. With the expansion of human communities in western Brazil, the geographic range of the *T. serrulatus* morph has increased considerably.

Selection of a parthenospecies by human environments: the case of the *Tityus serrulatus* morph

Several authors have noted that parthenogenetic animals tend to occur in habitats that are different from those of their related bisexuals (Vandel 1928; Udvardy 1969; Cuellar 1977; Glesener & Tilman 1978), a pattern for which Vandel (1928) coined the term 'geographic parthenogenesis'. According to White (1954, 1973), parthenogenetic animals are distinguished from their bisexual counterparts by two unique features, a greater dispersal ability and a much higher reproductive potential. Additionally, Cuellar (1977) has suggested that parthenogenetic animals can only evolve in areas devoid of bisexuals because fertilization would disrupt an all-female lineage, and competition would impede its successful establishment in nature.

The species *Tityus serrulatus* appears to conform with the above predictions. Although this species previously occupied a restricted area in the State of Minas Gerais in Brazil, it is today widely distributed throughout the southeast region of the country. The geographical expansion of this species was undoubtedly related to human colonization, which began about 300 years ago in the Atlantic coastal region and subsequently spread westward. Typically, newly erected towns are usually invaded by the scorpion within a few years after their foundation, although the surrounding natural areas remain virtually devoid of this species. The rapid expansion of *T. serrulatus* into human dwellings was recently demonstrated by its invasion of Brasilia, the capital of Brazil (Lourenço et al. 1994). Construction of Brasilia was initiated in 1956 and the city was completed during the 1970s. From 1971 to 1975, a precise inventory of the local scorpions and their densities was conducted in this region, yielding three species: *Tityus fasciolatus, Bothriurus araguayae* and *Ananteris balzanii* (Lourenço 1981). Of these, *T. fasciolatus,* represented 93% of the total population. This species is similar to *T. serrulatus* in several traits. Both average essentially the same adult size (65 versus 67 mm), the same brood size (16), the same period of embryonic development (2.5-3 months) and the same age to maturity (2.0 to 2.5 years). However, *T. fasciolatus* is bisexual, having a sex ratio of 1 male to 3 females. The two species also differ in their type of population regulation, that of *T. fasciolatus* being density dependent and that of *T. serrulatus* density independent. Consequently, the populations of *T. fasciolatus* have been stable for many years (Lourenço 1979b, 1995), whereas those of *T. serrulatus* have fluctuated widely (Lourenço et al. 1994). Since the introduction of *T. serrulatus* into Brasilia and the adjoining Federal District during the late 1980s and early 1990s, populations of *T. fasciolatus* have been rapidly declining. A new inventory conducted recently revealed that *T. serrulatus* now contitutes 70% of the total in this urbanized region (Lourenço et al. 1994), and is undoubtedly displacing the bisexual species.

The geographical expansion of *T. serrulatus* is undoubtdly due to its introduction into newly created cities and towns by human agency. The creation of new habitats suitable for colonization by *T. serrulatus* may be compared with natural clearings in dense primary forest (Blondel 1976). The cerrados of the Brazilian plateau may represent such forests and the new towns may be comparable to the clearings. In both cases, the new environments represent insular-type habitats which are now known to favor the establishment of parthenogenetic populations (Cuellar 1994). The new towns are, in many cases, separated by several hundred kilometers. Consequently, the countryside between them remains almost pristine, representing a formidable barrier to colonization. When parthenogenetic scorpions are transported by anthropogenic agents by road or rail, the process of colonization is greatly accelerated. This process is greatly facilitated by the higher prolificity and superior colonizing ability inherent in the parthenogenetic mode of reproduction, which favors the colonization of remote and unoccupied territory (Cuellar 1994). However, such territory need not be different, or in disclimax (disturbed environment), as is currently believed by many authors, but merely unoccupied (Hubbell & Norton 1978). In the case of parthenogenetic cave crickets, Hubbell and Norton (1978) found that the environments of the sexual and unisexual forms are 'equable, humid, undisturbed and highly predictable'. The essential factor for the establishment of parthenogenesis in this case is that remote and uninhabited caves 'must be colonized by sweepstake dispersal in which the chances for success are very small for any individual and overwhelmingly against the simultaneous arrival of both sexes'.

Advantages of sexual and parthenogenetic reproduction

Approximately 95% of all living species reproduce sexually. Yet the origin of sexual reproduction is not clear and it has probably evolved independently several times. Since sexual reproduction allows genetic recombination, it should also allow the rapid incorporation of favorable mutations. Muller (1932) was the first to propose that sex must accelerate evolution because two favorable mutations (A and B) are more likely to arise in different individuals of the same population than in a single individual. In asexual species, AB can only arise when two similar mutations occur simultaneously in the same individual (see also Williams 1975; Maynard Smith 1989).

A similar theory attempting to explain the advantages of sexual reproduction was formulated by Van Valen (1973). The environment of any given species is composed of two major factors, abiotic and biotic. Physical factors such as climate are abiotic, and biotic factors consist of other species in the environment, particularly closely related ones competing with each other for limited resources such as food. Any evolutionary modification adapting one species to the environment may be detrimental to the other, but their evolution does not usually influence the abiotic factors significantly. Therefore, each of the competing species must evolve constantly and rapidly in response to modifications of the other. Otherwise, the least alterable will ultimately be eliminated by the selective forces of competition. This is what Van Valen (1973) called 'the Red Queen hypothesis' (in analogy to Lewis Carol's book of Alice in Won-

derland, in which the Red Queen said to Alice 'Here you see, it takes all the running you can do to keep in the same place'). Therefore, in environments such as rain forests, where competition is extremely intense, sexual reproduction is not only advantageous, but a necessity.

The Red Queen hypothesis, however, does not seem to accord with the geographic distribution of parthenogenetic animals, the majority of which occur in remote habitats isolated from their bisexual congeners (Cuellar 1994). According to Cuellar, the major reason for the insular distribution of parthenogenetic species is the ability of single individuals to found a new colony without a member of the opposite sex being present. Assuming that parthenospecies are truly superior colonizers and have evolved in isolation from their bisexual progenitors (Cuellar 1977, 1994), then competition does not appear to have played an important role in their evolution. Aside from the potentially disruptive influence of competition on the establishment of unisexual clones, fertilization of the virgin females would also eliminate unisexual lineages by disrupting all-femaleness and the meiotic process. The latter regulates the constancy of ploidy and the integrity of the species (Cuellar 1977, 1987). Therefore, at least initially, parthenospecies must escape their bisexual counterparts in order to found new colonies. As long as they remain isolated from bisexuals, they can circumvent extinction as parthenogenetic.

Although sexuality is the predominant mode of reproduction among living organisms, it is not entirely devoid of costs, the most common of which are meiosis and the production of males (Williams 1975; Maynard Smith 1989). As emphasised by Mayr (1963), 'in parthenogenetic animals, all zygotes are egg-producing females that do not waste half of their eggs on males'. Neverless, sexual reproduction has the long-term advantage. This is undoubtely the reason why it has appeared so many times during evolutionary history and is the predominating reproductive mechanism in most organisms (Williams 1975; Maynard Smith 1978; Bell

1982). On the other hand, parthenogenesis is only advantageous under special environments (Cuellar 1994); it may not be very old on a geological scale (Bell 1982) and is considered to be an evolutionary blind alley (White 1954, 1973).

ACKNOWLEDGEMENTS
I am most grateful to the Organising Committee of the 19th European Colloquium of Arachnology and in particular to the Chairman Dr. Søren Toft for inviting me to present one of the key conferences of the meeting. I also express my thanks to the Danish Research Councils for financial support. In this paper I have benefited from the comments of Prof. J.L. Cloudsley-Thompson, London, to whom I express my gratitude.

REFERENCES
Anderson, D.T. 1973. *Embryology and phylogeny in annelids and arthropods.* Pergamon Press, Oxford.

Alexander, A.J. 1956. Mating in scorpions. *Nature* 178, 867-868.

Alexander, A.J. 1957. The courtship and mating of the scorpion *Opisthophthalmus latimanus. Proceedings of the Zoological Society of London* 128 (4), 529-544.

Alexander, A.J. 1959. Courtship and mating in the buthid scorpions. *Proceedings of the Zoological Society of London* 133 (1), 145-169.

Angermann, H. 1955. Indirekte Spermatophorenübertragung bei *Euscorpius italicus* (Herbst) (Scorpiones, Chactidae). *Naturwissenschaften* 42 (10), 303-306.

Auber, M. 1959. Observations sur le biotope et la biologie du scorpion aveugle: *Belisarius xambeui* E. Simon. *Vie et Milieu* 10, 160-167.

Auber, M. 1963. Reproduction et croissance de *Buthus occitanus* Amx. *Annales des Sciences Naturelles (Zoologie et Biologie Animale)* 12è ser., 5 (2), 273-285.

Bell, G. 1982. *The masterpiece of nature.* University of California Press, Berkeley.

Blondel, J. 1976. Stratégies démographiques et successions écologiques. *Bulletin de la Societé Zoologique de France* 101, 695-718.

Bull, J.J. 1983. *The evolution of sex chromosomes and sex determining mechanisms.* Benjamin & Cummings, Menlo Park, California.

Charnov, E.L. 1982. *The theory of sex allocation.* Princeton University Press, Princeton.

Cuellar, O. 1977. Animal parthenogenesis. *Science* 197, 837-843.

Cuellar, O. 1987. The evolution of parthenogenesis: a historical perspective. In: *Meiosis* (P.B. Moens ed.), pp. 43-104. Academic Press, New York.

Cuellar, O. 1994. Biogeography of parthenogenetic animals. *Biogeographica* 70 (1), 1-13.

Francke, O.F. 1982. Parturition in scorpions (Arachnida, Scorpiones). *Revue Arachnologique* 4, 27-37.

Fabre, J.H. 1907. *Souvenirs entomologiques.* Delagrave Editions, Paris.

Glesener, R.R. & Tilman, D. 1978. Sexuality and the components of environmental uncertainty: clues from geographic parthenogenesis in terrestrial animals. *American Naturalist* 112, 659-673.

Hubbell, T.H. & Norton, R.M. 1978. The systematics and Biology of the cave crickets of the North American tribe Hadenoecini (Orthoptera), Saltatoria, Ensifera, Rhaphidophoridae, Dolichopodinae. *Miscellaneous Publications – Museum of Zoology, University of Michigan* 156, 1-124.

Kahn, F. 1997. *Les palmiers de l'Eldorado.* Orstom Editions, Paris.

Kovoor, J., Lourenço W.R. & Muñoz-Cuevas, A. 1987. Conservation des spermatozoïdes dans les voies génitales des femelles et biologie de la reproduction des Scorpions (Chélicérates). *Comptes Rendus de l'Academie des Sciences, Paris* Ser. III, 304 (10), 259-264.

Laurie, M. 1890. The embryology of a scorpion (*Euscorpius italicus*). *Quarterly Journal of Microscopical Science* 31, 105-141.

Laurie, M. 1891. Some points in the development of *Scorpio fulvipes. Quarterly Journal of Microscopical Science* 32, 587-597.

Laurie, M. 1896a. Notes on the anatomy of some scorpions, and its bearing on the classification of the order. *Annals and Magazine of Natural History* Ser. 6, 17, 185-194.

Laurie, M. 1896b. Further notes on the anatomy and development of scorpions, and their bearing on the classification of the order. *Annals and Magazine of Natural History* Ser. 6, 18, 121-133.

Lourenço, W.R. 1979a. Le Scorpion Buthidae: *Tityus mattogrossensis* Borelli, 1901 (Morphologie, écologie, biologie et développement postembryonnaire). *Bulletin du Museum national d'Histoire naturelle, Paris* 4e ser., 1 (A1), 95-117.

Lourenço, W.R. 1979b. La biologie sexuelle et le développement postembryonnaire du Scorpion Buthidae: *Tityus trivittatus fasciolatus* Pessôa, 1935. *Revista Nordestina de Biologia* 2 (1-2), 49-96.

Lourenço, W.R. 1981. Sur l'écologie du scorpion Buthidae: *Tityus trivittatus fasciolatus* Pessôa, 1935. *Vie et Milieu* 31 (1), 71-76.

Lourenço, W.R. 1983. Contribution à la connaissance du Scorpion amazonien *Tityus metuendus* Pocock, 1897 (Buthidae). *Studies on Neotropical Fauna and Environment* 18 (4), 185-193.

Lourenço, W.R. 1985. Essai d'interprétation de la distribution du genre *Opisthacanthus* (Arachnida, Scorpiones, Ischnuridae) dans les régions néotropicale et afrotropicale. Étude taxinomique, biogéographique, évolutive et écologique. Thèse de Doctorat d'Etat, Universit, Pierre et Marie Curie.

Lourenço, W.R. 1991a. Biogéographie évolutive, écologie et les stratégies biodémographiques chez les Scorpions néotropicaux. *Compte Rendu des Seances de la Société de Biogéographie* 67 (4), 171-190.

Lourenço, W.R. 1991b. Parthenogenesis in the scorpion *Tityus columbianus* (Thorell) (Scorpiones, Buthidae). *Bulletin of the British Arachnological Society* 8 (9), 274-276.

Lourenço, W.R. 1991c. *Opisthacanthus* genre gondwanien défini comme groupe naturel. Caractérisation des sous-genres et des groupes d'espèces existant a l'intérieur du genre (Arachnida, Scorpiones). *Ihringia ser. Zoologia* 71, 5-42.

Lourenço, W.R. 1995. *Tityus fasciolatus* Pessôa,

Scorpion Buthidae à traits caractéristiques d'une espèce non-opportuniste. *Biogeographica* 71 (2), 69-74.

Lourenço, W.R. 1997. Additions à la faune de scorpions néotropicaux (Arachnida). *Revue Suisse de Zoologie* 104 (3), 587-604.

Lourenço, W.R. & Cloudsley-Thompson, J.L. 1996. Effects of human activities on the environment and the distribution of dangerous species of scorpions. In: *Envenomings and their treatments* (C. Bon & M. Goyffon eds.), pp. 49-60. Fondation Marcel Mérieux.

Lourenço, W.R. & Cloudsley-Thompson, J.L. 1999. Discovery of a sexual population of *Tityus serrulatus*, one of the morphs within the complex *Tityus stigmurus* (Scorpiones, Buthidae). *Journal of Arachnology* 27 (1),154-158.

Lourenço, W.R., Cloudsley-Thompson, J.L. & Cuellar, O. 2000. A review of Parthenogenesis in Scorpions with a description of postembryonic development in *Tityus metuendus* (Scorpiones, Buthidae) from Western Amazonia. *Zoologischer Anzeiger* 239, 267-276.

Lourenço, W.R. & Cuellar, O. 1994. Notes on the geography of parthenogenetic scorpions. *Biogeographica* 70 (1), 19-23.

Lourenço, W.R. & Cuellar, O. 1999. A new all-female scorpion and the first probable case of arrhenotoky in scorpions. *Journal of Arachnology* 27 (1), 149-153.

Lourenço, W.R., Cuellar, O. & Méndez de la Cruz, F. 1996. Variation of reproductive efforts between parthenogenetic and sexual populations of the scorpion *Tityus columbianus*. *Journal of Biogeography* 23, 681-686.

Lourenço, W.R., Knox, MB. & Yoshizawa, M.A. C. 1994. L'invasion d'une communaut, à le stade initial d'une succession secondaire par une espèce parthénogénétique de Scorpion. *Biogeographica* 70 (2), 77-91.

Lourenço, W.R., Kovoor, J. & Muñoz-Cuevas, A. 1986a. Modèle de la viviparit, chez les Scorpions. In: *Actas X Congreso Internacional de Aracnologia Jaca/España* (J.A. Barrientos ed.) Vol. 1, p. 62.

Lourenço, W.R., Kovoor, J. & Muñoz -Cuevas, A. 1986b. Morphogenèse des premiers stades embryonnaires chez des Scorpions. *Bolletino di Zoologia* 53 (suppl.), 105.

Maccary, M.A. 1810. *Mémoire sur le scorpion qui se trouve sur la montagne de Cette, département de l'Hérault*. Gabon Editions, Paris.

Mathew, A.P. 1956. Embryology of *Heterometrus scaber*. *Bulletin of the Centre of Research of the Institut (University of Travancore)* 1, 1-96.

Mathew, A.P. 1960. Embryonic nutrition in *Lychas tricarinatus*. *Journal of the Zoological Society of India* 12, 220-228.

Matthiesen, F.A. 1962. Parthenogenesis in scorpions. *Evolution* 16 (2), 255-256.

Matthiesen, F.A. 1969. Le développement post-embryonnaire du Scorpion Buthidae: *Tityus bahiensis* (Perty, 1834). *Bulletin du Museum national d'Histoire naturelle, Paris* 2e ser., 41, 1367-1370.

Maury, E.A. 1968. Aportes al conocimiento de los escorpiones de la Republica Argentina. I. Observaciones biologicas sobre *Urophonius brachycentrus* (Thorell, 1877) (Bothriuridae). *Physis* 27 (75), 131-139.

Maury, E.A. 1969. Observaciones sobre el ciclo reproductivo de *Urophonius brachycentrus* (Thorell, 1877) (Scorpiones, Bothriuridae). *Physis* 29 (78), 131-139.

Maynard Smith, J. 1978. *The evolution of sex*. Cambridge University Press, Cambridge.

Maynard Smith, J. 1989. *Evolutionary genetics*. Oxford University Press, Oxford.

Mayr, E. 1963. *Animal species and evolution*. Harvard University Press, Belknap, Cambridge, Mass.

Muller, H.J. 1932. Some genetic aspects of sex. *American Naturalist* 66, 118-138.

Nagelkerke, C.J. & Sabelis, M.W. 1991. Precise sex-ratio control in the pseudo-arrhenotokous phytoseiid mite *Typhlo-dromus occidentalis* Nesbitt. In: *The Acari. Reproduction, development and life-history strategies* (R. Schuster & P.W. Murphy eds.), pp. 193-207. Chapman & Hall, London.

Pavlovsky, E.N. 1924. On the morphology of the male genital apparatus in scorpions. *Travaux de la Société des Naturalistes de Leningrad* 53, 76-86.

Pavlovsky, E.N. 1925. Zur Morphologie des weiblichen Genitalapparatus und zur Embryologie der Skorpione. *Annuaire du Muséum de Zoologie de l'Académie des Sciences de l'U.R.S.S., Leningrad* 26, 137-205.

Pflugfelder, O. 1930. Zur Embryologie des Skorpions, *Hormurus australasiae* F. *Zeitschrift für wissenschaftliche Zoologie* 137, 1-23.

Polis, G.A. & Farley, R.D. 1979. Characteristics and environmental determinants of natality, growth and maturity in a natural population of the desert scorpion *Paruroctonus mesaensis* (Scorpionida, Vaejovidae). *Journal of Zoology* 187, 517-542.

Polis, G.A. & Farley, R.D. 1980. Population biology of a desert scorpion: Survivorship, microhabitat, and the evolution of life history strategy. *Ecology* 61 (3), 620-629.

Polis, G.A. & Lourenço, W.R. 1986. Sociality among scorpions. In: *Actas X Congreso Internacional de Aracnologia Jaca/España* (J.A. Barrientos ed.) Vol. 1, pp. 111-115.

Polis, G.A. & Sissom, W.D. 1990. Life History. In: *The Biology of scorpions* (G.A. Polis ed.), pp. 161-223. Stanford University Press, Stanford.

San Martin, P.R. & Gambardella, L.A. 1966. Nueva comprobacion de la partenogénesis en *Tityus serrulatus* Lutz y Mello-Campos, 1922 (Scorpionida, Buthidae). *Revista de la Sociedad entomologica argentina* 28 (1-4), 79-84.

Schultze, W. 1927. Biology of the large Philippine forest scorpion. *Philippine Journal of Science* 32 (3), 375-388.

Shulov, A. & Amitai, P. 1958. On mating habits of three scorpions: *Leiurus quinquestriatus* H & E., *Buthotus judaicus* E.S. and *Nebo hierochonticus* E.S. *Archives de l'Institut Pasteur d'Algérie* 36 (3), 351-369.

Shulov, A., Rosin, R. & Amitai, P. 1960. Parturition in scorpions. *Bulletin of the Research Council of Israel* B, 9 (1), 65-69.

Taylor, P.D. & Sauer, A. 1980. The selective advantage of sex-ratio homeostasis. *American Naturalist* 116, 305-310.

Udvardy, M.D.F. 1969. *Dynamic zoogeography. With special reference to land animals.* Van Nostrand Reinhold Company, New York.

Vandel, A. 1928. La parthénogenèse géographique: Contribution à l'étude biologique et cytologique de la parthénogenèse naturelle. *Bulletin de Biologie France Belgique* 62, 164-281.

Van Valen, L. 1973. A new evolutionary law. *Evolutionary Theory* 1, 1-30.

Varela, J.C. 1961. Gestacion, nacimiento y eclosion de *Bothriurus bonariensis* var. *bonariensis* (Koch, 1842) (Bothriuridae, Scorpiones). *Publicaciones del Departamento de Biologia y Genetica experimental de la Universidad de la Republica, Montevideo,* 24pp.

Waage, J.K. 1986. Family planning in parasitoids: adaptive patterns of progeny and sex allocation. In: *Insect parasitoids* (J.K. Waage & D. Greathead eds.), pp. 63-95. Academic Press, London.

White, M.J.D. 1954. *Animal cytology and evolution.* Cambridge University Press, Cambridge.

White, M.J.D. 1973. *Animal cytology and evolution.* Cambridge University Press, New York.

Williams, G.M. 1975. *Sex and evolution.* Princeton University Press, Princeton.

Williams, S. C. 1969. Birth activities of some North American scorpions. *Proceedings of the Californian Academy of Science* 4th series 37 (1), 1-24.

Yoshikura, M. 1975. Comparative embryology and phylogeny of Arachnida. *Kumamoto Journal of Science, Biology* 12 (2), 71-142.

European Arachnology 2000 (S. Toft & N. Scharff eds.), pp. 87-90.
© Aarhus University Press, Aarhus, 2002. ISBN 87 7934 001 6
(Proceedings of the 19th European Colloquium of Arachnology, Århus 17-22 July 2000)

Life history of *Caribetityus elii* (Armas & Marcano Fondeur, 1992) from the Dominican Republic (Scorpiones, Buthidae)

CHARLOTTE ROUAUD[1], DIETMAR HUBER[2], WILSON R. LOURENÇO[1]*

[1]*Laboratoire de Zoologie (Arthropodes), Muséum National d'Histoire Naturelle, 61 rue de Buffon 75005 Paris, France,* * correspondent author (arachne@mnhn.fr)
[2]*P.O. Box 27, A-6811 Göfis, Austria* (huber@gmx.net)

Abstract
The life cycle of *Caribetityus elii* (Scorpiones, Buthidae) was investigated. The duration of embryonic development in this species ranged 3-4 months, while the moults took place at average ages of 6, 155, 313, and 447 days. These developmental periods are only slightly greater than those recorded for several species of the genus *Tityus*. Also, the mean values of the growth rates observed between different instars are not significantly different from those observed for *Tityus*. However, *Caribetityus elii* completes its postembryonic development with just four moults, rather than the five or six observed in the *Tityus* species. This reduced number of moults had previously been observed only in species of the genus *Microtityus* Kjellesvig-Waering. A comparative analysis of the reproductive traits of both genera is presented.

Key words: Scorpion, *Caribetityus*, *Tityus*, Dominican Republic, life history

INTRODUCTION

The genus *Caribetityus* Lourenço was established for two species from the Dominican Republic, *Caribetityus quisqueyanus* (Armas) and *Caribetityus elii* (Armas & Marcano Fondeur), previously assigned to the genus *Tityus* C.L. Koch, 1836. *Tityus* was originally described from Brazil. This genus is the most numerous in terms of the species described (near to 160) and is very widely distributed in the Neotropical region. In two contributions to the fauna of the Caribbean region, Armas (1982) and Armas & Marcano Fondeur (1992) described two new species for the Dominican Republic, which they placed in the genus *Tityus*. In a recent publication, Lourenço (1999) reanalysed several morphological characters of these two species using scanning electron microscopy and transferred both to a new genus, *Caribetityus*.

During fieldwork in the Dominican Republic, one of the authors (DH) was able to collect several living specimens of *Caribetityus elii* (Armas & Marcano Fondeur) in the Province of La Vega, south of Santiago and east of the central cordillera. The scorpions were found at altitudes of 1400-1500 m in a rainforest formation. Living specimens were kept under laboratory conditions in Paris and Göfis. Mating was obtained for some females, which subsequently produced broods. Both the duration of embryonic development and the number of molts were recorded. In addition, a comparative analysis of reproductive traits of *Caribetityus elii* and several species of *Tityus* is presented.

MATERIAL AND METHODS

Scorpions were reared according to standard methods (Lourenço 1979a,b) using plastic terraria of different sizes. These contained a layer of soil, 2-3 cm in depth, as well as pieces of bark and a small Petri dish containing water. Food, consisting of crickets and spiders, was provided every 7-10 days. Temperatures ranged from 22-26 °C, but humidity was maintained at saturation level. After each molt, the exuviae were removed from the terraria and stored in boxes, one for each scorpion. Morphometric measures were taken from both dead specimens and exuviae. Three parameters were recorded: carapace length, metasomal segment V length, and movable finger length (Lourenço 1979a,b, 1991).

Voucher specimens are deposited in the Muséum National d'Histoire Naturelle, Paris, and in the Muséum d'Histoire Naturelle, Geneva.

RESULTS
Litter size
The observations made on *Caribetityus elii* show that the four broods obtained were composed of 10 to 12 (10, 11, 12, 12) individuals. These numbers are lower than those observed in most *Tityus* species, and only approximate values observed for some small species of the *Tityus clathratus* group (Lourenço 1991). In contrast, the size of pro-juveniles at birth is large. The first instar pro-juveniles are randomly positioned on the mother's back (Williams 1969).

Developmental period
The duration of embryonic development ranged from 3 to 4 months, while molts took place at average ages of 6, 155, 313, and 447 days. These developmental periods are not very different from those found in *Tityus* species or other genera of Buthidae, but the observed values of growth rates in the different instars are slightly greater than those of the other Buthidae that have been studied. Growth parameters, based on morphometric values, are shown in Fig. 1.

The adult lifespan of *Caribetityus elii* probably reaches 30 to 35 months, similar to that observed in some small species of the genus

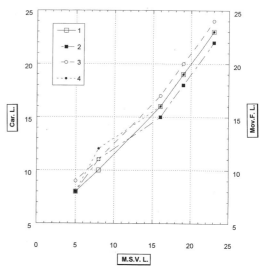

Fig. 1. Distribution of morphometric values (in mm) for juvenile and adult instars of *Caribetityus elii*. 1/3: Carapace length vs. Metasomal segment V length. 2/4: Movable finger length vs. Metasomal segment V length. 1-2: females; 3-4: males.

Tityus (Lourenço 1979; Lourenço & Cloudsley-Thompson 1998). Nevertheless, it is shorter than that observed in several other *Tityus* species (Lourenço 1991).

DISCUSSION
The analysis of the reproductive characteristics observed for *Caribetityus elii* shows that several traits are similar to those observed in species of *Tityus*. Closest similarities are found between *C. elii* and the smaller species of *Tityus*, belonging to the *T. clathratus* group. However, there are also some differences, which support the taxonomic position of *Caribetityus* as a lineage distinct from *Tityus*. Some aspects are listed below:

(a) - The number of molts necessary to reach adulthood in *C. elii* is only 4, compared with 5 or 6 observed in *Tityus* species (Lourenço 1979a,b, 1991; Lourenço & Eickestedt 1988; Lourenço & Cloudsley-Thompson 1998, 1999).

(b) - *Caribetityus elii* litters contain a smaller number of young: 10-12, compared with an average of 15–25 in most species of *Tityus*. Only 2 out of 3 species of the *T. clathratus* group have

Table 1. Life cycle parameters of *Caribetityus* and *Tityus* species. ++: parthenogenetic females.

	Gestation period (months)	Postembryonic development (months)	Litter size	Life span (years)	Adult size (mm)
C. elii	3	10-12	12	5	30-35
T. clathratus group					
T. bastosi	3	12	12	3	30-35
T. columbianus	3	12	13	3	25-30
T. mattogrossensis	3	13	12	2.5-3	30-35
T. bahiensis group					
T. bahiensis	3.5-4	10-14	15-25	3.5	55-75
T. fasciolatus	3.5-4	18-21	13-25	3.5-4	55-75
T. serrulatus++	3.5-4	15-16	15-25	3.5-4	55-70
T. strandi	4	12	12	3.5-4	50-70
T. trivittatus		15-18	15-24	3.5-4	55-70
T. asthenes group					
T. cambridgei	3.5-4	10-12	15-25	4-4.5	70-80
T. fuehrmanni	3	16	10-15	3.5-4	50-60
T. insignis	6-7	18-20	20-30	4-5	90-110
T. metuendus++	7-10	10	25-35	4-4.5	70-80

similarly low numbers. The initial body size of *C. elii* pro-juveniles at birth is slightly larger (in relation to mother's body size) than that observed in species of other genera, such as *Tityus* Koch, *Centruroides* Marx and *Rhopalurus* Thorell (Lourenço 1979, 1988, 1989). This more complete embryonic development may be correlated with the smaller number of postembryonic instars.

(c) - The average growth rates of the different instars observed for *C. elii* are similar to those observed for several species of *Tityus*. The theoretical morphometric growth rate defined by Dyar (1890) and Przibram and Megusar (1912) for the development of arthropods is 1.26. The values observed for species of *Tityus*, *Centruroides* and *Rhopalurus* vary from 1.22 to 1.33, depending on the parameter (segment) considered (Lourenço 1979, 1988, 1989). For males and females of *C. elii*, the growth rates were: carapace length 1.22 and 1.28; metasomal segment V length 1.28 and 1.29; movable finger length 1.24 and 1.26.

(d) - Several species of *Tityus*, *Centruroides* and *Isometrus* are able to store spermatozoa in the glandular tissue of the female genital tract (Kovoor et al. 1987). These females can therefore produce multiple broods from a single insemination. This capacity to store spermatozoa was not observed in *C. elii*.

ACKNOWLEDGEMENTS
We are very grateful to Mr Didier Geffard, Paris, for technical assistance and to Mark Judson, Paris, for comments on the manuscript.

REFERENCES
Armas, L.F. 1982. Adiciones a las escorpiofaunas (Arachnida: Scorpiones) de Puerto Rico y República Dominicana. *Poeyana* 237, 1-25.

Armas, L.F. & Marcano Fondeur, E. 1992. Nuevos alacranes de República Dominicana (Arachnida: Scorpiones). *Poeyana* 420, 1-36.

Dyar, H. 1890. The number of molts in Lepidopterous larvae. *Psyche* 5, 420-422.

Kovoor, J., Lourenço, W.R. & Muñoz-Cuevas, A. 1987. Conservation des spermatozoïdes dans les voies génitales des femelles et biologie de la reproduction des Scorpions (Chélicérates). *Comptes Rendus de l'Académie des Sciences, Paris* 304, sér. III, 10, 259-264.

Lourenço, W.R. 1979a. Le scorpion Buthidae: *Tityus mattogrossensis* Borelli, 1901 (morphologie, écologie, biologie et développement postembryonnaire). *Bulletin du Muséum national d'Histoire naturelle, Paris* 4e sér. 1 (A1), 95-117.

Lourenço, W.R. 1979b. La biologie sexuelle et développement postembryonnaire du scorpion Buthidae: *Tityus trivittatus fasciolatus* Pessôa, 1935. *Revista Nordestina de Biologia* 2 (1-2), 49-96.

Lourenço, W.R. 1988. Le développement postembryonnaire de *Centruroides pococki* Sissom & Francke, 1983 (Buthidae) et de *Didymocentrus lesueurii* (Gervais, 1844) (Diplocentridae) (Arachnida, Scorpiones). *Revue Arachnologique* 7 (5), 213-222.

Lourenço, W.R. 1989. Le développement postembryonnaire de *Rhopalurus princeps* (Karsch, 1879) (Scorpiones, Buthidae). *Revista Brasileira de Biologia* 49 (3), 743-747.

Lourenço, W.R. 1991. Biogéographie évolutive, écologie et les stratégies biodémographiques chez les Scorpions néotropicaux. *Compte Rendu des Séances de la Société de Biogéographie* 67 (4), 171-190.

Lourenço, W.R. 1999. Origines et affinités des scorpions des Grandes Antilles: Le cas particulier des éléments de la famille des Buthidae. *Biogeographica* 75 (3), 131-144.

Lourenço, W.R. & Cloudsley-Thompson, J.L. 1998. A note on the postembryonic development of the scorpion *Tityus bastosi*

Lourenço, 1984. *Newsletter British Arachnological Society* 83, 6-7.

Lourenço, W.R. & Cloudsley-Thompson, J.L. 1999. Notes on the ecology and postembryonic development of *Tityus insignis* (Pocock, 1889) (Scorpiones, Buthidae) from the Island of St. Lucia in the Lesser Antilles. *Biogeographica* 75 (1), 35-40.

Lourenço, W.R. & von Eickstedt, V.R.D. 1988. Notes sur le développement postembryonnaire de *Tityus strandi* (Scorpiones, Buthidae). *Journal of Arachnology* 16, 392-393.

Przibram, H. & Megusâr, F. 1912. Wachstummessungen an *Sphodromantis bioculata* Burm. 1. Länge und Masse. *Archiv für Entwicklungsmechanik der Organismen* (Wilhelm Roux) 34, 680-741.

Williams, S.C. 1969. Birth activities of some North American scorpions. *Proceedings of the Californian Academy of Science* 4th sér., 37 (1), 1-24.

European Arachnology 2000 (S. Toft & N. Scharff eds.), pp. 91-96.
© Aarhus University Press, Aarhus, 2002. ISBN 87 7934 001 6
(Proceedings of the 19th European Colloquium of Arachnology, Århus 17-22 July 2000)

Structure of the ovariuterus of the scorpion *Euscorpius carpathicus* (L.) (Euscorpiidae) before fertilization

L. SORANZO, R. STOCKMANN, N. LAUTIE & C. FAYET
URREDAT. Université P. et M. Curie, 7 quai St Bernard, F-72252 Paris Cedex 05, boîte 4, France
(Roland.Stockmann@snv.jussieu.fr)

Abstract

The ultrastructure of the ovariuterus of *Euscorpius carpathicus* (Scorpiones, Euscorpiidae) before fertilization was studied. Its wall presents a smooth muscular tunica and a pseudostratified columnar epithelium. At the level of the oocytes, the muscular wall is interrupted. The structure of the cells of the oocyte peduncle, issued from the epithelium, undergoes change. The vitellogenesis is divided into four steps, which are characterized by the size of the oocyte and by differences in the number and the structure of the cell organelles.

Key words: *Euscorpius carpathicus*, genital apparatus, ovariuterus, ultrastructure, vitellogenesis

INTRODUCTION

The different embryonic patterns of development allow the grouping of families in Scorpions. Three main patterns of development have been reported: within the tubules of ovariuterus (apoïkogenic), within specialized diverticula (katoikogenic) (Laurie 1890; Pawlowsky 1924, 1926) or in an intermediate and complex form of apoïkogenic ovariuterus (Farley 1996, 1998, 1999). *Euscorpius carpathicus* presents an apoïkogenic development. It is characterized by the oocytes, which are filled up with yolk, and by the development of embryos within the ovariuterine tubules. In this work, we study the ovariuterus ultrastructure of *Euscorpius carpathicus* females before fertilization.

MATERIAL AND METHODS

Adult female scorpions were caught in southern France. They were kept singly in bottles containing moistened leaf-mould at 26°C, and were fed young crickets.

Adult females were anaesthetized with chloroform. The genital tract was dissected and fixed with 2.5% glutaraldehyde in 0.2M cacodylate buffer (pH 7.3), containing 2.45% saccharose. Samples of the genital tract were rinsed in this buffer, then post-fixed with 1% osmium tetroxide and dehydrated. They were then embedded in epoxy resin. Semi-thin sections were stained with toluidine blue. Thin sections were stained with uranyl acetate and lead citrate and observed with a Philips electron microscope.

RESULTS

Anatomical study

The female reproductive system of *Euscorpius carpathicus* is composed of a tubular ovariuterus organized like rungs of a ladder, with three longitudinal tubules interconnecting with four pairs of transverse tubules (Pawlowsky 1926). It is prolonged at each side by a short oviduct and a seminal receptacle. The genital aperture is situated on the second opisthosomal segment.

The earliest immature germ cells are located

Fig. 1. Semi-thin section of the ovariuterus. EC: Epithelial Cells; MT: Muscular Tunica; Pd: Peduncle; Oo: Oocyte.
Fig. 2. Smooth muscular tunica. BC: Basal Cell; BL: Basal Lamina; MC: Muscular Cell; SF: Smooth fibers.
Fig. 3. Pseudostratified epithelium. d: degenerating epithelial nucleus; EC: Epithelial Cells; L: Lumen.
Fig. 4. Apical side of epithelial cells. d: degenerating epithelial nucleus; CE: Cytoplasmic Expansions; L: Lumen; SJ: Septate Junctions.

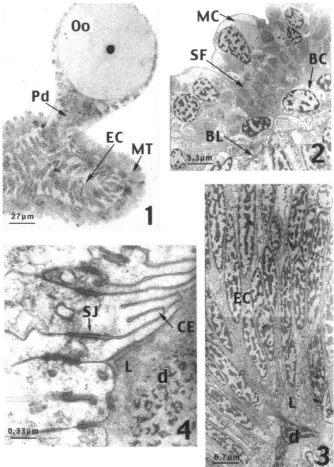

inside the wall of the ovariuterus (see histological study). Afterwards, they pass on the periphery of the ovariuterus and start to grow. During this period they are held by a peduncle (Pd, Fig. 1).

After fertilization, the embryos pass into the tubules and complete their development there in the normal manner for apoïkogenic scorpions.

Ultrastructural study

The ovariuterus

The ovariuterus is circular in transverse section. Its wall is composed of a muscular tunica (MT) organized into several layers of smooth muscular cells (MC) (Figs. 1 and 2) and a pseudostratified epithelium (EC), lying on a basal lamina (BL). Connective tissue is well devel-

oped between muscular cells. The epithelium consists of very high columnar cells (EC), whose nuclei are elongated with a concentration of chromatin (Fig. 3).

At the apical end of these cells are cytoplasmic expansions (CE) and septate junctions (SJ) (Fig. 4). The lumen (L) of the ovariuterus is narrow and often occupied by a flocculent material and degenerating epithelial nuclei (d).

The replacement of cells occurs from undifferentiated basal cells (BC). These lie on a thick basal lamina, which is often irregular. At the basal side of the columnar cells, numerous indentations promote cohesion without obvious junctions.

Female germ cells are incorporated in the wall of ovariuterus and are noticeable by their round nuclei and clear cytoplasm, thus appear-

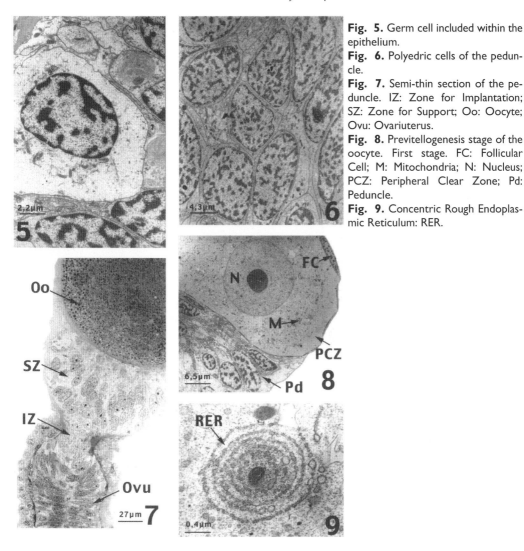

Fig. 5. Germ cell included within the epithelium.
Fig. 6. Polyedric cells of the peduncle.
Fig. 7. Semi-thin section of the peduncle. IZ: Zone for Implantation; SZ: Zone for Support; Oo: Oocyte; Ovu: Ovariuterus.
Fig. 8. Previtellogenesis stage of the oocyte. First stage. FC: Follicular Cell; M: Mitochondria; N: Nucleus; PCZ: Peripheral Clear Zone; Pd: Peduncle.
Fig. 9. Concentric Rough Endoplasmic Reticulum: RER.

ing very different from the columnar cells (Fig. 5).

The peduncle

The cells of the peduncle undergo a specific differentiation from ovariuterine columnar cells. These cells alter their direction and become polyedric, showing an important change in their nucleus structure: the nuclei become round and take on a highly granular appearance (Fig. 6). These modifications may indicate an increased activity of the cells.

The cells nearest to the oocyte infold their membrane and become filled with dense inclu-sions (local synthesis in these cells with transfer to oocyte?).

The peduncle differentiates later in two dif-ferent parts: a zone for implantation (IZ) and a zone for support (SZ), which may have a nutri-tive function (Fig. 7).

The oocytes

Four stages of oocyte development are observ-ed during oogenesis.

First stage of previtellogenesis. All the oocytes, with a size less than 100 μm, show a granular cytoplasm with many ribosomes. Two parts are

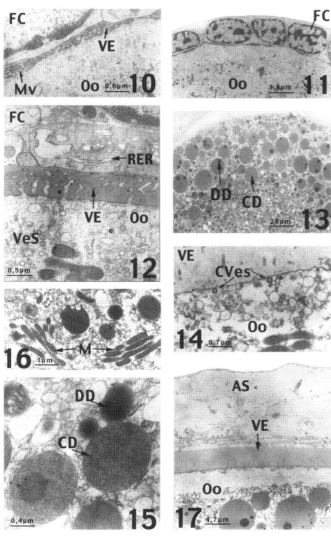

Fig. 10. Vitelline Envelope (VE). FC: Follicular cell; Mv: Microvilli; Oo: Oocyte.

Fig. 11. Cubic Follicular Cells (FC). Oo: Oocyte.

Fig. 12. Vesicular System (VeS) during the second stage of previtellogenesis. FC: Follicular cell; Oo: oocyte; RER: Rough endoplasmic reticulum; VE: Vitelline Envelope.

Fig. 13. Semi-thin section of the oocyte during the primary vitellogenesis stage. DD: Dark-staining Droplets; CD: Clear staining Droplets.

Fig. 14. Coated vesicles (CVes) associated with zone of Vitelline Envelope (VE); Oo: Oocyte.

Fig. 15. Fused yolk droplets during the secondary vitellogenesis stage. CD: Clear staining Droplets; DD: Dark staining Droplets.

Fig. 16. Groups of mitochondria (M).

Fig. 17. Thickened Vitelline Envelope (VE). AS: Nearly-Amorphous Substance; Oo: Oocyte.

currently discriminative: a peripheral clear zone (PCZ, Fig. 8) without mitochondria and a central part with many mitochondria (M) and other organelles: Golgi complex, rough endoplasmic reticulum (RER) organized in concentric structures (Fig. 9).

The formation of the vitelline envelope (VE) initially discontinuous, takes place at the level of the oocyte cytoplasmic membrane. It presents microvilli which soon close (Fig. 10).

Follicular cells (FC) form a coat created by the forward migration of some peduncular cells. They are first elongated and later become cubic (Fig. 11).

Second stage of previtellogenesis. The oocytes (100 µm to 200 µm) present similar characteristics to those of the first stage: the clear zone under the cytoplasmic membrane is still present. The cytoplasmic granulations decrease with the decreasing number of ribosomes.

The vesicular system (VeS) consists of single round or curved vesicles and of large united vesicles. Some of them are homogenous dark-staining and the others contain a granular material. The concentric reticulum is no longer observed (Fig. 12).

The continuous vitelline envelope, whose thickness increases from 0.3 to 0.8 µm, shows microvilli at wide intervals in a parallel direction.

Follicular cells overlap one another and the cellular contacts between these cells are not strong, so peculiar junctions are not observed. This membranous network could allow a permeability, which may facilitate exchanges between the haemolymph and oocyte through the vitelline envelope.

The follicular cells are now in place, very numerous and have more organelles. They could be involved in the synthesis of the oocyte reserves.

Primary vitellogenesis stage. This stage is characterized by the accumulation of yolk reserves. The density and complexity of the intracytoplasmic granules allow the recognition of two steps for storage during the growth of the oocyte.

The number of ribosomes decreases, while the cytoplasmic vesicles increase (perhaps through endocytosis). This interpretation would involve an exogenous means of establishing yolk reserves.

Dark-staining droplets (DD) appear first, followed by clearer and granular droplets (CD, Fig. 13).

The vitelline envelope can reach several micrometers (about 3 μm) in thickness and associated coated vesicles (CVes) are often observed (Fig. 14).

The follicular coat breaks up and at last we observe few fragments of cells with few organelles. Sometimes, single nuclei remain in a nearly amorphous substance. A thick basal lamina surrounds this substance.

Secondary vitellogenesis stage. The oocyte reaches its largest size and takes on an ovoid form (360 μm x 270 μm). The end of growth is characterized by the fusion of yolk droplets (Fig. 15). The more strongly-staining ones, of small size, coalesce with the clearer ones for organizing reserves into heterogeneous droplets (about 2 μm in diameter). They then form vitelline platelets.

Rare mitochondria are set side by side in characteristic groups (Fig. 16).

The vitelline envelope (VE), is very thick (about 7 μm) and no longer shows microvilli at this final stage (Fig. 17).

The follicular coat is entirely changed at the end of oogenesis, with a great development of the nearly amorphous substance (AS) of 8 to 16 μm and forms a protective layer around the oocyte. Together they protrude into the haemolymphatic cavity of the scorpion.

DISCUSSION

This is the first time that an ultrastructural study of the ovariuterus and of oocyte maturation has been carried out in *Euscorpius carpathicus*.

The germ cells are localized in the wall of the ovariuterus. But the question of whether they are able to divide or not as in Buthidae (Warburg et al. 1992, 1995) remains unanswered.

During oogenesis we can characterize 2 stages: previtellogenesis and vitellogenesis. From cytological observations, these stages are subdivided into 4 steps. During previtellogenesis, we observe many ribosomes, the Golgi apparatus, the development of a vitelline envelope around the oocyte and the follicular coat. During vitellogenesis, we observe numerous microvilli at the level of the vitelline envelope and the accumulation of yolk. This cycle is very similar to that observed in Crustacean oocytes (*Orchestia gammarella*) (Zerbib 1975; Charniaux-Cotton 1980). However, a few differences should be noted such as the membrane modification, with macrovilli that bear microvilli, and the formation of microtubules in the Crustacean.

The cytological arguments, ribosomes and RER, on the one hand, and pinocytosis, on the other hand, seem to indicate two origins: exogenous and endogenous of the oocyte reserves. If this hypothesis is correct it will be necessary to corroborate it by biochemical tests.

CONCLUSION

The development of the oocyte is similar in its main stages to that of Crustaceans (pre-vitellogenesis, vitellogenesis). The existence of follicular cells, which are formed from ovari-uterus taking place around the ovocyte, and then beginning to degenerate, is proved. Contrary to all other Arachnid groups (Kaufman 1999) the oocyte of this scorpion is surrounded by follicular cells. Our results corroborate those of Laurie (1890), who also observed follicular cells .

The peduncle, which is also formed from the ovariuterine cells, undergoes important structural modifications before ovulation.

This study will be continued by investigating the changes which occur during later development, and obviousness of vitellogenins and their incorporation.

ACKNOWLEDGMENTS

The authors thank A. Hafdi for assistance in breeding scorpions, and IFR 2062 – Biologie Integrative for use of the electron microscope. We thank M. Judson for revising our text.

REFERENCES

Charniaux-Cotton, H. 1980. Experimental studies of reproduction in Malacostracan crustaceans. Description of vitellogenesis and its endocrine control. In: *Advances in invertebrate reproduction.* (W.R Clark, Jr. & T.S . Adams eds.), pp. 177-186. Elsevier, North Holland.

Farley, R.D. 1996. Formation of maternal trophic structures for embryos of *Paruroctonus mesaeensis* (Scorpionida: Vejovidae). *Revue Suisse de Zoologie* vol. hors série 1, 189-202.

Farley, R.D. 1998. Matotrophic adaptations and early stages of embryogenesis in the Desert Scorpion *Paruroctonus mesaensis* (Vaejovidae). *Journal of Morphology* 237, 187-211.

Farley, R.D. 1999. Scorpiones. In: *Microscopic anatomy of invertebrates, Vol. 8A Chelicerate Arthropoda* (F.W. Harrison & R.F. Foelix ed.), pp 117-222. Wiley-Liss, New-York.

Kaufman, W.R. 1999. Chelicerate Arthropods. In: *Encyclopedia of reproduction* Vol. 1 (E. Knobil & J.D. Neill eds.), pp. 564-571. Academic Press, New York.

Laurie, M. 1890. The embryology of the scorpion *Euscorpius italicus*. *Quarterly Journal of Microscopical Science* 31, 105-141.

Pawlowsky, E.N. 1924. Zur Morphologie der weibliche Genitalorgane der Skorpione. *Zoologische Jahrbücher für Anatomie* 46, 473-506.

Pawlowsky, E.N. 1926. Zur Morphologie des weiblichen Genitalapparates und zur Embryologie der Skorpione. *Annales du Musée Zoologique de l'Académie des Sciences Leningrad* 26 (1925), 137-205.

Warburg, M.R. & Rosenberg, M. 1992. The reproductive system of female *Buthotus judaicus* (Scorpiones, Buthidae). *Biological Structures and Morphogenesis* 4 (1), 33-37.

Warburg, M.R., Elias, R. & Rosenberg, M. 1995. Ovariuterus and oocyte dimensions in the female buthid scorpion *Leiurus quinquestratus* H. & E. (Scorpiones, Buthidae) and the effect of higher temperature. *Invertebrate Reproduction and Development* 27 (1), 21-28.

Zerbib, C. 1975. Contribution à l'étude ultrastructurale de l'ovocyte chez le Crustacé Amphipode *Orchestia gammarella* Pallas. *Comptes-Rendus de l'Académie des Sciences Paris* ser. D 277, 1209-1212.

SPIDER WEBS

SILK

European Arachnology 2000 (S. Toft & N. Scharff eds.), pp. 99-106.
© Aarhus University Press, Aarhus, 2002. ISBN 87 7934 001 6
(Proceedings of the 19th European Colloquium of Arachnology, Århus 17-22 July 2000)

Form and function of the orb-web

SAMUEL ZSCHOKKE

Department of Integrative Biology, Section of Conservation Biology (NLU), University of Basel, St. Johanns-Vorstadt 10, CH-4056 Basel, Switzerland (samuel.zschokke@unibas.ch)

Abstract

In the present article, I review the physical and biological constraints that spiders face when they construct and use orb-webs, and I show how these constraints influence the form of the orb-web. Using the orb-web of the common garden cross spider *Araneus diadematus* as an example, I illustrate and explain a number of features of the orb-web and show alternatives employed by other spider species in their webs. In particular, I discuss why the orb-web generally is planar and vertical, why it has a radial structure with concentric loops of sticky silk, producing a regular meshwork, and why it has a top-down asymmetry. Furthermore, I discuss possible reasons for the increasing distance between the sticky silk loops from the centre to the periphery of the web and the function of the secondary frame threads. I conclude that the shape of the orb-web is a logical consequence of various constraints and optimisations and can therefore not be taken as evidence of a monophyletic origin of the orb-web.

Key words: Biomechanics, evolution, orb-web, web construction, web design

INTRODUCTION

Has the orb-web a monophyletic or a polyphyletic origin? This old question still remains unanswered (for a review of the history of the controversy, see Coddington 1986). Many have argued in favour of a monophyletic origin, mainly because most orb-webs share several key features: all orb-webs have a sticky spiral placed on radial threads which converge at a central location, the hub; typical orb-webs are highly regular structures which are more or less circular and planar; during the construction of orb-webs, an auxiliary spiral is built; etc. (Thorell 1886; Wiehle 1931; Coddington 1986). However, all these features may be or may not be an indication of a common origin. The physical constraints of webs have to be analysed to reveal whether these features may be an adaptation to the function of the finished web, or an adaptation to the web construction process, or whether orb-webs share them because of common ancestry. The aim of the present article is to discuss the features of the well-known orb-web, using the well known ecribellate orb-web of *Araneus diadematus* Clerck as a model. Only those features that cannot be shown to be adaptive to the function or the construction of the web may serve as isolated evidence for a common origin.

THE BASIC SHAPE OF THE ORB-WEB

Spiders make their living by catching insects. To catch flying insects, many spiders have evolved the ability to build traps in the air; the orb-web is one kind of such aerial traps. Since natural selection favours structures that are efficient, orb-webs should cover the largest area possible with a limited amount of material, which is best achieved with a planar web (Wainwright et al. 1976; Opell 1999a; Zschokke

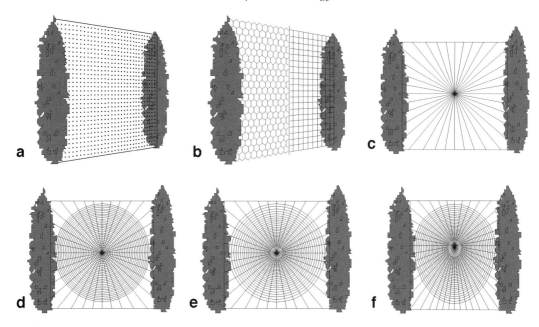

Fig. 1. Hypothetical webs. **(a)** planar vertical web with undefined structure; **(b)** web with regular mesh-work (two alternative meshworks are shown); **(c)** radial structure that allows quick alerting of the spider; **(d)** radial structure with loops of sticky silk; **(e)** same as d) but with increasing distance between sticky silk loops; **(f)** basic hypothetical 'ideal' web.

& Vollrath 2000). As insects tend to fly more or less horizontally in most habitats, vertical webs are best suited to intercept their flight paths (Chacón & Eberhard 1980; Eberhard 1989). Vertical webs have the additional advantage that insects struggling to get away and dropping down are held back by a lower part of the web, whereas they completely drop out of horizontal webs (Eberhard 1990b). Fig. 1a shows such a hypothetical, vertical web, suspended between two bushes or trees for support.

Alternative: In some habitats, for instance above water surfaces, insects primarily fly up and down. Spiders specialised for these habitats, e.g. *Tetragnatha* spp., consequently build horizontal webs. Cribellate orb-weavers (e.g. *Uloborus* spp.) also build horizontal orb-webs. However, their sticky silk and web engineering differ quite strongly from that of the ecribellates (Peters 1987; Köhler & Vollrath 1995; Opell 1997) which makes direct comparisons difficult.

Traps must not be recognised as such by the potential prey. This can be achieved either by camouflage or by making the web invisible; orb-webs try to be invisible (Rypstra 1982; Craig 1986). Since the silk threads themselves are not transparent, they must be as thin as possible and there must be as few threads as possible. Under these constraints, a large area is most efficiently covered using a regular meshwork, similar to a fishing net (von Frisch 1974; Thompson 1992) (Fig. 1b).

When an insect hits the web, the web must fulfil two physical requirements: it must stop and retain the insect (Eberhard 1990b). In other words: prey should neither fly through the web, nor should it bounce back as it would from a trampoline. A solution to this problem is to have two different, specially adapted kinds of thread (Lin et al. 1995). To stop the prey - especially the large ones - and to keep the sticky silk in place, strong and rather stiff threads are used. To retain the prey, flexible sticky silk is used which can absorb the energy of the struggling prey without breaking

(Eberhard 1986; Opell 1999b). In orb-webs, the two kinds of threads are placed in a grid, with one kind of thread running in one direction and the other kind of thread placed perpendicularly on them. The mesh size of this grid differs between spider species and also changes with spider size, spider condition, environmental condition and with the available prey spectrum (Wiehle 1927; Uetz et al. 1978; Sandoval 1994; Vollrath et al. 1997; Zschokke 1997; Schneider & Vollrath 1998).

Alternative: Some spiders have separated the two functions completely; with a tangle of so-called knock-down threads to stop the prey at the top and - spatially separated - a sheet below the knock-down threads onto which the insects fall and can then be captured by the spider. These kinds of webs are built by agelenid, linyphiid spiders and by *Cyrtophora* spp. (Bristowe 1958; Lubin 1973). At the same time, these webs - together with theridiid space webs - usually last longer (typically several weeks) than orb-webs (which last a few days at most). Even though their construction requires more silk than that of orb-webs, they are therefore probably as economical as orb-webs.

Once an insect has hit the web, the spider wants to be quickly alerted to its presence before it escapes. This is best achieved with direct, rather stiff threads running from the different areas of the web to the spider (Masters 1984; Eberhard 1990b). In other words, a structure with radial threads is best suited (Fig. 1c). In orb-webs, these radial threads are also the ones that stop the prey and keep the sticky silk in place. To obtain a regular meshwork based on a radial structure, the sticky threads to retain the prey are best placed in concentric circles on the radial threads (Fig. 1d).

ADDITIONAL CONSIDERATIONS

The structure shown in Figure 1d shows strong similarities to an orb-web. It has non-sticky, stiff radii joining up at one point at the centre of the web, the hub, and loops of sticky, flexible silk. However, orb-webs are more sophisticated than this simple structure.

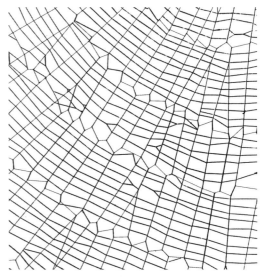

Fig. 2. Part of *Nephila* web below the hub. The vertical threads are the radii. The straight horizontal threads are part of the sticky spiral and the horizonal zigzag threads are part of the auxiliary spiral.

The two functions of stopping and retaining the prey need to be matched in all areas of the web, otherwise the spider might risk that a large insect can break through the web if there are not enough stopping threads, or that an insect may bounce back if there are not enough retaining threads (Eberhard 1981). Since the radii which stop the prey are further apart at the periphery of the web, the loops with sticky silk also have to be spaced further apart at the periphery of the web. Peters (1939; 1947; 1954) even proposed the 'segment rule' which suggests that the distances between the loops of the sticky silk are proportional to the distances between the radii. This precise relationship has since been disputed (ap Rhisiart & Vollrath 1994). However, the basic relationship that the sticky silk is spaced further apart where the radii are further apart is true (Witt 1952; ap Rhisiart & Vollrath 1994; Heiling & Herberstein 1998; own observations).

Alternative: To keep the distance between neighbouring radii approximately equal, some spiders (e.g. *Nephila* spp.) have adopted the strategy of using subsidiary radii (Fig. 2)

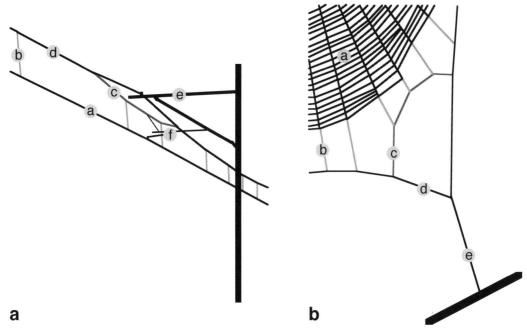

a   **b**

Fig. 3. Similarity between supports of high speed railway overhead contact lines **(a)** and of secondary frames in orb-webs **(b)**. a: contact wire/capture area; b: droppers/outer part of radii; c: auxiliary catenary (Y-dropper)/secondary frame (Y-frame); d: messenger wire (main catenary)/primary frame; e: cantilever and mast/anchor thread and environmental support (e.g. branch of bush); f: steady arm (lateral stabiliser). In both cases, there is no direct connection between the rigid structure (e) and the supported parts (a).

Subsidiary radii are those radii that do not start at the hub but somewhere further out (Zschokke 1999). In webs with subsidiary radii, the distance between the sticky silk loops is consequently roughly the same in the whole web; there is no increase from the centre to the periphery as it is in other orb-webs (Peters 1953).

When an insect has hit the web, the spider must be able to reach it quickly, before it escapes. Since spiders can run faster downwards than upwards, the area that they can cover within a certain time is larger below the hub than above the hub. Consequently, orb-webs have a top-down asymmetry, with the larger part below the hub (Masters & Moffat 1983; ap Rhisiart & Vollrath 1994; Herberstein & Heiling 1999). The top-down asymmetry requires several modifications. In the first place, the radii above the hub are shorter than those below the hub (Krieger 1992; ap Rhisiart & Vollrath 1994). Additionally, the spiders adjust the shape of the sticky silk loops (Zschokke 1993) and insert additional sticky silk threads in the lower half of the web (Mayer 1952; Witt et al. 1968). The angles between radii are smaller in the lower part of the web (Peters 1937; Tilquin 1942; Mayer 1952; Krieger 1992), presumably to adjust towards the ideal length of sticky silk between two radii. This seems to be more relevant than the weight of the spider sitting on the hub, which one could expect to require more radii in the upper part of the web (Langer 1969). Fig. 1f shows a hypothetical web with all the features mentioned so far.

Sitting in the centre of the web is advantageous for catching prey. However, sitting in the centre also has disadvantages. Since the spider can easily be seen there, it may attract predators or deter prey. Many spiders therefore hide in a retreat somewhere at the edge of the web during the day (Tilquin 1942). Some spiders (e. g. *Zygiella* spp.) even have a specialised signal thread running from the centre of the web to the retreat. Unfortunately, when an insect has hit the web, the spider loses valuable time because it first has to rush to the hub, and only when it has arrived there it can locate and reach the prey. To minimise these costs, the spider builds the web in such a way that the hub is close to the retreat (Le Guelte 1967).

Alternative: An alternative to hiding is camouflage. Some spiders (e.g. *Cyclosa* spp.) achieve camouflage by adding a stabilimentum that looks like a twig (Rovner 1976; Neet 1990). Other spiders (e.g. *Arachnura* spp.) achieve camouflage by making themselves look like something completely different, for instance like a small dead leaf.

The tensions in each loop of the sticky silk produce a centripetal force on the radii, resulting in an increase in tension along each radius from the centre of the web to the periphery (Wirth & Barth 1992). A few species (e.g. *Zilla diodia*) adapt the structure of their radii accordingly by building radii that are single stranded near the centre of the web (where the tensions are lower) and double stranded at the periphery of the web (Zschokke 2000).

On top of all these considerations, the orb-web should be able to withstand environmental stresses, e.g. wind or impact of large insects. The connection between the web and the anchor threads (e in Fig. 3b) which are attached to the rather rigid environment must therefore be very flexible; in particular, the spider should avoid connecting a radius directly to an anchor thread. Humans face similar problems when they build railway overhead lines. Rigid masts support the contact line which needs to be flexible to avoid temporary contact loss resulting in electric arcs. For high-speed railways, this problem has been solved with so-called auxiliary catenaries (Bauer & Kießling 1986), a solution which is remarkably similar to the secondary frames employed by spiders (Fig. 3).

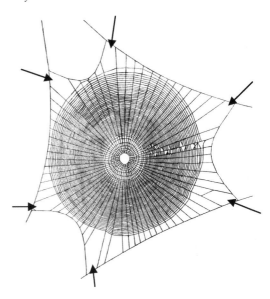

Fig. 4. Web of *Micrathena triangularis*. Note the larger distance between two neighbouring radii near the anchor points (arrows). Redrawn, with permission, after Eberhard (1986).

Alternative: Some spiders (e.g. *Micrathena triangularis* and some *Uloborus* species), achieve a flexible connection between anchor threads and web without a secondary frame. They avoid connecting a radius to the frame near an anchor thread by enlarging the distance between neighbouring radii around the anchor threads (Fig. 4).

CONSTRAINTS OF THE CONSTRUCTION

Despite their ingenuity, orb-webs do not last forever, since the sticky silk dries out or glues together, threads are torn or debris falls into the web (Wiehle 1927; Opell 1999a). Araneid orb-weavers therefore rebuild their webs usually every night or every other night (Wiehle 1927; Breed et al. 1964; own observations). This requires the construction of the web to be economical and therefore several deviations from the 'ideal' orb-web outlined above can be found. First, the sticky thread is laid down in a spiral and not in circles. Since the spider turns around to insert additional sticky threads in the lower half of the web, the spiral is not a continuous one. Comparing the angles between neighbouring radii, some angles are considerably larger or smaller than their neighbouring angles. The most divergent angles are usually the ones around the radii that the spider constructed last, where the spider was constrained by the previously laid radii (Mayer 1952).

CONCLUSION

All features of the macro-structure of the orb-webs shown can be explained by mechanical and biological constraints. The fact that most orb-webs share these features can therefore not be taken as a proof of a monophyletic origin of the orb-web. It also provides no support for the hypothesis of dual or multiple origin of the orb-web. To learn more about the phylogeny of the orb-webs, comparative studies on the web-construction behaviour, especially the early stages (Eberhard 1982; 1990a; 1996; Zschokke & Vollrath 1995) and on the fine structure of the webs (Jackson 1971; Benjamin et al. 2002) are needed.

ACKNOWLEDGEMENTS

I thank Rolf Locher, Søren Toft and two anonymous referees for helpful comments on the manuscript, and Joachim Zschokke for providing the terminology for various parts of the railway overhead line. I am also grateful to Fritz Vollrath for providing me with *Nephila* spiders.

REFERENCES

ap Rhisiart, A. & Vollrath, F. 1994. Design features of the orb web of the spider, *Araneus diadematus*. *Behavioral Ecology* 5, 280-287.

Bauer, K.H. & Kießling, F. 1986. Oberleitungen für den Schienen-Schnellverkehr. *Glasers Annalen: Zeitschrift für Eisenbahnwesen und Verkehrstechnik* 110, 367-373.

Benjamin, S.P., Düggelin, M. & Zschokke, S. 2002. Fine structure of sheet-webs of *Linyphia triangularis* (Clerck) and *Microlinyphia pusilla* (Sundevall), with remarks on the presence of viscid silk. *Acta Zoologica* 83, 49-59.

Breed, A.L., Levine, V.D., Peakall, D.B. & Witt, P.N. 1964. The fate of the intact orb web of the spider *Araneus diadematus* Cl. *Behaviour* 23, 43-60.

Bristowe, W.S. 1958. *The world of spiders.* Collins, London.

Chacón, P. & Eberhard, W.G. 1980. Factors affecting numbers and kinds of prey caught in artificial spider webs, with consideration of how orb webs trap prey. *Bulletin of the British Arachnological Society* 5, 29-38.

Coddington, J.A. 1986. The monophyletic origin of the orb web. In: *Spiders - webs, behavior and evolution.* (W.A. Shear ed.), pp. 319-363. Stanford University Press, Stanford.

Craig, C.L. 1986. Orb-web visibility: the influence of insect flight behaviour and visual physiology on the evolution of web designs within the Araneoidea. *Animal Behaviour* 34, 54-68.

Eberhard, W.G. 1981. Construction behaviour and the distribution of tensions in orb webs. *Bulletin of the British Arachnological Society* 5, 189-204.

Eberhard, W.G. 1982. Behavioral characters for the higher classification of orb-weaving spiders. *Evolution* 36, 1067-1095.

Eberhard, W.G. 1986. Effects of orb-web geometry on prey interception and retention. In: *Spiders - webs, behavior, and evolution.* (W. A. Shear ed.), pp. 70-100. Stanford University Press, Stanford.

Eberhard, W.G. 1989. Effects of orb web orientation and spider size on prey retention. *Bulletin of the British Arachnological Society* 8, 45-48.

Eberhard, W.G. 1990a. Early stages of orb construction by *Philoponella vicina, Leucauge mariana*, and *Nephila clavipes* (Araneae, Uloboridae and Tetragnathidae), and their phylogenetic implications. *Journal of Arachnology* 18, 205-234.

Eberhard, W.G. 1990b. Function and phylogeny of spider webs. *Annual Review of Ecology and Systematics* 21, 341-372.

Heiling, A.M. & Herberstein, M.E. 1998. The web of *Nuctenea sclopetaria* (Araneae, Araneidae): Relationship between body size and web design. *Journal of Arachnology* 26, 91-96.

Herberstein, M.E. & Heiling, A.M. 1999. Asymmetry in spider orb-webs: a result of physical constraints? *Animal Behaviour* 58, 1241-1246.

Jackson, R.R. 1971. Fine structure of the thread connections in the orb web of *Araneus diadematus. Psyche* 78, 12-31.

Köhler, T. & Vollrath, F. 1995. Thread biomechanics in the two orb-weaving spiders *Araneus diadematus* (Araneae, Araneidae) and *Uloborus walckenaerius* (Araneae, Uloboridae). *Journal of Experimental Zoology* 271, 1-17.

Krieger, M. 1992. Radienbau im Netz der Radnetzspinne. Diplomarbeit, Universität Basel.

Langer, R.M. 1969. Elementary physics and spider webs. *American Zoologist* 9, 81-89.

Le Guelte, L. 1967. La structure de la toile et les facteurs externes modifiant le comportement de *Zygiella-x-notata* Cl. (Araignées, Argiopidae). *Revue du Comportement Animal* 1, 23-70.

Lin, L., Edmonds, D. & Vollrath, F. 1995. Structural engineering of a spider's web. *Nature* 373, 146-148.

Lubin, Y.D. 1973. Web structure and function: the non-adhesive orb-web of *Cyrtophora moluccensis* (Doleschall) (Araneae, Araneidae). *Forma et Functio* 6, 337-358.

Masters, M.W. 1984. Vibrations in the orbwebs of *Nuctenea sclopetaria* (Araneidae). I. Transmission through the web. *Behavioral Ecology and Sociobiology* 15, 207-215.

Masters, M.W. & Moffat, A. 1983. A functional explanation of top-bottom asymmetry in vertical orbwebs. *Animal Behaviour* 31, 1043-1046.

Mayer, G. 1952. Untersuchungen über Herstellung und Struktur des Radnetzes von *Aranea diadema* und *Zilla x-notata* mit besonderer Berücksichtigung des Unterschiedes von Jugend- und Altersnetzen. *Zeitschrift für Tierpsychologie* 9, 337-362.

Neet, C.R. 1990. Function and structural variability of the stabilimenta of *Cyclosa insulana* (Costa) (Araneae, Araneidae). *Bulletin of the British Arachnological Society* 8, 161-164.

Opell, B.D. 1997. The material cost and stickiness of capture threads and the evolution of orb-weaving spiders. *Biological Journal of the Linnean Society* 62, 443-458.

Opell, B.D. 1999a. Changes in spinning anatomy and thread stickiness associated with the origin of orb-weaving spiders. *Biological Journal of the Linnean Society* 68, 593-612.

Opell, B.D. 1999b. Redesigning spider webs: stickiness, capture area and the evolution of modern orb-webs. *Evolutionary Ecology Research* 1, 503-516.

Peters, H.M. 1937. Studien am Netz der Kreuzspinne (*Aranea diadema*). I. Die Grundstruktur des Netzes und Beziehungen zum Bauplan des Spinnenkörpers. *Zeitschrift für Morphologie und Ökologie der Tiere* 32, 613-649.

Peters, H.M. 1939. Über das Kreuzspinnennetz und seine Probleme. *Naturwissenschaften* 27, 777-786.

Peters, H.M. 1947. Zur Geometrie des Spinnen-Netzes. *Zeitschrift für Naturforschung* 2b, 227-232.

Peters, H.M. 1953. Weitere Untersuchungen über den strukturellen Aufbau des Radnetzes der Spinnen. *Zeitschrift für Naturforschung* 8b, 355-370.

Peters, H.M. 1954. Worauf beruht die Ordnung im Spinnen-Netz? *Umschau in Wissenschaft und Technik* 54, 368-370.

Peters, H.M. 1987. Fine structure and function of capture threads. In: *Ecophysiology of spiders* (W. Nentwig ed.), pp. 187-202. Springer, Berlin.

Rovner, J.S. 1976. Detritus stabilimenta on the webs of *Cyclosa turbinata* (Araneae, Araneidae). *Journal of Arachnology* 4, 215-216.

Rypstra, A.L. 1982. Building a better insect trap: an experimental investigation of prey capture in a variety of spider webs. *Oecologia* 52, 31-36.

Sandoval, C.P. 1994. Plasticity in web design in the spider *Parawixia bistriata*: a response to variable prey type. *Functional Ecology* 8, 701-707.

Schneider, J.M. & Vollrath, F. 1998. The effect of

prey type on the geometry of the capture web of *Araneus diadematus*. *Naturwissenschaften* 85, 391-394.

Thompson, D.A.W. 1992. *On growth and form.* Cambridge University Press, Cambridge.

Thorell, T. 1886. On Dr. Bertkau's classification of the order Araneae, or spiders. *The Annals and Magazine of Natural History, London* 5, 301-326.

Tilquin, A. 1942. *La toile géométrique des araignées.* Presses Universitaires de France, Paris.

Uetz, G.W., Johnson, A.D. & Schemske, D.W. 1978. Web placement, structure, and prey capture in orb-weaving spiders. *Bulletin of the British Arachnological Society* 4, 141-148.

Vollrath, F., Downes, M. & Krackow, S. 1997. Design variability in web geometry of an orb-weaving spider. *Physiology & Behavior* 62, 735-743.

von Frisch, K. 1974. *Animal architecture.* Harcourt Brace Jovanovich, New York.

Wainwright, S.A., Biggs, W.D., Currey, J.D. & Gosline, J.M. 1976. *Mechanical design in organisms.* Edward Arnold, London.

Wiehle, H. 1927. Beiträge zur Kenntnis des Radnetzbaues der Epeiriden, Tetragnathiden und Uloboriden. *Zeitschrift für Morphologie und Ökologie der Tiere* 8, 468-537.

Wiehle, H. 1931. Neue Beiträge zur Kenntnis des Fanggewebes der Spinnen aus den Familien Argiopidae, Uloboridae und Theridiidae. *Zeitschrift für Morphologie und Ökologie der Tiere* 22, 349-400.

Wirth, E. & Barth, F.G. 1992. Forces in the spider orb web. *Journal of Comparative Physiology A* 171, 359-371.

Witt, P.N. 1952. Ein einfaches Prinzip zur Deutung einiger Proportionen im Spinnennetz. *Behaviour* 4, 172-189.

Witt, P.N., Reed, C.F. & Peakall, D.B. 1968. *A spider's web: problems in regulatory biology.* Springer, Berlin.

Zschokke, S. 1993. The influence of the auxiliary spiral on the capture spiral in *Araneus diadematus* Clerck (Araneidae). *Bulletin of the British Arachnological Society* 9, 169-173.

Zschokke, S. 1996. Early stages of web construction in *Araneus diadematus* Clerck. *Revue Suisse de Zoologie* hors série 2, 709-720.

Zschokke, S. 1997. Factors influencing the size of the orb web in *Araneus diadematus*. In: *Proceedings of the 16th European Colloquium of Arachnology* (M. Żabka ed.), pp. 329-334. Wyższa Szkoła Rolniczo—Pedagogizna, Siedlce, Poland.

Zschokke, S. 1999. Nomenclature of the orb-web. *Journal of Arachnology* 27, 542-546.

Zschokke, S. 2000. Radius construction and structure in the orb-web of *Zilla diodia* (Araneidae). *Journal of Comparative Physiology A* 186, 999-1005.

Zschokke, S. & Vollrath, F. 1995. Web construction patterns in a range of orb-weaving spiders (Araneae). *European Journal of Entomology* 92, 523-541.

Zschokke, S. & Vollrath, F. 2000. Planarity and size of orb-webs built by *Araneus diadematus* (Araneae: Araneidae) under natural and experimental conditions. *Ekológia (Bratislava)* 19 Suppl. 3, 307-318.

European Arachnology 2000 (S. Toft & N. Scharff eds.), pp. 107-116.
© Aarhus University Press, Aarhus, 2002. ISBN 87 7934 001 6
(Proceedings of the 19th European Colloquium of Arachnology, Århus 17-22 July 2000)

Radius orientation in the cross spider *Araneus diadematus*

FRITZ VOLLRATH[1,2,3], THOMAS NØRGAARD[3,4] & MICHAEL KRIEGER[1,2,5]

[1]*Department of Zoology, South Parks Rd, Oxford OX1 3PS, UK (fritz.vollrath@zoology.oxford.ac.uk)*
[2]*Zoologisches Institut, Rheinsprung 11, CH-4051 Basel, Switzerland*
[3]*Department of Zoology, Universitetsparken B135, DK-8000 Århus C, Denmark*
[4]*Department of Zoology, Winterthurerstrasse 190, CH-8057 Zürich, Switzerland*
[5]*Laboratoire de Micro-informatique, Swiss Federal Institute of Technology, CH-1015 Lausanne, Switzerland*

Abstract
We studied radius building in the orb-weaving spider *Araneus diadematus*. Distorting webs during construction did not affect radius placement indicating that tensions were not a major factor in orientation. However, locally and specifically displacing part of the spider's walking thread did affect radius placement and led the spider to shift its radius attachment point predictably, and thus the hub angle. We conclude that path integration could be a mechanism by which the spider determines a radius attachment point on the web frame.

Key words: Orb web, orientation, path integration, web construction, web tension

INTRODUCTION

Orb webs like that of the garden cross spider *Araneus diadematus* are examples of animal design that can be studied in great detail, both for ontogeny (building behaviour) and final morphology (web architecture). The construction of the frame (the 'rim' in analogy with a wheel) and radius threads (the 'spokes') are crucial parts of the building behaviour since here the animal determines the working platform on which to fashion the rest of the web (Peters 1937a, 1939; Tilquin 1942; Eberhard 1990; Vollrath 1992; Zschokke 1996). This stage can be divided into two principal phases: (1) construction of the frame threads, guy lines and primary radii, and (2) filling in the remaining radii following a characteristic pattern (Peters 1937a, 1937b; König 1951; Mayer 1953; Reed 1969a; Eberhard 1982; Zschokke & Vollrath 1995a). Web-building proper starts from a star of

threads (Mayer's 'Fadenstern' 1953) or proto-web (Eberhard 1972; Zschokke 1996) consisting of a proto-hub with 3-7 radiating threads (Petrusewiczowa 1938; Mayer 1953). These proto-radii will become the original guy threads of the future web and define it's plane (Zschokke & Vollrath 2000). Only when the spider has completed this proto-web will it commence constructing the frame together with the primary radii (Zschokke 1999). This is followed by filling in the orb-space with the remaining radii which are often placed in opposing directions, although in this pattern there is much irregularity on top of species-specific traits (McCook 1889; Peters 1937b; Petrusewiczowa 1938; Mayer 1953; Reed 1969b; Zschokke & Vollrath 1995b; Eberhard 1972, 1981; Zschokke 1996, 1999, 2000).

When constructing a secondary radius, *Araneus diadematus* clambers along existing radius and frame threads (Fig. 1 A, B). In the early stages, existing radii can be cut and replaced; and in the process the radius wheel (such as it is) can be greatly distorted and the hub be moved several times (Mayer 1953; Zschokke 1996; Zschokke & Vollrath 2000). In the later phases, radii are only added, never removed, and there is no visible distortion. However, the possibility remains that underneath the visible radius wheel there lies — invisible to us but highly tangible to the spider — a wheel of tensions that is altered by the spider adding radii. Radii under unusual tension are replaced or restrung (Dahl 1885; Nielsen 1932) and cut radii replaced with tensions taken into account (Wiehle 1927; Le Guelte 1969). Wiehle (1927) and Le Guelte (1969) interpret the replacement of a cut radius as evidence that the underlying guiding principle of radius construction is tension. Eberhard noted that in certain orb webs tensions are equilibrated (Eberhard 1972, 1981) but he also demonstrated that they are unlikely to be used as guide in orientation (Eberhard 1988). We have confirmed this in our experiments with *Araneus diadematus*.

Tensions would provide local landmarks for orientation during web construction. Hans Peters (1937b) took a very different view and hypothesised that the spider might use some form of spatial map. To test this idea Peters (1937b) displaced a radius thread with a match stick in a way that enlarged the gap between two adjacent radii. When the spider, after circling the hub, arrived at this new 'oversized' opening, it built a new radius thread to fill this gap, resulting again in a radius interval typical of this web segment. Peters (1937b) only did a few pilot experiments and interpreted his results as evidence that the spider may orient not by tension but by path integration, i.e. by constantly recalculating the vector pointing home. There is strong evidence that spiders use medium range orientation by path integration, e.g. funnel spiders hunting in their webs (for reviews see Görner & Glaas 1985; Mittelstaedt 1985), jumping spiders hunting in a bush (Hill 1979; Tarsitano & Jackson 1994), egg-sac

searching in wolf spiders (Görner & Zeppenfeld 1980), return to lost prey in ctenid spiders (Seyfarth et al. 1980) and orb spiders during prey capture (Peters 1932). There is additional evidence that orb spiders use it in the short range (millimeters) during spiral construction (Peters 1937b; Eberhard 1988). Our study investigated the possibility that such a mechanism might also be used by orb weavers in the medium range (10-15 cm) during radius placement.

MATERIAL AND METHODS
We used juvenile (ca. 20 mg) *Araneus diadematus* garden cross spiders acclimatised to our laboratory conditions (45-55% rH, 24 ± 2 °C); and we used the same set of individuals for repeated measures on subsequent webs. The spiders were kept in open ended PVC frames (30x30x5 cm); they were watered daily and fed a fruit fly after web construction. The web-building behaviour was recorded on videotape from its early beginning and analysed frame by frame (25 fps). Webs were also photographed and digitised using our standard lab procedures (Vollrath et al. 1997). We conducted three experiments using either 3 or 5 spiders for series of repeated measurements:

(i) In the first experiment, on the effect of local tensions, we changed the local structure of tension by displacing a single vertical radius in the lower part of the web sideways by 1 cm. For the displacement we used a fine metal pin (1 mm diameter and 15 cm long) attached to a micro manipulator.

(ii) In the second experiment (on the effect of global tensions) we used standard frames hinged in the corners that could be sheared by 30° which distorts the entire web anchored to the sides of the frame to such an extend that untensed radii completely relaxed and sagged.

(iii) The third experiment (on the issue of detour integration) built on an analysis of natural detours which the spider did in the corners of the web. In the actual experiment we shifted the spider's out-radius using a fine metal rod, just before the animal turned from this radius

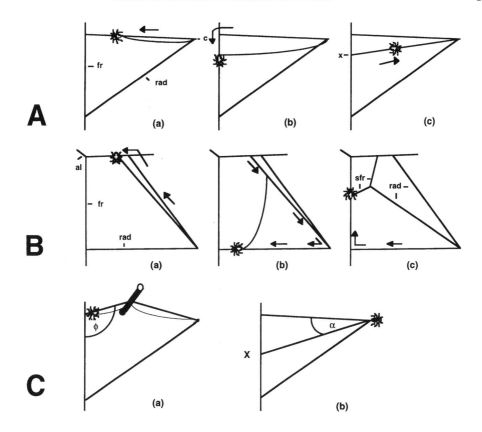

Fig. 1 Attachment of a radius thread to a frame thread in the orb of *Araneus*. **A**. Radius construction without detour. **(a)** The spider walks on the higher radius thread to the periphery. **(b)** It attaches the trailing line to the frame. **(c)** The spider returns to the centre on this line. **B**. Radius construction with detour. **(a)** The spider walks on the higher radius thread to the periphery and attaches the thread. **(b)** On the way back to the centre, the spider connects a second thread to the new radius **(c)** The spider walks now on the lower radius again to the frame. There, it first tightens and then attaches the trailing line (al: anchor line, c: centre, fr: frame, rad: radius, sfr: secondary frame, x: attachment point of new radius). **C**. Radius construction during experimental manipulation. The spider walks along the existing out-radius dragging along the new radius thread (stippled). **(a)** The out-radius is shifted carefully a few millimeters upwards behind the spider with a small metal rod altering angle ϕ. **(b)** The spider attaches the new radius at point X, resulting in the angle α at the hub.

onto the frame thread. We shifted it only a few millimeters (to minimise affecting tensions) and always upwards (Fig. 1C). Since the spider always moves downwards on the frame thread, this served to increase the spider's detour angle (ϕ in Fig. 4). On the whole this manipulation did not seem to irritate the spiders. However, since the spider moves fast, one had to react quickly which introduced a variable degree of error. This crucial experiment was repeated twice. In our first set of trials (n = 21) photos were taken and analysed, and the expected hub angle was calculated from the neighbouring angles; our analysis of radius geometry showed that this was possible. In our second trials (n = 70) the behaviour was filmed on video (25 f.p.s.) and analysed frame by frame, and all angles measured on the screen (mag. 2 times) using a customised analysis program. In this set of trials the last radius put into place by the spider was cut and the experiment performed when the spider replaced it. The angle pairs compared in this case were (i) the experimental hub angle (which was the result of the manipulation) and (ii) an angle predicted from a possible algorithm that models path integration (Fig. 2).

Unless otherwise noted, all comparisons were made with 1-factor ANOVAs. If the degree of freedom was greater than one (df > 1), then a multiple comparison procedure (SNK, α = 0.05) was additionally used. The data on experimental radius displacement (first set) were analysed by first noting the sign of the angle (smaller, larger, indifferent to the calculated normal angle, i.e. the local set angle); thus the detour to the set value was calculated for each web using the average angle for that sector. We then performed sign tests both for all data pooled (with and without the confidence limits given by the standard deviation) and for the measurements repeated during one web-construction process averaged. The data on experimental radius displacement (second set), the control, and experimental angle pairs were analysed using a t-test after normality had been established by probit analysis.

RESULTS

To establish baseline data on the radii, we measured and analysed radius geometry and examined the sequence in which radii were laid down. To study the role of tensions we determined experimentally how accurately the spider could replace a radius that had been removed, both in control webs and in webs with distorted tensions. To examine the possibility of path integration we compared angular regularity between radii that were built with, and without, the spider walking a natural as well as an experimental detour.

Detailed description of radius construction

The spider begins adding radii to the initial radial cross by circling the web's centre, holding on to the hub with the legs on the inward-facing side of its body and measuring the angle between two neighbouring radii with its first and second leg on the outward-facing body

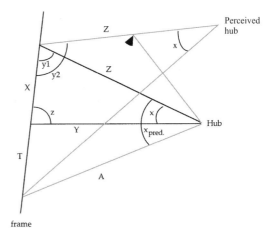

frame

Fig. 2. Sides and angles used in the calculation of the predicted angle at the hub. In the equations small letters x, y and z refer to angles and capital letters X, Y, Z, T and A refer to sides (line segments) in the web. The arrow illustrates the point behind the spider where the out-radius and the new radius thread was lifted during the experiments. The algorithm used to calculate the predicted hub angle: (1) $x_{predicted}$ = x + (arc sin T * sin (180 - z) / A) where (2) T = (sin x / sin (180 - y2 - x) * Z) - X and (3) A = (T2 + Y2 - 2TY * cos(180 - z))$^{1/2}$.

side. If this radius angle exceeds the spider's set value it clambers out along a radius towards the frame trailed by its omnipresent drag line, clambering always on the upper (with respect to gravity) of the two radii (Fig. 1). It walks along the frame and, at one point X, attaches a thread to become the new radius. It reels in its drag line until this is taut, and returns straight to the hub on this line, thus drawing the new radius from the point X of attachment on the frame to the centre, where it is attached to the thread circling the hub. The point X on the frame thread determines the angle at the hub between the new radius and the out-going radius. Thus point X is reached by a detour from the hub along a neighbouring radius to the frame and along the frame to this attachment point. The experiment manipulated the spacial relationship of the radii and frame threads and thus affected the various outer angles between the radii and frame threads as well as the inner angles between the radii at the hub (Fig. 2).

Radius geometry

We found that individual spiders (N = 5) building webs under our controlled conditions made webs (n = 96) with a fairly predictable radius geometry. Radius number was 32.1 ± 0.5 with significant differences (F = 13.66, P < 0.0001, df 4/91) between spiders. Inter-radius angles (the angles between adjacent radii) varied within each web according to their position: the north (above the hub) had significantly larger angles (= fewer radii) than the south (below the hub), east and west (either side of the hub) lay between. Again there were significant individual differences (2-factor ANOVA, F = 187.28, orientation P < 0.0001, spiders P < 0.0001, interaction P = 0.09). We could normalise for the individual differences by expressing the number of radii per sector as percent of all radii in a web. The up/down asymmetry and left/right symmetry of the web's radius numbers and angles was repeated in the length of the radii in the different sectors: longest in the south and shortest in the north. The difference between the sectors was significant, and there were significant differences between spiders (2-factor ANOVA, F = 63.80, orientation P < 0.001, spiders P < 0.0001, interaction P = 0.40).

Radius construction behaviour

In three spiders we filmed and analysed radius construction (n = 12, 10, 12 webs). Here we observed between 5 and 8 primary radii, with spider specific means of 6 ± 0.74, 6.8 ± 0.68 and 5.9 ± 0.58 . There were no significant differences between individual spiders (F = 1.36, df = 2, P = 0.27). However, there was a significant difference (F = 20.14, P < 0.0001) in the numbers of primary radii between north and south sectors (north = 2.2 ± 0.64, south = 1.71 ± 0.8) but none between the east and west sectors (east = 1.09 ± 0.67, west = 1.15 ± 0.61). We analysed the construction sequence of the remaining radii, using a 'coefficient of radius construction' which is composed of the rank of each radius in the sequence of construction (the radius index) divided by the total number of all radii (to normalise for radius numbers between webs). We

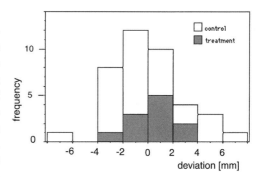

Fig. 3. Displacement of radial attachment points under normal (control) conditions and under conditions of slight distortion.

observed significant differences (F = 59.32, P < 0.0001) in the coefficients between different sectors: radii in the lower sector (south) were built later than those in the other sectors (north, east and west). The coefficients of radius construction were (n = 30) : north = 0.41 (n = 113), south = 0.71 (n = 226), east = 0.41 (n = 185) and west = 0.46 (n = 196).

Local tension experiment

If the spiders were to use tensions in order to determine the placement of a radius, then distorting the tensions in a web should lead to a different placement of a focal radius when compared with normal conditions. We allowed a spider to build its last radius before it would have proceeded to spiral construction. We then continued to cut this radius until the spider stopped replacing it and began auxiliary spiral construction. As long as the spider did replace this radius (N = 7 webs, n = 38 cuts), it did so accurately, attaching it within a range of a few mm (1.9 ± 1.5 mm), in a normal distribution around the original (control) attachment point (Fig. 3).

When we distorted the web slightly with a tripod-mounted micro manipulator pulling the frame sideways, this radius was still always replaced in the typical fashion. Attachment points in such distorted webs were not placed differently (F = 0.30, P = 0.59) from the control attachments that had been made previously while the web was undisturbed.

Global tension experiment

In the 15 cases where we distorted the web grossly by shearing its holding frame by 30º, the spider either kept on building (9 times) or it stopped (6 times) for an average of 5 minutes (4'52" ± 4'65"). This kind of extreme distortion had some radii stretched by 20% and others relaxed so much that they sagged. In the nine cases where the spider continued without interruption it either did nothing about the situation (6 times) or else it connected two adjacent threads (3 times). On the six occasions where it interrupted its behaviour, the spider always readjusted the tensions by laying what appeared to be a single reinforcement radius in the area of highest distortion.

Detour analysis

We assumed that the spider attempts to place the attachment point on the frame thread in such a way that each radius receives its particular hub angle specific to each sector. If the spider uses path integration to determine the placement of a radius, then we would expect that the regularity of the radius angle at the centre would be negatively affected by the length of the detour. *Araneus diadematus* builds two types of radii: those that involve only a minor detour along the outgoing radius and the frame, and those where a secondary piece of frame is laid down (Fig. 1). We analysed 21 webs of two spiders (n = 10, 11) for the different web parameters (Fig. 2). The difference between the two spiders was significant, therefore they were treated separately in our statistics. The variances of the angle in both spiders was always significantly larger (F = 1.73 resp. F = 1.57, P < 0.05) when the animals had walked a detour (n = 34 resp. n = 45) than when they had not done so (n = 207 resp. n = 184).

Displacement experiments

If the spider uses path integration for radius placement then changing one parameter of the path in a specific way would affect the point of attachment X in a predictable manner (Fig. 4). We increased the angle ϕ between the out-going radius and the frame from $\phi 1$ to $\phi 2$ by slightly lifting the out-going radius upward just before the spider walked over this junction A. We observed that this resulted in the spider shortening its path section A-X on the frame thread which led to a decreased hub angle α.

For the first set of trials we measured this angle and compared it to the average set-angle for this particular web and web section. These experimental angles were on average smaller, sometimes marginally, sometimes considerably. Since we had a variable number of repeated measures for each spider and web, and since the experimental setup had some uncontrollable variables (exact repeatability of displacement, spider's detection of our interference), we decided to use a conservative measure of impact. Therefore, for each spider we only used one trial (i.e. the first) of a sequence. We tested the distribution of differences between the observed angle α and the expected angle α for normality using the Kolmogorov-Smirnov-Lilliefors test (Legendre & Vaudor 1991) with the result that the null-hypothesis of normality could not be rejected. A non-directional 1-sample t-test was then performed on the distribution of differences with P = 0.031, indicating (at the 5% level) that our manipulation might have had an effect. We likewise examined the data for interactions between the observed angles α (at the hub) and ϕ (at the frame) and again found a weak positive correlation.

For the second set of trials we modified our analysis because we filmed the experiment and behaviour. This allowed us to do pairwise comparisons of a calculated angle (computed from the control angle and the size of the out-radius displacement) and the actual experimental angle (built in the experiment); the data were obtained from digitised measurements of single-frame video pictures. From the 70 superficially good trials (where the spider only stopped briefly) we took the 46 best trials (where the animal continued without a stop and where the image was so good that we could clearly distinguish all threads). For this

Fig. 4. Path integration experiment. The new radius thread being actually or hypothetically drawn is shown as a fine grey line. **(a)** From the hub H the spider exits via the out-radius and when it encounters the turns the frame thread at A, it moves towards the intersection of an already existing radius at B. Before reaching this point it places a thread at point X on the frame which gives a typical angle α at the hub. In our experiments, when the spider was just before A, the out-radius behind it was displaced a given amount (using a small metal rod) at S. Such radius displacement resulted in changes in the radius-frame angle ϕ from $\phi 1$ to $\phi 2$ which in turn affected the position of X moving it from X1 to X2 with resulting changes of the hub angle α from $\alpha 1$ to $\alpha 2$. The experiment has different predictive outcomes depending on whether the original angle ϕ at A is smaller than 90° **(b)** or whether it is larger than 90° **(c)**. The dotted line shows the spider's dragline as it links the points of experimental displacement S and frame-fixation X2.

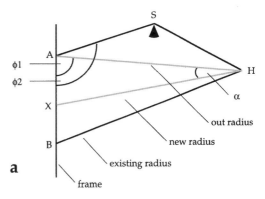

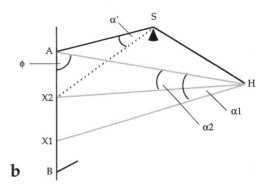

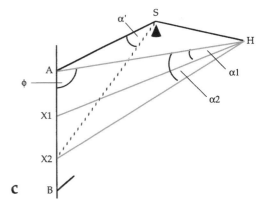

data set, our results showed that the estimated angle deviated from the calculated angle by an average of +1.46 degrees (variance estimate: 11.45). Thus, with a probability of $P < 0.0017$ (t = 3.45), we have to reject the null-hypothesis that there was no difference between the experimental and the calculated angle. We then reanalysed our data after taking out all values where the deviation was greater than 5 degrees. The reasoning behind this adjustment was the thought that, to make such large 'mistakes', these spiders perhaps were disturbed imperceptibly by our measurements (we could only detect the stopping of building). In this potentially more conservative dataset of 41 trials the average angle was +0.68 degrees larger than the calculated angle (variance estimate: 3.78) which was significant at $P < 0.0303$ (t = 2.246).

DISCUSSION

Orb spider web-building behaviour is fine-tuned orientation coupled to manipulation of threads. During web-construction orb weavers might conceivably use path integration (or its leg-positioning equivalent) over the short distances spanned during capture spiral construction (Eberhard 1988). During radial construction the distances over which the spider would

have to integrate are considerably larger indeed, they are over 10 times the spider's length. If here path integration were used then this could radically alter our interpretation of all other phases of site-exploration and orb-construction. This would have serious repercussions for the assumptions presently made in interpretations of orb web-construction (Eberhard 1981, 1982, 1988; Coddington 1986; Vollrath 1992; Hormiga et al. 1995), and in computer simulations where, so far, memory — but not path integration — has to be invoked to model realistic web-geometries (Eberhard 1969; Gotts & Vollrath 1991, 1992; Krink & Vollrath 1997, 1999, 2000).

Our detour analysis, tension experiments, and displacement experiments were designed to determine the likelihood that an orb spider might use such a mechanism as opposed to tensions during radius construction. We show that web parameters like radius positioning can be normalised and that radius angles can be predicted for each sector; this allowed us to interpret our perturbation experiments.

We conclude that tensions are not used as major landmarks during web construction. The first experiment showed that slightly altering the tension had no effect. Even when the web was grossly distorted and a reaction could be observed (9 out of 15 cases), in each web only one single radius thread was adjusted (or in the case of the connection of two threads, two). The other radii were left untouched whether they were highly tensed or slack to the degree of flapping. The spider simply continued laying new radii. If tensions were a major guiding principle, we would expect a more notable adjustment after our perturbance of the distribution of tensions in the whole web. Note that in windy conditions in nature webs do flap about without apparently perturbing the spider. Spider silk is a viscoelastic material (Denny 1980) and the initial tension in a thread will decrease rapidly after installation in the web. This would make it difficult for the spider to use thread tensions as a reliable aid during web construction. However, for web-engineering and prey capture, tensions are of course important (Craig 1987; Wirth & Barth 1992; Linn et al. 1995). The orb web spider *Zilla diodia*, for example, seems to adjust radius morphology to balance tensions by doubling the outer part of many, if not all, radii (Zschokke 2000).

Our data support the use of path integration during web building although it falls short

of proving it. The aim of these experiments was to alter spatial parameters of the path walked by *A. diadematus* during the placement of new radius threads. This was done by lifting the existing radius thread (along with the silk dragline) at a point just behind the building spider and before it turned from this radius thread to the frame thread. However, manipulation of the existing radius in this manner would increase the spider's silk payout as well as its detour angle. And the spider might use silk payout as a measure of distance walked.

The outcome of the first set of trials was in favour of path integration. According to the initial predictions, the angle α at the hub should be smaller than expected if angle ϕ were equal to, or smaller than, 90° and larger than expected if angle ϕ were greater than 90° (Fig. 4 b,c). However, we must consider that the spider's dragline was lifted together with the existing radius thread - which should have resulted in the spider shortening the path section at the frame and hence reducing angle α accordingly. If this effect was fairly consistent, we would still expect a correlation (if only a weak one); this we see.

In the second set of trials we actually calculated the expected angle assuming path-integration. The outcome was a significant shift of angular dimensions in the right direction. If (e.g., by measuring its payout of silk) the spider used the distance it actually travelled on the lifted radius to determine the cut angle, then this could explain a certain mismatch between calculated and observed angles.

On a note of caution, we want to point out that our experimental interference will certainly have been sensed by the spider, and we do not know whether it has affected its behaviour. Nevertheless, we deduce from the results of all our experiments and from comparison with orientation behaviour in other spiders where path integration has been shown to exist (Görner & Glaas 1985; Mittelstaedt 1985) that, as Hans Peters had thought (Peters 1937b), the use of path integration during radius construction in orb weavers is now a likely possibility.

ACKNOWLEDGEMENTS

This paper is dedicated to the lasting memory of the grand old master of web-building behaviour, the late Professor Hans Peters. We thank Peter Görner, Thiemo Krink and two anonymous experts for pertinent comments. Finally, we are greatly indebted to Professor Horst Mittelstaedt for his invaluable help, support, modelling skills and many interesting and important suggestions for future experiments.

REFERENCES

Coddington, J. 1986. The monophyletic origin of the orb web. In: *Spiders - webs, behaviour and evolution* (W.A. Shear ed.), pp. 319-363. Stanford University Press, Stanford.

Craig, C. L. 1987. The ecological and evolutionary interdependence between web architecture and web silk spun by orb web-weaving spiders. *Biological Journal of the Linnean Society* 30, 135-162.

Dahl, F. 1885. Versuch einer Darstellung der Psychischen Vorgänge in den Spinnen II. *Vierteljahrsschrift für wissenschaftliche Philosophie* 9, 162-190.

Denny, M.W. 1980. Silks - their properties and functions. In: *The Mechanical properties of biological materials* (J.F.V. Vincent & J.D. Currey eds.), pp. 245-271. Cambridge University Press, Cambridge UK.

Eberhard, W. G. 1969. Computer simulation of orb-web construction. *American Zoologist* 9, 229-238.

Eberhard, W.G. 1972. The web of *Uloborus diversus* (Araneae: Uloboridae) *Journal of Zoology* 166, 417-465.

Eberhard, W. G. 1981. Construction behaviour and distribution of tensions in orb webs. *Bulletin of the British Arachnological Society* 5, 189-204.

Eberhard, W. G. 1982. Behavioural characters for the higher classification of orb-weaving spiders. *Evolution* 36, 1067-1095.

Eberhard, W. G. 1988. Memory of distances and directions moved as cues during temporary spiral construction in the spider Leucauge

mariana (Araneae:Araneidae). *Journal of Insect Behavior* 1, 51-66.

Eberhard, W.G. 1990. Function and phylogeny of spider webs. *Annual Review of Ecology and Systematics* 21, 341-372.

Gotts, N.M. & Vollrath, F. 1991. Artificial intelligence modelling of web-building in the garden cross spider. *Journal of Theoretical Biology* 152, 485-511.

Gotts, N.M & Vollrath, F. 1992. Physical and theoretical features in the simulation of animal behavior: The spider's web. *Cybernetics and Systems* 23, 41-65.

Görner, P. & Class, B. 1985. Homing behaviour and orientation in the funnel-web spider, *Agelena labyrithica* Clerck. In: *Neurobiology of arachnids*. (F. Barth ed.), pp. 275-298. Springer, Berlin.

Görner, P. & Zeppenfeld, C. 1980. The runs of *Pardosa amentata* (Araneae, Lycosidae) after removing its cocoon. In: *Proceedings of the 8th International Congress of Arachnology, Vienna* (J. Gruber ed.), pp. 243 - 248. Verlag H. Egermann, Wien.

Hill, D. 1979. Orientation by jumping spiders of the genus *Phidippus* (Araneae: Salticidae) during the pursuit of prey. *Behavioral Ecology and Sociobiology* 5, 301-322.

Hormiga, G., Eberhard, W.G. & Coddington, J. A. 1995. Web-construction behaviour in Australian *Phonognatha* and the phylogeny of nephiline and tetragnatid spiders (Araneae: Tetragnatidae) *Australian Journal of Zoology* 43, 313-364.

Krink, T. & Vollrath, F. 1997. Analysing spider web-building behaviour with rule-based simulations and genetic algorithms. *Journal of Theoretical Biology* 185, 321-331.

Krink, T. & Vollrath, F. 1999. Virtual spiders guide robotic control design. *IEEE Intelligent Systems* 14, 77-84.

Krink, T & Vollrath, F. 2000 Optimal area use in orb webs of the garden cross spider. *Naturwissenschaften* 87, 90-93.

König, M. 1951. Beiträge zur Kenntnis des Netzbaus orbiteler Spinnen. *Zeitschrift für Tierpsychologie* 8, 462- 492.

Legendre, P. & Vaudor, A. 1991. *The R package: multidimensional analysis, spatial analysis.* Dept. de Sciences Biologiques, Univ. of Montreal, Montreal.

Le Guelte, L. 1969. Comportement de construction de l'araignee et tension des premiers rayons de sa toile. *Revue du Comportement Animal* 3 (1), 27-32.

Lin, L., Edmonds, D. & Vollrath, F. 1995. Structural engineering of a spider's web. *Nature* 373, 146-148.

Mayer, G. 1953. Untersuchungen über Herstellung und Struktur des Radnetzes von *Aranea diadema* und *Zilla x-notata* mit besonderer Berücksichtigung des Unterschiedes von Jugend- und Altersnetzen. *Zeitschrift für Tierpsychologie* 9, 337-362.

McCook, H. 1889-1893. *American spiders and their spinning work. I - III.* Philadelphia.

Mittelstaedt, H. 1985. Analytical cybernetics of spider navigation. In: *Neurobiology of arachnids.* (F. Barth ed.), pp. 298-318. Springer, Berlin.

Nielsen, E. 1932. *The biology of spiders I.* Levin and Munksgaard, Copenhagen.

Peters, H.M. 1932. Experimente über die Orientierung der Kreuzspinne *Epeira diade-matus* Cl. im Netz. *Zoologische Jahrbücher der Abteilung für Allgemeine Zoologie und Physiologie der Tiere* 51, 239-288.

Peters, H.M. 1937a. Studien am Netz der Kreuzspinne (*Aranea diadema* L). I. Die Grundstructur des Netzes und Beziehungen zum Bauplan des Spinnenkörpers. *Zeitschrift für Morphologie und Ökologie der Tiere* 32, 613-649.

Peters, H. 1937b. Studien am Netz der Kreuzspinne (*Aranea diadema* L). II. Über die Herstellung des Rahmens, der Radialfäden und der Hilfspirale. *Zeitschrift für Mor-phologie und Ökologie der Tiere* 33, 128-150.

Peters, H.M. 1939. Über das Kreuzspinne und seine probleme. *Naturwissenschaften* 47, 777-786.

Petrusewiczowa, E. 1938. Obserwacje budowania sieci przez Pajaka Krzycaka. *Prace Towarzystwa przyjaciol nauk w Wilnie* 13, 1-24.

Reed, C.F. 1969a. Order of radius construction in the orb web. *Bulletin du Museum National d'Histoire Naturelle, 2nd series* 41, 85-87.

Reed, C.F. 1969b. Cues in the web-building process, *American Zoologist* 9, 211-221.

Tarsitano, M.S. & Jackson, R.R. 1994. Jumping spiders make predatory detours requiring movement away from prey. *Behaviour* 131, 65-73.

Tilquin, A. 1942. *La toile geometrique des araignees.* Presses Universitaires de France, Paris.

Vollrath, F. 1992. Analysis and interpretation of orb spider exploration and web-building behavior. *Advances in the Study of Behavior* 21, 147-199.

Vollrath, F., Downes, M. & Krackow, S. 1997. Design variability in web geometry of an orb-weaving spider. *Physiology & Behavior* 62, 735-743.

Wiehle, H. 1927. Beiträge zur Kenntnis des Radnetzbaues der Epeiriden, Tetragnathiden und Uloboriden. *Zeitschrift für Morphologie und Ökologie der Tiere* 8, 468 - 537.

Wirth, E. & Barth, F. G. 1992. Forces in the spider orb web. *Journal of Comparative Physiology A* 171, 359-371.

Zschokke, S. & Vollrath, F. 1995a. Unfreezing the behaviour of two orb web-weaving spiders. *Physiology & Behavior* 58, 1167-1173.

Zschokke, S. & Vollrath, F. 1995b. Web construction patterns in a range of orb-weaving spiders (Araneae). *European Journal of Entomology* 92, 523-541.

Zschokke, S. & Vollrath, F. 2000. Planarity and size of orb-webs built by *Araneus diadematus* (Araneae: Araneidae) under natural and experimental conditions. *Ekólogia (Bratislava)* 19, 307-318.

Zschokke, S. 1999. Nomenclature of the orb-web. *Journal of Arachnology* 27, 542-546

Zschokke, S. 2000. Radius construction and structure in the orb-web of *Zilla diodia* (Araneidae). *Journal of Comparative Physiology A* 186, 999-1005.

European Arachnology 2000 (S. Toft & N. Scharff eds.), pp. 117-122.
© Aarhus University Press, Aarhus, 2002. ISBN 87 7934 001 6
(Proceedings of the 19th European Colloquium of Arachnology, Århus 17-22 July 2000)

A computerised method to observe spider web building behaviour in a semi-natural light environment

SURESH P. BENJAMIN & SAMUEL ZSCHOKKE

Department of Integrative Biology, Section of Conservation Biology (NLU), University of Basel, St. Johanns-Vorstadt 10, CH-4056 Basel, Switzerland (Suresh.Benjamin@unibas.ch)

Abstract

Spider webs are a record of the application of a series of behavioural patterns. Web building behaviour is of great interest to ethologists and taxonomists studying the evolutionary relationships of spiders. However, due to the inability of the researcher to observe the spider around the clock during web building, many details of the behavioural patterns remain undetected. To overcome this problem we developed a novel, computerised method to continually observe the spider during web building. The spider is kept in a temperature controlled room, on a reversed light cycle, confined to an observation arena placed in front of an infrared illuminated background. An infrared sensitive digital video camera is used to capture live images which are transferred to a computer where they are analysed in real time. A separate program allows a detailed study of the recorded movements, including various spatial and temporal analyses. It also allows for the export of movement patterns. The method of observation and data analysis developed by us, enables the detailed study of the web building behaviour of nocturnal spiders and eliminates most constraints encountered to date. Due to the inaccuracy of human observation of long chains of behavioural events and the stereotypic nature of web building behaviour, computerised observation systems are preferable.

Key words: Araneae, spiders, web building behaviour, computerised observation, computerised data analysis

INTRODUCTION

Spider webs are a semi-permanent record of the application of a series of behavioural patterns. The web building behaviour and the finished web of a spider are of immense interest to ethologist and taxonomist studying the evolutionary relationships within different spider taxa (Eberhard 1982).

However, due to the inability of the researcher to observe the spider continuously during web building, many details of the behavioural patterns can remain undetected. For instance, the earlier stages of web construction, which are considered important for deducing taxonomic relationships, are the least studied. The complicated movement patterns of the spider during this stage of construction make it difficult for the observer to keep an accurate frame of reference (Eberhard 1990; Zschokke 1996). The time of initiation is very unpredictable. An observer would have to continuously keep an eye on the spider for a minimum of 24 hours to detect the first steps of the web building process. As discussed in detail in Eberhard (1990), these constraints have led to many irregularities in the description of web building behaviour.

Probably to avoid predation, most spiders

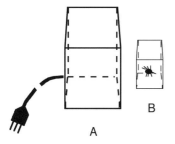
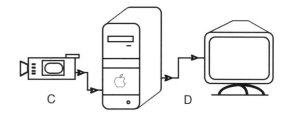

Fig. 1. Basic laboratory layout, which consists of the infrared illuminated background (A), a perspex box in which the spider is kept (B), an infrared capable digital video camera (C) used to capture live images and the computer (D).

build their webs during the night and many are disturbed even by very small amounts of light, typically leading to an interruption of web building. Consequently it is rather difficult to observe undisturbed web construction behaviour of these spiders.

These constraints can be best overcome with computerised observation using infrared light. We developed and tested a novel method, using an infrared sensitive digital video camera and a high end computer with real time analysis of the video frames, to observe the spider during web building.

MATERIALS AND OBSERVATION METHOD

The study animals are kept in 8*8*16 cm perspex boxes in a lab with a reverse 12L:12D light cycle. As day light sources we use 'Osram Daylight ' (Osram GmbH, Hellabruner Strasse 1, D-81536, München) fluorescent bulbs.

To record the spider's movements, we place the perspex box with the spider in front of an infrared illuminated background (Fig. 1, *A*). An infrared sensitive digital video camera (Fig. 1, *C*) is used to capture live images which are transferred to a computer where they are analysed in real time. The computer records the position of the spider at a maximal rate of 14 frames per second.

Background light sources

To overcome the problem that some spiders are disturbed in their web construction by small

amounts of normal light, we use infrared light. Background infrared illumination consists of a PVC box 30.5*30.5*15 cm in size, containing 16 light bulbs arranged in 4 rows of 4 bulbs each. Light bulbs with a capacity of 24.0 Volt, 50 mA and 1200 mW each were used. The front side of the box is covered by an infrared transmitting filter (Farnell AG, Postfach 675, CH-8027 Zurich) and a diffuser. The inner side of the box is painted with a reflective paint.

The use of normal light bulbs enables the construction of a intensive but inexpensive infrared background. An intense light source is necessary to achieve sufficient contrast between the spider and the background. To the human eye and presumably also to the spider's eye (Yamashita 1985), web construction takes place in complete darkness.

Observation arena

The spiders are kept in transparent enclosures, the observation arenas. The size of the observation arena depends on the size of the spiders' webs in nature, but should not exceed the size of the infrared background. In our present studies where we are recording the web construction of small theridiid spiders, we keep them in 8*8*16 cm perspex boxes. Alternatively, U-shaped supporting structures similar to structures used by Zschokke (1994) for the recording of orb-webs can be used. All structures and enclosures are constructed with transparent material. In the case of material with a smooth surface, scratches with a sharp

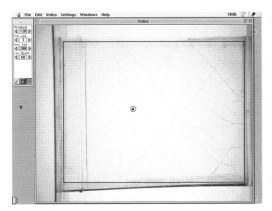

Fig. 2. Screen shot of the recording program. The small window on the left is the control window where the user can set the intensity threshold, minimum and the maximum size of the spider and the minimal distance the spider is required to move before a new position is recorded. The larger window on the right shows the most recent video frame (rotated by 90° to optimise the use of the available area). The black rectangle indicates the observation area in which the objects are detected, the black pixels indicate all pixels that are darker than the defined intensity threshold and the circle highlights the position of the spider as it is detected by the program.

tool were made to enable the spider to crawl about freely.

Video camera

Sony digital video cameras (DCR-TRV10E and DCR-TRV6E) with high infrared sensitivity ('NightShot' function) are used. The 'NightShot' function enables the recording of the spiders' movements in the dark using infrared light instead of normal light. The built-in infrared lamp of the camera is turned off to achieve maximum contrast. The boxes we used were more high than wide (16 cm and 8 cm), whereas the video frames were more wide than high (720 pixels wide and 576 pixels high). To optimise the use of the available area we rotated the camera by 90°. This rotation is compensated for during data recording.

The video camera is connected to the computer using a digital video (DV) link, also known as FireWire (FireWire is a registered trademark of Apple Inc.) or i.LINK (i.LINK is a

registered trademark of Sony Corporation). With digital-to-digital connection, video signals are transmitted at high quality, with a vastly reduced amount of noise compared to analogue video signals. This improved image quality greatly, and decreased disturbances which may cause problems during image analysis (see below).

Data recording

The program to record the data continuously grabs single video frames and analyses them. The following is a simplified schematic outline of the algorithm. Once a video frame has been grabbed, it is scanned within the observation area (user definable) to find all pixels that are darker than a user defined intensity threshold (Fig. 2). Ideally, these dark pixels include just the spider, but typically they may also include some smaller objects (e.g. dirt particles) within the observation area. In a second step, objects are detected by clustering all contiguous dark pixels, using the fill-algorithm described in Zschokke (1990). All objects that are smaller or larger than the user definable minimum and maximum sizes are then discarded. This eliminates most noise created when dirt particles adhere to the perspex boxes. In most cases (if the settings are appropriate), the program should then end up with a single object, the spider. If multiple objects are detected, the object whose position is closest to the previous position of the spider is considered. The program has now successfully detected the current position of the spider. This position is then compared with the previously recorded position. If these two positions differ more than a user defined distance, i.e. the spider has moved, this new position is recorded, together with the exact time. Without delay, the next video frame is then grabbed and analysed. We use an Apple PowerMac G4 computer running at a speed of 400 MHz which allows acquisition and analysis of 14 video frames per second. At any time, the user can stop the recording and save the recorded data on disk.

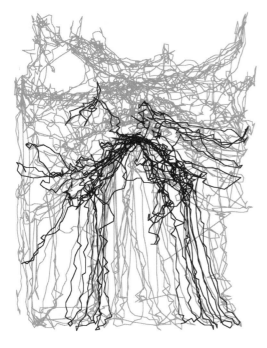

Fig. 3. Example of a recorded track of the theridiid *Achaearanea tepidariorum* (C.L. Koch, 1841) during construction of its tangle web. The construction of the supporting structure is shown in grey, whereas the construction of some gumfooted lines is highlighted in black.

Fig. 4. Gumfooted line construction of the web shown in Fig. 3 with speed codes. Lighter lines denote slower movements and darker lines denote faster movements. It can easily be seen that the spider slowed down wherever it attached a thread.

DATA ANALYSIS

The recordings are analysed with a separate program, typically running on another computer. This program can draw the track of all or parts of the recording. It also allows the colouring of different parts of the track to, for example, highlight a certain stage of web construction (Fig. 3). It is furthermore possible to replay the movement of the spider at the original speed or a multiple thereof allowing rapid visual analysis of movement patterns.

Different numerical analyses are also possible. Probably the most important one is the activity pattern which plots speed against time (cf. Zschokke & Vollrath 1995a,b), or against distance. To visualise the speed of the spider during different parts of the web construction, the movements of the spider can be drawn in a colour corresponding to the speed of the movement (Fig. 4). This allows for rapid localisation of those parts of the web construction, where the spider moved slowly or hastily. In some cases, the speed of the spider can be used to determine the kind of silk it produces (Zschokke & Vollrath 1995b) or to locate the position where a thread was attached or cut (Zschokke 2000). It is furthermore possible to identify areas where the spider was most active or spent most of the time with the so-called position pattern (Fig. 5). For this, the observation area is divided into squares with a user definable resolution, and the program then counts either the time the spider spent (Fig. 5A) or how often the spider entered each square (Fig. 5B). With this analysis, the observer can find the focal point or focal points in the web that are probably important for the spider. This is especially important in the study of non orb-weaving spiders where the existence of such a focal point (possibly equivalent to the hub in orb-webs) is of great interest. All analyses can be done on the entire or just a part of the web construction. In all windows, the user can select single or multiple elements and the program

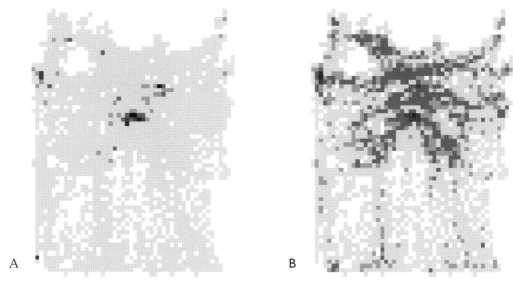

Fig. 5. Position patterns of the web construction shown in Fig. 3. **(A)** shows where the spider spent most of the time during the construction of its web, darkness proportional to the cumulated time spent in each square. **(B)** shows where the spider was most active, darkness proportional to the number of visits to each square.

will then display various data (like position or speed) pertaining to the selected element.

CONCLUSIONS
Our computerised observation method requires high contrast between the spider and background. This makes it difficult to offer the spider a structured environment. It also limits operation of the system in the field. This is not unique to our system but is a general problem in automated vision systems (Bakchine-Huber et al. 1992; Baatrup & Bayley 1993; Zschokke 1994). Nevertheless, due to the inaccuracy of human observation of long chains of behavioural events and the stereotypic nature of web building behaviour, computerised observation systems are preferable.

Since the observation method determines the sequence of spider positions in two dimensions only, all movements in the third direction are ignored.

This observation method and data analysis enable detailed study of the behaviour of nocturnal animals and eliminates most constraints encountered to date. Although this observation method was developed for the observation of spider web building behaviour, its use is not limited to the study of spiders. It is also ideal for the recording and analysis of movement patterns of any nocturnal animal or animals that are disturbed by the presence of visible light, provided that they can be studied under laboratory conditions.

ACKNOWLEDGEMENTS
We thank the Swiss National Science Foundation (Grant no. 31-55617.98) for financial support. We thank Dr. Michel Wurtz for invaluable help in setting up the laboratory, the workshop of the Biozentrum for building the infrared background, and Christin Zschokke, Søren Toft and two anonymous referees for comments on the manuscript.

REFERENCES
Baatrup, E. & Bayley, M. 1993. Quantitative analysis of spider locomotion employing automated video tracking. *Physiology & Behavior* 54, 83-90.

Bakchine-Huber, E., Marion-Poll, F., Pham-Delegue, M.H. & Masson, C. 1992. Real-time detection and analysis of the exploratory behavior of small animals. *Naturwissenschaften* 79, 39-42.

Eberhard, W.G. 1982. Behavioral characters for the higher classification of orb-weaving spiders. *Evolution* 36, 1067-1095.

Eberhard, W.G. 1990. Early stages of orb construction by *Philoponella vicina, Leucauge mariana*, and *Nephila clavipes* (Araneae, Uloboridae and Tetragnathidae), and their phylogenetic implications. *Journal of Arachnology* 18, 205-234.

Yamashita, S. 1985. Photoreceptor cells in the spider eye: spectral sensitivity and efferent control. In: *Neurobiology of arachnids*. (F.G. Barth ed.), pp. 103-117. Springer, Berlin.

Zschokke, S. 1990. PainT - Editor for Digitized Pictures. Diploma Thesis, ETH Zürich.

Zschokke, S. 1994. Web construction behaviour of the orb weaving spider *Araneus diadematus* Cl. PhD Thesis, Universität Basel.

Zschokke, S. 1996. Early stages of web construction in *Araneus diadematus* Clerck. *Revue Suisse de Zoologie* Hors série 2, 709-720.

Zschokke, S. 2000. Radius construction and structure in the orb-web of *Zilla diodia* (Araneidae). *Journal of Comparative Physiology A* 186, 999-1005.

Zschokke, S. & Vollrath, F. 1995a. Unfreezing the behaviour of two orb spiders. *Physiology & Behavior* 58, 1167-1173.

Zschokke, S. & Vollrath, F. 1995b. Web construction patterns in a range of orb-weaving spiders (Araneae). *European Journal of Entomology* 92, 523-541.

European Arachnology 2000 (S. Toft & N. Scharff eds.), pp. 123-126.
© Aarhus University Press, Aarhus, 2002. ISBN 87 7934 001 6
(Proceedings of the 19th European Colloquium of Arachnology, Århus 17-22 July 2000)

Linear and circular dichroism can help us to understand the molecular nature of spider silk

JOHN M. KENNEY[1], DAVID P. KNIGHT[2], CEDRIC DICKO[1,3] & FRITZ VOLLRATH[2,3]
[1]*Institute for Storage Ring Facilities, University of Aarhus, DK-8000 Århus C, Denmark*
(kenney@ifa.au.dk)
[2]*Department of Zoology, University of Oxford, Oxford OX1 3PS, UK*
[3]*Department of Zoology, University of Aarhus, DK-8000 Århus C, Denmark*

Abstract

Tougher than steel, synthetic spider silk would have a revolutionary impact in the material sciences with applications ranging from bullet-proof vests to bone implants. A first step towards the development of synthetic spider silk is understanding how the spider produces it. In addition to detailing the morphology and chemistry of the silk production pathway in the spider, it is vital to characterise the molecular protein (Spiroin) constituents of spider silk in both the soluble (in the gland) and insoluble (in the fibre) states. Here we describe the techniques of Circular Dichroism (CD) and Linear Dichroism (LD) spectroscopy for the structural characterisation of spider silk.

Key words: Spider silk, circular dicroism, linear dichroism

INTRODUCTION

The production of silk in the spider depends on the controlled conversion of soluble proteins (members of the Spidroin family) to form insoluble silk fibres (Knight & Vollrath 1999). This process seems to be based on the refolding of the Spidroin proteins (Gosline et al. 1999; Simmons et al. 1996). Spiders take advantage of the low viscosity in the liquid crystalline regime (Magoshi et al. 1985) and extrude an aqueous solution of silk protein to spin molecules into oriented fibres. During the process of spinning, the silk protein solution undergoes an irreversible phase change to a water insoluble solid fibre with high breaking strength and large extensibility (Willcox et al. 1996). The resultant silk fibre is a composite consisting of protein crystals embedded in a protein matrix (Simmons et al. 1996). By varying the relative proportions of crystals to matrix, as well as the length and width of the crystals, the spider can alter the properties of its silk as required (Knight & Vollrath 1999). Previous X-ray diffraction (Becker et al. 1994) and liquid crystalline (Knight & Vollrath 1999; Kerkam et al. 1991) investigations have shown that the silk fibres contain a β-sheet crystalline component and that the fluid secretions exhibit a large-scale helical organisation. The mechanical properties of the final fibre exceed those of many synthetic fibres, which imply a high degree of molecular alignment (Becker et al. 1994). The complexity of the extrusion process suggests that a number of different factors, along the production pathway, contribute to the extraordinary toughness of spider silk. Indeed, silk's extraordinary physical properties (O'Brien et al. 1998) are a consequence of both the composition and molecular architecture of a fibrous protein, Spidroin (Vollrath & Knight 1999; Vollrath et al. 1998).

To understand the formation of silk fibres

in the spider requires an understanding of the Spidroin structure and its conversion on the molecular level. This can help us to understand and predict its function, stability, and intermolecular associations. The structure of Spidroin, like all protein molecules, is organised at different levels. The amino-acid sequence is known as its primary structure. The next level of organisation is called the secondary structure. It is at this level that the three-dimensional structure is first revealed. The secondary structure is defined as the local spatial arrangement of the amino acids. Secondary structural elements form the building blocks that define the total structure of the protein molecule. Several common secondary structures in proteins are the α-helix, β-sheet, and random coil. Some of these can be associated with particular functions whereas others have no specific biological function alone but are part of larger structural and functional assemblies.

CIRCULAR AND LINEAR DICHROISM SPECTROSCOPY

The secondary structure of soluble Spidroin in the gland can be described using Circular Dichroism (CD), while the insoluble Spidroin in the fibre can be analysed using Linear Dichroism (LD) spectroscopy (Rodger & Nordén 1997). CD and LD applications can be largely grouped into (i) monitoring conformational changes; and (ii) determining the secondary structural content. Understanding the secondary structure of a protein can help to clarify the assembly of large molecular organisations of silk fibres. Besides silk, another example of the importance of secondary structure in determining protein assembly and function is the amyloid fibril-forming proteins of pathogenic diseases, such as Creutzfeld-Jacob disease (Jackson et al. 1999). CD and LD are ideally suited to the study not only of the structure of Spidroin, but also its solution-to-fibre conversion.

CD spectroscopy measures the difference in the absorption of left-handed circularly-polarised light versus right-handed circularly-polarised light as a function of wavelength, typically in the visible to ultraviolet region. A preference for one over another only occurs when a symmetric chromophore is located in an asymmetric environment or in the presence of chiral molecules. The absence of regular structure results in a zero CD intensity, while an ordered structure results in a spectrum, which contains both negative and positive signals. CD has a number of advantages over other structural biology techniques. Only a small amount (micrograms) of protein sample is necessary. Indeed, a single silk gland yields enough material for a set of experiments. Also, the protein concentration and buffering can be physiological. Finally, the protein is studied under soluble conditions - which is not possible with x-ray crystallography.

STUDIES OF SPIDER SILK

Already work has been done to begin to characterise the Spidroin conversion in the major ampullate gland of *Nephila edulis* using CD. The gland consists of a sac or ampulla with an apical 'tail', which produces some part of the silk feedstock, and a funnel that leads to the duct. The secretory portion of the gland ampulla has two morphologically distinct zones. Upstream in the silk production pathway the A zone secretes material that makes the silk fibre core and downstream the B zone secretes a coating (Vollrath & Knight 1999). These secretions are passed through a 'funnel' into a long duct. The duct is folded back to itself in an elongated 'S' to give three limbs, which progressively narrow to form a hyperbolic die. The raw solute rich in spider silk protein was carefully extracted from the major ampullate glands of *Nephila edulis* spiders kept under identical conditions. Samples were prepared of raw solute and solute gently diluted (≈1:4) in Schartau's Ringer solution pH 7.4 (Schartau & Leidersher 1983). A small quantity (≈10 µl) of sample was loaded into a 0.01 mm light path quartz cell (Hellma 124-QS). The loaded sample was examined by polarised light microscopy to confirm that it had not suffered from mechanical shear-induced polymerisation.

The results of this first study show that the soluble Spidroin in the gland undergoes a dramatic structural conversion (Kenney et al. subm.). Material isolated from different parts of the gland (upstream in the A zone and downstream in the B zone) exhibit two completely different CD spectra. Thus, during passage in the gland, Spidroin is converted from one conformational state to another. The suggestion is that the spider takes advantage of these two states to control the production, transportation and conversion of Spidroin in the gland so that Spidrion can be kept in a soluble state at high concentration ($\approx 50\%$) just prior to spinning the insoluble silk fibre. These results illuminate three important aspects of the conversion of the soluble Spidroin: 1) there is a refolding of the protein structure in the gland, 2) the refolding can be induced, and 3) that the converted state is stable and irreversible. Further CD measurements will be performed on silk molecules in solution to determine as a function of location in the gland: (i) whether the proteins are folded, and if so, characterise their secondary structure; (ii) the conformational stability of the proteins under stress; (iii) environmental effects (e.g., pH); and finally, (iv) whether protein-protein interactions alter the conformation of protein. The CD experiments are ongoing.

The LD principle (optical anisotropy) is the same as the CD except that linearly polarised light is used instead of circularly polarised light and one measures differences in the absorption of light linearly polarised parallel and perpendicular to an orientation axis of the fibre. CD is not an optimal technique for studying the insoluble, linearly organised structures of silk fibres. LD, on the other hand, is ideal for this. LD will be used to access: (i) information about how the chromophores are oriented in anisotropic media; and (ii), information about the mobility and binding of particular centres; and finally, (iii) assignment of transition moment directions. The collected data can be used to understand the organisation of oriented structures, such as crystalline domain within the fibre and their effect on the fibre macroscopic properties.

The extremely high brightness and wide spectral range (from near infrared to vacuum ultra violet) synchrotron radiation (SR) has been recognised as a valuable source for CD (Sutherland 1996). SR CD available on the photobiology facility on the ASRID storage ring, ISA, has recently been developed. This is the only CD facility using a synchrotron radiation source (SRCD) in continental Europe, and one of three worldwide. This new SR CD facility allows CD and LD data to be collected with high spatial resolution and low noise background, which is not possible with a conventional commercial machine. Development of CD and LD on the synchrotron is especially interesting in that one can access shorter wavelength (down to 130 nm), and thereby extend classical analysis of spectra in order to obtain more accurate structural features.

PERSPECTIVES

A more complete understanding of spider silk offered by CD and LD, especially on the molecular level, will not only have a strong impact on building new materials but should also have implications in the understanding of the structural changes associated with biological fibre formation in general. This may be particularly interesting in medicine for pathogenic amyloid fibrils such as α-synuclein in Parkinson's disease (Jensen et al. 2000). Amyloid fibrils are insoluble aggregates that result from the self-assembly of partially unfolded proteins (Cohen & Skinner 1990; Zhang & Rich 1997). We conclude that the application of CD and LD techniques to silk and other spider products (e.g., venom) is expected to provide us with a better understanding, and perhaps leading to successful mimicking, of the self-assembly of silk and other biological products found in nature.

REFERENCES

Becker, M.A., Mahonney, D.V., Lenhert, P.G.,

Eby, R.K., Kaplan, D. & Adams, W.W. 1994. X-ray moduli of silk fibers from *Nephila clavipes* and *Bombyx mori*. *Silk Polymers* ACS Symp. Ser. 544, 185-195.

Cohen, A.S. & Skinner, M. 1990. New frontiers in the study of amyloidosis. *New England Journal of Medicine* 323, 542-543.

Gosline, J.M., Guerette, P.A., Ortlepp, C.S. & Savage, K.N. 1999. The mechanical design of spider silks: from fibroin sequence to mechanical function. *Journal of Experimental Biology* 202, 3295-3303.

Jackson, G.S., Hosszu, L.L.P., Power, A., Hill, A.F., Kenney, J., Saibil, H., Craven, C.J., Waltho, J.P., Clarke, A.R. & Collinge, J. 1999. Reversible conversion of monomeric human prion protein between native and fibrilogenic conformations. *Science* 283, 1935-1937.

Jensen, P.H., Islam, K., Kenney, J.M., Nielsen, M.S., Power, J. & Gai, W.P. 2000. Microtubule-associated Protein 1B is a Component of Cortical Lewy Bodies and Binds α-synuclein Filaments. *Journal of Biological Chemistry* 275, 21500-21507.

Kenney, J.M., Wise, M., Knight, D. & Vollrath, F. Submitted. Amyloidogenic nature of spider silk.

Kerkam, K., Viney, C., Kaplan, D. & Lombardi, S. 1991. Liquid crystallinity of natural silk secretions. *Nature* 349, 596-598.

Knight, D.P. & Vollrath, F. 1999. Liquid crystals and flow elongation in a spider's silk production line. *Proceedings of the Royal Society of London - Biological Sciences* 266, 519-523.

Magoshi, J., Magoshi, Y. & Nakamura, S. 1985. Crystallization, liquid-crystal, and fiber formation of silk fibroin. *Journal of Applied Polymer Science: Applied Polymer Symposium* 41, 187-204.

O'Brien, J.P., Fahnestock, S.R., Termonia, Y. & Gardner, K.C.H. 1998. Nylons from nature: Synthetic analogs to spider silk. *Advanced Materials* 10, 1185-1195.

Rodger, A. & Nordén, B. 1997. *Circular dichroism and linear dichroism*. Oxford Chemistry Masters Series, Oxford Univ. Press.

Schartau, W. & Leidersher, T. 1983. Composition of the hemolymph of the tarantula *Eurypelma californicum*. *Journal of Comparative Physiology* 152, 73-77.

Simmons, A.H., Michal, C.A. & Jelinsky, L.W. 1996. Molecular-orientation and 2-component nature of the crystalline fraction of spider dragline silk. *Science* 271, 84-87.

Sutherland, J.C. 1996. *Circular dichroism using synchrotron radiation* in *circular dichroism and the conformational analysis of biomolecules* (G. D. Fasman ed.). Plenum Press, New-York.

Vollrath, F. & Knight, D. 1999. Structure and function of the silk production pathway in the spider. *International Journal of Biological Macromolecules* 24, 243-249.

Vollrath, F., Knight, D. & Hu, X.W. 1998. Silk production in a spider involves acid bath treatment. *Proceedings of the Royal Society of London - Biological Sciences* 263, 817-820.

Willcox, P.J., Gido, S.P., Moller, W. & Kaplan, D. 1996. Evidence of a cholesteric liquid crystalline phase in natural silk spinning processes. *Macromolecules* 29, 5106-5110.

Zhang, S. & Rich, A. 1997. Direct conversion of an oligopeptide from a beta-sheet to an alpha-helix: A model for amyloid formation. *Proceedings of the National Academy of Science USA* 94, 23-28.

SEXUAL SELECTION

SPERM COMPETITION

European Arachnology 2000 (S. Toft & N. Scharff eds.), pp. 129-137.
© Aarhus University Press, Aarhus, 2001. ISBN 87 7934 001 6
(Proceedings of the 19th European Colloquium of Arachnology, Århus 17-22 July 2000)

Sexual selection in the drumming wolf spider *Hygrolycosa rubrofaciata*

JARI J. AHTIAINEN, RAUNO V. ALATALO, JANNE S. KOTIAHO, JOHANNA MAPPES, SILJA PARRI & LAURA VERTAINEN
University of Jyväskylä, Department of Biological and Environmental Science, P.O. Box 35, FIN-40351 Jyväskylä, Finland (jjahti@dodo.jyu.fi)

Abstract

We have studied the significance of courtship and agonistic signalling in the wolf spider *Hygrolycosa rubrofasciata* from the viewpoint of sexual selection theory. According to the 'good genes' sexual selection theory, females base their choice of mates on traits indicating heritable viability differences and are expected to benefit through better offspring survival. During the brief mating season in early spring, males of the wolf spider *H. rubrofasciata* move around their habitat (bog, meadow or abandoned field) searching for females and drum their abdomen on dry leaves to produce short audible drummings. There are two distinct signalling types: courtship drumming in intersexual signalling and agonistic drumming in intrasexual encounters. The courtship drumming of males increases their metabolic rate and tends to increase the risk of predation. As increased courtship drumming also reduces survival, courtship drumming can be seen as an evolutionarily significant cost. However, in unmanipulated conditions there is a positive correlation between survival and courtship drumming rate. The survival cost of increased courtship drumming is condition-dependent, such that males manipulated to be in a good condition tolerate the cost of drumming better than males manipulated to be in a poor condition. The offspring of sires with high courtship drumming rate has greater viability than offspring of sires with low courtship drumming rate, and females basing their choice on male drumming benefit in terms of increased offspring viability. Therefore, male courtship drumming is a honest viability indicator for choosy females in *H. rubrofasciata*. We have also found that males with higher agonistic drumming rate have better fighting ability in male-male encounters than males with lower agonistic drumming rate, but the role of agonistic interactions in sexual selection of *H. rubrofasciata* is likely to be minor compared to that of courtship drumming. Although body mass is one of the most common sexually selected male traits among animal taxa, body mass in *H. rubrofasciata* is not correlated with male mating success or drumming rate. However, differences in body mass increase the winning probability of the larger male in agonistic male-male interactions.

Key words: drumming, honesty of signalling, viability, intersexual selection, intrasexual selection, wolf spider

INTRODUCTION

Although Charles Darwin pointed out the importance of sexual selection already in 1871 in his *The Descent of Man and Selection in Relation to Sex*, it took over a century for scientists to ultimately discover its power in explaining the diversity and function of secondary sexual characters. Sexual selection can be divided into two components: intra- and intersexual selection (Andersson 1994). Intrasexual selection relates to the interactions between individuals of the same sex, and it is manifested as, for example, weapons or status signals. In intersexual selection the choosier sex, usually the female, is choosing either resources or secondary sexual traits of the opposite sex. While resource-based female choice is easy to understand due to the direct effects on female reproductive success, indirect selection affecting through offspring-performance has been debated. There are three main ideas of intersexual selection, the 'Fisherian self-reinforcing' hypothesis, the 'sensory bias' hypothesis and the 'good genes' hypothesis, that all focus on the evolution of sexual preferences for secondary sexual traits. The 'Fisherian self-reinforcing' hypothesis is based on the idea that when females prefer males with higher life-time reproductive success, their preference genes will be united in their offspring with the male's genes for higher attractiveness (Kirkpatrick 1982; Pomiankowski & Iwasa 1993). In this way both the female preference genes and the male attractiveness genes will then spread in the population. The preferred male secondary sexual traits can be neutral or even disadvantageous, the only thing that matters is that these traits must be carried by males with higher fitness. The 'sensory bias' hypothesis argues that there are pre-existing sexual preferences ('sensory bias') for certain male secondary sexual traits that females inherit from their ancestors (Ryan 1985). The preferred secondary sexual traits will spread in the population by simply utilizing this pre-existing sensory bias. According to the 'good genes' or 'handicap' hypothesis, females should prefer males displaying honest, costly secondary sexual traits that signal genes for survival (Andersson 1986, 1994; Johnstone 1995). The 'good genes' hypothesis states that as the size of a secondary sexual trait increases, male fitness increases in terms of mating success, but this is traded-off with viability which decreases as ornament size increases. Males vary in their ability to withstand the costs of secondary sexual characters so that males in poor condition have lower op-

timum level of the ornament expression than males in good condition. Thus, males in good condition have better mating success and fitness, and females mating with them will enjoy increased viability for their offspring inherited from their fathers.

The sexual signalling of the wolf spider *Hygrolycosa rubrofasciata* (Ohlert), i.e. drumming, offers a convenient way of studying both inter- and intrasexual selection. Sexual signalling rate of *H. rubrofasciata* males can be manipulated without a risk of manipulation itself having side effects. This can be accomplished by not manipulating males themselves, but manipulating only their environment; when males are placed in the vicinity of a virgin female they spontaneously increase their drumming rate. However, males of *H. rubrofasciata* produce drumming signals even without the presence of females. The signals that males of *H. rubrofasciata* produce travel as vibrations both in the substrate and in the air. Thus, it can be directly tested whether there is any difference in female response behaviour to male signals depending on how they receive the signals. Also, males of *H. rubrofasciata* produce agonistic drumming signals while encountering other males. The characteristics of tape-recorded male drumming can be manipulated to test for the importance of different components of the drum in sexual selection.

ECOLOGY AND LIFE-HISTORY

Hygrolycosa rubrofasciata is a ground-dwelling wolf spider (Lycosidae) which can be found in patchily located populations widely distributed over northern Europe. It is found along coastal regions as well as inland in south-eastern and northern parts of Finland. *H. rubrofasciata* basically inhabits two kinds of habitats: abandoned fields and other meadow habitats and half-open bogs with deciduous trees. Although there may be tens of *Hygrolycosa rubrofasciata* in a square meter of the high-density habitat, they are not readily visible, since they usually hide under leaf litter. These two habitats differ from each other in many ecological characters, e.g. in soil moisture, pH, vegetation, soil fauna, and soil litter thickness.

Male drumming

In southern Finland most matings of *H. rubrofasciata* take place during sunny days immediately after the snow has melted in late April or early May. All males die during or shortly after the mating season which lasts a few weeks. During the mating season, males of *H. rubrofasciata* produce drumming signals by hitting their abdomen on dry leaves or other suitable substratum to court females (Kronestedt 1984, 1996). One drumming consists of ca. 30-40 separate pulses, lasting ca. 1 second (Rivero et al. 2000), and it is audible to the human ear up to a distance of several meters. Female spiders receive these signals both as audible airborne signals and as substrate borne vibrations. In the experiment conducted by Parri et al. (submitted) investigating the effect of the manipulation of the substrate contact between males and females on female response behavior, the proportion of trials in which females responded to male drumming did not differ between the contact and non-contact set-ups. This means that females were equally choosy for airborne signals as in situations where vibrations via the substrate were present. Thus, the acoustic component of male drumming is truly important for female choice in *H. rubrofasciata*.

Based on both mark-recapture method and direct observations in field and laboratory, Kotiaho et al. (2000) have found that males of *H. rubrofasciata* are not distributed randomly among the habitat: fewer males are found in areas that have high sedge cover and low elevation, and males spend more time on dry leaf substrate than on other substrates. Drumming rate in the field is positively correlated with dry leaf cover, and males clearly prefer dry leaves as a drumming substrate. Male drumming rate and mobility are positively correlated with temperature. Therefore, males may be sexually selected to optimize their signalling habitat by active microhabitat choice. Interest-ingly, on forested bog habitats the availability of dry leaf litter is considerably higher than on more open meadow habitats that are dominated by grass vegetation. Male drumming is more frequent in bogs than in meadows (Vertainen et al. unpublished data). Thus, adaptive differences in male mating tactics may have evolved between these two habitat types. During the mating season, males are actively searching for receptive females, and while searching engage in agonistic encounters with other males. In these encounters males use a different type of drumming signal (Kronestedt 1984). This agonistic drumming is also audible to the human ear, but it is shorter and more intense than courtship drumming. There is no possibility for observers to confuse agonistic signal with courtship signal.

Life-history

Males of *H. rubrofasciata* drum while wandering around the habitat searching for receptive females. Female draglines seem to affect males, and thus chemical substances in them might be used for sexual communication. When a male encounters a female he stops and drums several times. If the female is willing to copulate with the male, she responds by vibrating her body. This response is given immediately after the male drum, and it is clearly visible like the male drummings, although female percussions are usually not sufficiently intense to produce audible sounds. The existence of such a clear response allows the experimenter to determine explicitly which male or male signal the female prefers. After an initial female response male and female, while approaching each other, drum several times before the copulation begins (the so-called duetting, Kronestedt 1996).

Mated females of *H. rubrofasciata* produce an egg sac, which they carry attached to their spinnerets (silk glands in their abdomen). After approximately three to four weeks the offspring will emerge from the egg sac. The offspring usually remain on the female's abdomen or on top of the empty egg sac for a day to chitinise their exoskeleton, after which they disperse. The development of *H. rubrofasciata* to maturation typically lasts 2-3 years in Southern Finland depending on the prevailing environmental conditions. Male spiderlings resemble cryptic brownish female spiderlings, and only after the final moult does the typical blackish appearance of males emerge. The final moult of maturing individuals happens in autumn preceding the mating season the following spring. Each adult male cohort reproduces only during one mating season, and males die during or immediately after it. However, we have observed individually marked females surviving until the next mating season.

GENERAL METHODOLOGY

Hygrolycosa rubrofasciata can easily be collected alive by pitfall traps and by hand picking. Unmated virgin females can be collected while there is still some snow on the ground. After collection, the spiders are placed individually into small plastic containers (film jars) with some moist moss (*Sphagnum* spp.) until they are brought to the laboratory. Before experiments spiders are prevented from reaching the active sexual phase by keeping them at a cool temperature (+3-7 °C). In the laboratory *H. rubrofasciata* are fed with fruit flies (*Drosophila* spp.) and springtails.

The characteristics of male drumming can be directly manipulated to test for the importance of different components of the drum in female choice (Parri et al. 1997; Parri 1999; Rivero et al. 2000). Repeatability and variability of the acoustic signal characteristics have been used to categorize signal components into static and dynamic parts (Gerhardt 1991; Castellano & Giacoma 1998). Static characteristics are highly repeatable within individuals with a low degree of variability among them, whereas dynamic characteristics are described as being repeatable within and variable among individuals. Such studies are essential to understand the significance of signals as honest handicaps, arbitrary Fisherian traits or species recognition traits. If signal components are repeatable, highly variable among individuals,

and correlated with fitness, they may convey information about male quality. Repeatable but less variable components might serve for species recognition purposes or they might be more of the Fisherian type of traits.

SEXUAL SELECTION
Intrasexual selection
Kotiaho et al. (1997) showed that large differences in body mass and courtship drumming rate between the two rivals seem to increase independently the winning probability of the larger or more active male *H. rubrofasciata*. Kotiaho et al. (1999a) found that agonistic drumming rate indicates male fighting ability, and that relative size asymmetry and motivation to fight both contribute to fighting ability. They also found that male-male competition decreases the courtship drumming rate of subdominant males, suggesting that male-male competition limits the opportunities for female choice. While courtship and agonistic drumming rate, and body mass in particular, affect the fighting success of males, male-male interactions may, however, be of relatively minor importance in sexual selection of *H. rubrofasciata*.

Intersexual selection
In the wolf spider *Hygrolycosa rubrofasciata* females choose males with the highest drumming rates as mating partners (Kotiaho et al. 1996). Females disproportionately choose more males with higher drumming rate than would be expected if females exhibited passive choice, i.e. if females were randomly choosing males in direct proportion to drumming rates. The sexual signalling, i.e. drumming rate, is thus under directional female choice. Drumming rate is repeatable within males and highly variable among males (i.e. dynamic) (Kotiaho et al. 1996; Rivero et al. 2000). Female *H. rubrofasciata* also prefer higher volume of the signal and longer drumming signals of males (Rivero et al. 2000). In addition, females respond more quickly to playbacks of male signals with a higher drumming rate and volume (Parri et al.

1997). This suggests that females may use a threshold level when responding to male courtship signals, and that they are prepared to suffer costs of waiting for an opportunity to choose between different males. In addition to drumming rate, signal length is also repeatable within males and has high variability among males (Rivero et al. 2000). Likewise, signal volume exhibits moderate within-male repeatability, but in reality the differences in distance and substrate confound female perception of volume differences. Both signal length and volume are positively correlated with the rate at which males produce drumming signals. Two dynamic characteristics of the male acoustic signal, i.e. drumming rate and length, are thus most likely to operate as indicators informing the females about the quality of wolf spider males.

In contrast, pulse rate (number of pulses divided by signal length where signal length is the total duration of the signal in milliseconds), symmetry of the signal (peak time/signal length, where peak time is the section of the signal at which the maximum intensity occurs), or peak frequency (the frequency at which the highest amplitude occurs, in Hz) have not been found to be targets of female choice in *H. rubrofasciata* (Parri 1999; Rivero et al. 2000). They have low repeatability and/or small among-male variability, and are not related to any other male trait (e.g. body mass, drumming rate, or mobility). Pulse rate might operate as a static characteristic related to species recognition. In several studies on insects pulse rate has indeed turned out to be under stabilizing rather than directional selection (e.g. Ritchie et al. 1994). Indeed, many spiders use substrate-borne vibrations presumably for species recognition (e.g. Schüch & Barth 1990; Barth & Schmitt 1991). While in other animal groups the peak frequency of sexual signals may convey honest information on male size, in this spider species peak frequency is not related to male size and has low repeatability (Rivero et al. 2000).

In *H. rubrofasciata*, there is no correlation between male drumming rate and body mass (e.g. Kotiaho et al. 1996; Mappes et al. 1996), and the body size of males does not seem to be intersexually selected (Kotiaho et al. 1996). This is surprising because body size is one of the most common sexually selected male traits among other animal taxa (Andersson 1994). In *H. rubrofasciata*, male viability is independent of body mass, even when males are stressed, large males have a small but significant survival advantage compared with smaller males (Mappes et al. 1996). There is no detectable effect of male body mass on overwinter survival (Kotiaho et al. 1999b). Interestingly, however, larger males lose proportionally more mass than smaller males. This suggests that larger males may have an additional energy expenditure because of their large body mass while overwintering, but also that large males can afford to expend more energy than smaller males.

Honesty of male courtship drumming

The sexual signalling of *H. rubrofasciata* is a particularly well-tested signalling system where drumming rate has really proved to be an honest viability indicator. There are three conditions that must be met before a signal can be classified as an honest signal: (I) there must be within-individual repeatability and among-individual variability in the signal, which must have an effect on mate choice; (II) there must be substantial costs of signalling in terms of increased mortality or reduced subsequent mating success; (III) there must be condition-dependence of signalling in a way that there are differential costs between individuals in different condition. Then, in a given signalling level individuals in poor condition pay higher costs than individuals in better condition.

I. Variation and repeatability. In *H. rubrofasciata*, there are considerable within-individual repeatability and among-male variability in drumming rate (Kotiaho et al. 1996), which have an effect on mate choice. If there was no phenotypic variability in sexual signalling, there could not be any selective pressure

Within-individual repeatability is also important, since otherwise a signal cannot be used as a reliable source of information about male quality.

II. Costs. Every new signal incurs costs of production. Those costs may be physiological such as increased energy expenditure, or direct costs such as increased risk of predation. These fitness costs have rarely been found even though they play an important role for the evolution of secondary sexual traits.

a. Physiological costs. Drumming of *H. rubrofasciata* is energetically highly demanding to its bearer (Kotiaho et al. 1998a): During drumming, metabolic rate is 22 times higher than at rest and four times higher than when males are actively moving. Metabolic rate per unit mass is positively related to absolute body mass during sexual signalling but not during other activities. Indeed, it seems that the largest males can drum only 12 times per minute before reaching the maximum sustainable metabolic rate, whereas the smallest males may drum up to 39 times per minute. However, there is no relationship between body mass and drumming rate, indicating that larger males are able to compensate for the higher cost of drumming (Kotiaho et al. 1998a; Kotiaho 2000).

Physiological costs are ultimately realized as an increased probability of mortality. When Mappes et al. (1996) induced a set of male *H. rubrofasciata* to increase their drumming rate by presenting females in their proximity, these males suffered higher mortality and lost significantly more weight than control males, directly verifying that drumming is costly. Within the female induced treatment group males that drummed most actively survived better than less active males.

b. Predation costs. In addition to physiological costs of male drumming, higher mate searching activity clearly impairs male *H. rubrofasciata* by causing a direct increase in predation risk (Kotiaho et al. 1998b). Also, there is a tendency that more actively drumming males have higher risk of predation. Increased predation risk while drumming has lead to a counter-adaptation: male drumming rate decreases drastically in the presence of a predator (Kotiaho et al. 1998b). In *H. rubrofasciata* both increased mate searching activity and drumming benefit males through sexual selection, but at the same time natural selection provokes direct balancing costs on the same traits.

III. Condition-dependence and differential costs. In *H. rubrofasciata*, there is a clear positive correlation between male drumming rate and survival (Kotiaho et al. 1996, 1999c; Mappes et al. 1996). When Mappes et al. (1996) manipulated the phenotypic condition of males, males in a high food level treatment maintained their drumming rate at a high level, while males with intermediate and low food levels exhibited a reduction in drumming rates. Thus, males vary in their ability to bear the costs of drumming. Kotiaho (2000) showed that when the presence of females induced males to increase their drumming rate, large males manipulated to be in good condition survived better than large males manipulated to be in poor condition, that is, there is a three-way interaction between trait expression, condition and size on survival (see also Kotiaho et al. 1996, 1999c). Therefore, survival costs of male drumming are condition-dependent, being manifested in decreased viability of males in poorer condition.

Indirect genetic benefits for offspring
In *H. rubrofasciata* females prefer males with high drumming rate. As male drumming is not known to indicate any direct benefits affecting female fecundity or survival, there may be indirect genetic benefits for offspring. Indeed, by choosing males with the highest drumming rates, females of *H. rubrofasciata* benefit through better offspring survival (Alatalo et al. 1998). The estimated correlated response in offspring viability was rather small (0.12). However, it may be sufficiently large if the costs of being choosy are small. In fact, females mate with

better-than-average males just by responding passively to a random drumming signal, since this allows more actively drumming males to be observed more frequently. The active choice by females seems to increase this benefit only slightly. In many mating systems, females obtain better-than-average males as a consequence of intense male-male competition or because of the extraordinary variance in male signalling. This may be the general solution to the lek paradox, which stems from the fact that in polygamous species females appear to copulate with a small subset of available males. Such strong directional selection is predicted to deplete additive genetic variance in the preferred male traits (Fisher 1958; Falconer 1989; Roff 1997), yet females continue to mate selectively, thus generating the lek paradox (Borgia 1979; Taylor & Williams 1982; Kirkpatrick & Ryan 1991). Most studies only report small genetic fitness benefits of female choice (review in Møller & Alatalo 1999). Our results indicate that the costs of any additional choice may be so minor that female choice for honestly signalling males may evolve even with small benefits in offspring viability.

To conclude, male drumming rate is an honest signal which indicates heritable viability for females, thus reflecting phenotypic and genetic quality of males. These results provide evidence that a good-genes process based on the costly male drumming trait, reflecting heritable viability and thus genetic benefits for offspring, is involved in the evolution of sexual selection through female choice in the wolf spider *H. rubrofasciata*.

FUTURE PROSPECTS

Recently, there has been much interest in estimating fluctuating asymmetry (FA) of morphological traits as a short-cut measure of overall individual quality (review in Møller & Thornhill 1998). FA deals with small differences and thus measurement error is often relatively large. However, repeated measurements and large sample sizes allow reliable estimates of fluctuating asymmetry that can be corrected for errors. As an example, Ahtiainen et al. estimated how much fluctuating asymmetry predicts male quality in the wolf spider *H. rubrofasciata*. Pedipalps as the most repeatable of all bilateral traits in this species were measured (n = 804) with drumming rate and mobility as fitness-related references. Our data showed a weak negative relationship between an estimate of male sexual performance and pedipalp asymmetry ($r = -0.100$, $P = 0.007$), which inevitably underestimates the true relationship, given the measurement error. It is possible to estimate the unbiased relationship by correcting the above correlation coefficient with effective reliability estimates of FA and sexual performance, and thus r will be -0.181. In conclusion, fluctuating asymmetry as a measure of variation in male quality at the population level is suitable only if the sample size (and/or the number of repeated measurements) is large enough to overcome the masking effect of measurement error.

Our recent interest is in the study of adaptive variation in isolated populations of *Hygrolycosa rubrofasciata* with the particular goal of conservation applications. However, this research will also have the ambition to understand evolution of adaptive variation in general. Gene flow between populations, even those located within a few hundred meters from each other, seems to be highly restricted, and we have observed adaptive genetic differences between populations on a very small geographical scale (Vertainen et al. unpublished data). We are paying particular attention to the interaction between sexual signalling and the viability of small isolated populations. These studies utilize modern quantitative genetic techniques.

ACKNOWLEDGEMENTS
The work was financed by the Academy of Finland.

REFERENCES
Ahtiainen, J.J., Alatalo, R.V., Mappes, J. & Vertainen, L. (submitted). Fluctuating asymme-

try and sexual performance in the drumming wolf spider *Hygrolycosa rubrofasciata*.

Alatalo, R.V., Kotiaho, J., Mappes, J. & Parri, S. 1998. Mate choice for offspring performance: major benefits or minor costs? *Proceedings of the Royal Society of London B* 265, 2297-2301.

Andersson, M. 1986. Evolution of condition-dependent sex ornaments and mating preferences: sexual selection based on viability differences. *Evolution* 40, 804-816.

Andersson, M. 1994. *Sexual selection*. Princeton University Press, Princeton.

Barth, F.G. & Schmitt, A. 1991. Species recognition and species isolation in wandering spiders (*Cupiennius* spp.; Ctenidae). *Behavioral Ecology and Sociobiology* 29, 333-339.

Borgia, G. 1979. Sexual selection and the evolution of mating systems. In: *Sexual selection and reproductive competition in insects.* (M.S. Blum & N.A. Blum eds.), pp. 19-80. Academic Press, New York.

Castellano, S. & Giacoma, C. 1998. Stabilizing and directional female choice for male calls in the European green toad. *Animal Behaviour* 56, 275-287.

Darwin, C.R. 1871. *The descent of man and selection in relation to sex.* John Murray, London.

Falconer, D.S. 1989. *Introduction to quantitative genetics.* 3rd edition. Longman, New York.

Fisher, R.A. 1958. *The genetical theory of natural selection.* Oxford University Press, Oxford.

Gerhardt, H.C. 1991. Female mate choice in treefrogs: static and dynamic criteria. *Animal Behaviour* 42, 615-635.

Johnstone, R.A. 1995. Sexual selection, honest advertisement and the handicap principle: reviewing the evidence. *Biological Reviews* 70, 1-65.

Kirkpatrick, M. 1982. Sexual selection and the evolution of female choice. *Evolution* 36, 1-12.

Kirkpatrick, M. & Ryan, M.J. 1991. The evolution of mating preferences and the paradox of the lek. *Nature* 350, 33-38.

Kotiaho, J.S. 2000. Testing the assumptions of conditional handicap theory: costs and condition dependence of a sexually selected trait. *Behavioral Ecology and Sociobiology* 48, 188-194.

Kotiaho, J., Alatalo, R.V., Mappes, J. & Parri, S. 1996. Sexual selection in a wolf spider: male drumming activity, body size and viability. *Evolution* 50, 1977-1981.

Kotiaho, J., Alatalo, R.V., Mappes, J. & Parri, S. 1997. Fighting success in relation to body mass and drumming activity in the male wolf spider *Hygrolycosa rubrofasciata*. *Canadian Journal of Zoology* 75, 1532-1535.

Kotiaho, J.S., Alatalo, R.V., Mappes, J., Nielsen, M.G., Parri, S. & Rivero, A. 1998*a*. Energetic cost of size and sexual signalling in a wolf spider. *Proceedings of the Royal Society of London B* 265, 2203-2209.

Kotiaho, J., Alatalo, R.V., Mappes, J., Parri, S. & Rivero, A. 1998*b*. Male mating success and risk of predation in a wolf spider: a balance between sexual and natural selection? *Journal of Animal Ecology* 67, 287-291.

Kotiaho, J.S., Alatalo, R.V., Mappes, J. & Parri, S. 1999*a*. Honesty of agonistic signalling and effects of size and motivation asymmetry in contests. *Acta Ethologica* 2, 13-21.

Kotiaho, J.S., Alatalo, R.V., Mappes, J. & Parri, S. 1999*b*. Overwintering survival in relation to body mass in a field population of the wolf spider (*Hygrolycosa rubrofasciata*). *Journal of Zoology* 248, 270-272.

Kotiaho, J.S., Alatalo, R.V., Mappes, J. & Parri, S. 1999*c*. Sexual signalling and viability in a wolf spider (*Hygrolycosa rubrofasciata*): measurements under laboratory and field conditions. *Behavioral Ecology and Sociobiology* 46, 123-128.

Kotiaho, J.S., Alatalo, R.V., Mappes, J. & Parri, S. 2000. Microhabitat selection and audible sexual signalling in the wolf spider *Hygrolycosa rubrofasciata* (Araneae; Lycosidae). *Acta Ethologica* 2, 123-128.

Kronestedt, T. 1984. Ljudalstring hos vargspindeln *Hygrolycosa rubrofasciata. Fauna och Flora* 79, 97-107.

Kronestedt, T. 1996. Vibratory communication in the wolf spider *Hygrolycosa rubrofasciata*

(Araneae, Lycosidae). *Revue Suisse de Zoologie* Hors série 2, 341-354.

Mappes, J., Alatalo, R.V., Kotiaho, J. & Parri, S. 1996. Viability costs of condition-dependent sexual male display in a drumming wolf spider. *Proceedings of the Royal Society of London B* 263, 785-789.

Møller, A.P. & Thornhill, R. 1998. Bilateral symmetry and sexual selection: a meta-analysis. *American Naturalist* 151, 174-192.

Møller, A.P. & Alatalo, R.V. 1999. Good-genes effects in sexual selection. *Proceedings of the Royal Society of London B* 266, 85-91.

Parri, S. 1999. Female choice for male drumming characteristics in the wolf spider *Hygrolycosa rubrofasciata* (dissertation). *Biological Research Reports from the University of Jyväskylä*, 108 p.

Parri, S., Alatalo, R.V., Kotiaho, J. & Mappes, J. 1997. Female choice for male drumming in the wolf spider *Hygrolycosa rubrofasciata*. *Animal Behaviour* 53, 305-312.

Parri, S., Alatalo, R.V., Kotiaho, J.S., Mappes, J. & Rivero, A. (submitted). Sexual selection in the wolf spider *Hygrolycosa rubrofasciata*: female preference for drum length and pulse rate.

Pomiankowski, A. & Iwasa, Y. 1993. Evolution of multiple sexual preferences by Fisher's runaway process of sexual selection. *Proceedings of the Royal Society of London B* 253, 173-181.

Ritchie, M.G., Yate, V.H. & Kyriacou, C.P. 1994. Genetic variability of the interpulse interval of courtship song among some European populations of *Drosophila melanogaster*. *Heredity* 72, 459-464.

Rivero, A., Alatalo, R.V., Kotiaho, J.S., Mappes, J. & Parri, S. 2000. Acoustic signalling in a wolf spider: can signal characteristics predict male quality? *Animal Behaviour* 60, 187-194.

Roff, D. 1997. *Evolutionary quantitative genetics.* Chapman & Hall, New York.

Ryan, M. 1985. *The tungara frog: A study in sexual selection and communication.* University of Chicago Press, Chicago.

Schüch, W. & Barth, F.G. 1990. Vibratory communication in a spider: female responses to synthetic male vibrations. *Journal of Comparative Physiology* 166, 817-826.

Taylor, P.D. & Williams, G.C. 1982. The lek paradox is not resolved. *Theoretical Population Biology* 22, 392-409.

European Arachnology 2000 (S. Toft & N. Scharff eds.), pp. 139-144.
© Aarhus University Press, Aarhus, 2002. ISBN 87 7934 001 6
(Proceedings of the 19th European Colloquium of Arachnology, Århus 17-22 July 2000)

Copulation and emasculation in *Echinotheridion gibberosum* (Kulczynski, 1899) (Araneae, Theridiidae)

BARBARA KNOFLACH

Institute of Zoology und Limnology, University of Innsbruck, Technikerstrasse 25, A-6020 Innsbruck, Austria
(konrad.thaler@uibk.ac.at)

Abstract

In *Echinotheridion gibberosum* (Kulczynski, 1899) from the Canary Islands and Madeira copulation regularly ends in emasculation and sexual cannibalism, which is unusual among spiders. The male dies already at the beginning of insertion. After ca. 4 min of insertion, the female amputates the single male palp by circling. The disconnected gonopod remains fastened to the epigynum for about 5 hours, probably acting as a non-permanent mating plug. During this period, the palpless male is consumed by the female. In a new species of the closely related genus *Tidarren* similar behaviour has already been observed.

Key words: Copulatory behaviour, emasculation, sexual cannibalism, *Echinotheridion*, *Tidarren*, Theridiidae, Canary Islands.

INTRODUCTION

The genus *Echinotheridion* Levi, 1963 is considered to be closely related to *Tidarren* Chamberlin & Ivie, 1934 (Wunderlich 1992). In both genera males are dwarf and possess only one palp (Fig. 1), which is unique among spiders. Generic separation is mainly based on a female character, spurs on the fourth coxae, which are typical of *Echinotheridion*. In *Tidarren* outstanding behavioural traits were found. Males amputate one of their palps after the penultimate moult. The single-palp copulation routinely ends in sexual cannibalism (Branch 1942; Knoflach & van Harten 2000). *Tidarren argo* Knoflach & van Harten, 2001 (from Yemen) even synchronises copulation and sexual cannibalism by amputation of the male palp, which then remains attached to the epiygnum as an ephemeral mating plug, while the female feeds on the emasculated male (Knoflach & van Harten 2001). Does similar behaviour occur also in *Echinotheridion*? To answer this question, it was of interest to investigate more closely the

only Old World representative of this genus, *E. gibberosum* (Kulczynski 1899) from the Canary Islands and Madeira, and to re-evaluate a first casual copulatory observation on this species by Schmidt (1980). Self-removal of a palp was already suggested for *Echinotheridion* by Levi (1980)

Fig. I. *Echinotheridion gibberosum.* Copulation, late phase, right palp inserted; female already has started to turn around.

and Ramirez & Gonzalez (1999), but has not been observed hitherto. Even emasculation might be expected, as once a female was found with the male palp fixed to the epigynum (Ramirez & Gonzalez 1999).

MATERIAL AND METHODS

Specimens of *Echinotheridion gibberosum* were collected in pine and laurel forest on Tenerife, Canary Islands, above Orotava/Aguamansa and in mts. Anaga, near Las Mercedes, February 2000. They were maintained in plastic boxes at room temperature. Behaviour was observed with a stereo microscope with horizontal objective body (Nikon SMZ-2B), magnification up to x50. Five copulations were videotaped with a SONY DXC-325P. Altogether 25 successful copulations were observed, all referring to virgin pairs. There are differences in the numbers given for each copulatory element, as it was not possible to record all

traits with the same accuracy in each observation. Functional contact of genitalia was analysed in 8 females freezed after emasculation, with the aid of a stereo- and compound microscope.

RESULTS
Morphology

Dimensions of male / female [mean (min-max), n = 10]: body length 1.7 (1.4 - 1.9) / 3.5 (3.3 - 3.7), prosoma length 0.7 (0.6 - 0.7) / 1.4 (1.3 - 1.6), prosoma width 0.6 (0.5 - 0.6) / 1.1 (1.0 - 1.2), length femur I 1.3 (1.2 - 1.4) / 3.0 (2.7 - 3.4), tibia I 0.8 (0.8 - 1.0) / 1.9 (1.8 - 2.2) mm; femur of male palp ca. 0.3 mm long.

All adult males possess only one palp. Out of 45 males investigated, 20 had the left and 25 the right palp. The male palp is pincershaped, owing to the strongly protruding cymbium and conductor (Figs. 2, 3, 9, 10), see also Schmidt (1981; sub *Tidarren pseudogibberosum* Schmidt, 1973). The cymbium ends in a hook-like, apical process and

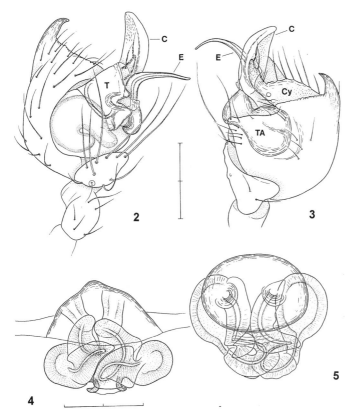

Figs 2 - 5. *Echinotheridion gibberosum*: **(2)** Right male palp, prolateral. **(3)** retrolateral-apical. **(4)** epigynum/vulva, aboral. **(5)** ventral. Scale lines: 0.2 mm. Abbreviations: C: conductor, Cy: cymbium, E: embolus, T: tegulum, TA: tegular apophysis.

Fig. 6. *Echinotheridion gibberosum.* **(a)** Ecdysis into subadult stage of male. **(b)** Moulting completed, legs bent, both palps present. **(c)** Early phase of amputation, circling with left palp raised. **(d)** Another specimen, right palp just amputated.

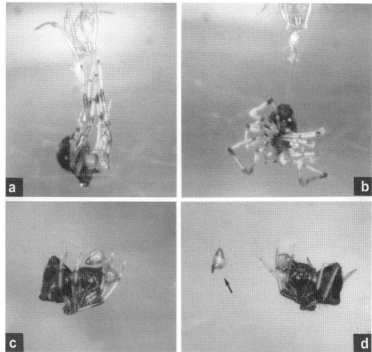

in a hairy, slender lobe, which bears the tarsal organ and which adjoins the broad furrow of the conductor. Base of embolus is slender.

Females have a spur on the posterior base of coxae IV. The claw of the female palp is fan-like. The epigynum is a broadly rounded, sclerotised protuberance (Figs. 4, 5, 9, 10). On its anterior declivity open the separate introductory orifices. The copulatory ducts are heavily sclerotised, coiled and folded, with narrow lumen.

Genitalia coupling (Figs. 9, 10): Holdfast structures of the palp are sclerites, not modified haematodochae. The basal haematodocha is large, the distal one rudimentary. The hook-like apex of the cymbium locks behind the epigynal protuberance in a fold of the integument, whereas the conductor adjoins the anterior side of the epigynum. The coxal spurs on legs IV of the female are not necessarily used for genitalia coupling. During insertion, contact between the male palp and the coxal spurs sometimes was present and sometimes not.

Palp-amputation

Loss of one palp is achieved by self-amputation also in *E. gibberosum* males. For the first time in this genus two amputations were observed. Two (three) hours after the penultimate moult (Figs 6a, b) the subadult male raised one palp (Fig. 6c) and started to rotate until the palp became fixed to the threads of the web (see Knoflach & van Harten 2000). Then the male circled around its own appendage, thereby twisting off the palp (Fig. 6d). The process lasted 6 (11) minutes and involved 16 (17) rotations. The detached palp was sucked out. Another two freshly moulted males did not perform the amputation during 7 hours of observation.

Copulatory behaviour

Sperm induction obviously takes place in the period before copulation, but was not observed. For courtship the male walks around the female, attaching draglines and moving his abdomen in jerks from time to time. He regularly approaches the female and palpates her forelegs. The female vibrates her body throughout the whole courtship, regularly moves her first legs up and down and also cleans them with the chelicerae. She faces towards the male, adjusting her position

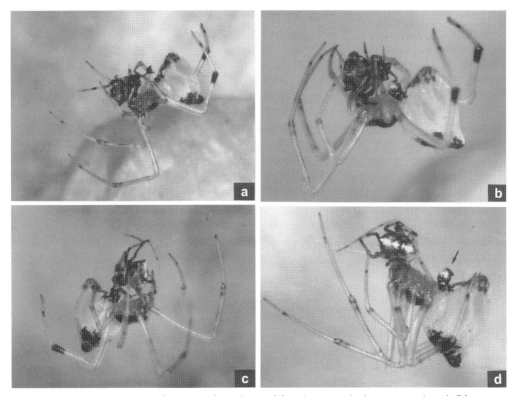

Fig. 7. *Echinotheridion gibberosum*. Sequence of copulation: **(a)** male approach along mating thread; **(b)** insertion of left palp; **(c)** emasculation by female circling; **(d)** emasculation completed, female starts to feed on male . Arrow points to dismembered palp, breaking point between trochanter and femur.

relative to him. The male then spins the mating thread, which may be reinforced several times and is 2 - 5 cm long. This thread is intensively pulled by the male 1 - 19 times. With each pulling sequence the female becomes strongly shaken and approaches along the thread. Finally, the male also approaches her for insertion with his palp extended forwards (Fig. 7a). Courtship lasts about 12 minutes ($\bar{x}$ = 12.2 ± 4.2 s.e.; range 1.4 - 56.0; n = 14). It should be mentioned, that not every pairing ended in successful copulation. An additional 12 virgin pairings were negative.

Copulation takes place on the mating thread. Insertion was successful at once in 12 out of 18 copulations; further two copulations with 1, two with 2, one with 3, and one with 5 unsuccessful insertion attempts. During insertion both partners are completely motionless (Figs. 1, 7b). The male dies from fatigue as soon as genitalia contact is achieved. He remains passively coupled to the female, his legs being contracted as in dead spiders. The female is cataleptic. After about 4 minutes ($\bar{x}$ = 3.9 ± 0.9 s.e.; range 0.3 - 15.4; n = 21) the female awakes from her catalepsy, entangles the male and starts to turn around in circles (Fig. 7c) by using her legs. After 3-20 rotations the palp becomes amputated (Figs. 7d, 8). Some females started several times until they succeeded in breaking off the palp. Emasculation is completed after ca. 3 minutes ($\bar{x}$ = 3.4 ± 0.8 s.e.; range 0.5 - 12.0; n = 17). The palp broke off between tibia and tarsus (Fig. 8) in 23 cases, between femur and trochanter in two cases (Fig. 7d). The detached palp then remains on the epigynum for the next five hours on average ($\bar{x}$ = 5.4 ± 1.0 s.e.; range 1.0 - 11.4; n = 9), until it is removed by the female. Synchronously the female feeds on the palpless male (for 2-5 h; Figs. 7d, 8). One female was observed to suck out the removed male palp, after she had pushed it away from the epigynum with her

Fig. 8. *Echinotheridion gibberosum.* Sexual cannibalism after emasculation; palpless male is sucked out, detached palp (arrow) fastened to epigynum. Arrow points to dismembered palp, breaking point between tibia and tarsus.

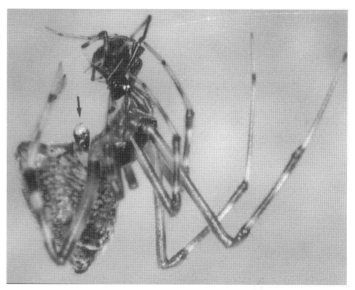

hindlegs. Three females were allowed to remate on the next day. However, instead of vibrating their body they pulled threads and no copulation followed (observation time from 7 hours up to 3 days).

Anomalous observations: One female performed 55 rotations for successful emasculation, changing five times direction of rotation. Another female started to feed on the male already during insertion when she still was circling. 90 minutes later she completed emasculation. A third female did not amputate the male palp nor did she feed on him during observation. For about 3 hours the

dead male was observed to be just connected to the epigynum. On the following day I found him emasculated and sucked out.

DISCUSSION

Echinotheridion gibberosum exhibits similar behavioural traits as the two *Tidarren* species hitherto studied. Palp-amputation of the subadult male follows the same pattern (Branch 1942; Knoflach & van Harten 2000). Also courtship and copulation correspond to *T. cuneolatum* (Tullgren, 1910) and in particular to *T. argo* from Yemen: Courtship proceeds via a mating thread; copulation

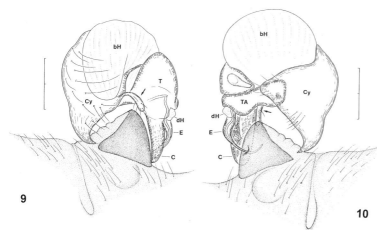

9

10

Figs 9 - 10. *Echinotheridion gibberosum.* Functional contact of genitalia after emasculation, epigynum with left palp, **(9)** from the left side and **(10)** from the right side, female punctured. Abbreviations: bh: basal haematodocha, dh: distal haematodocha; C: conductor, Cy: cymbium, E: embolus, T: tegulum, TA: tegular apophysis; arrow points to lobe of cymbium. Scale lines: 0.2 mm.

takes place on the mating thread, involves single usage of the male palp and ends in exhaustion of the male, in emasculation and subsequent sexual cannibalism (Knoflach & van Harten, 2001). The courtship repertoire of *E. gibberosum* females lacks twanging with second legs, which is a typical element of *Tidarren* spp. Instead, the first legs are used. Remarkably, copulatory behaviour of *E. gibberosum* was already basically observed by Schmidt (1980), with reference to palp-disconnection and fatigue of the male. However, this observation was mistaken as a heterospecific copulation between two genera (Schmidt 1980, 1981).

In *Tidarren argo* from Yemen the male is emasculated immediately after application of the male palp. The separated palp remains fixed to the epigynum for several hours, supported by special haematodochal holdfast structures. Unlike this new *Tidarren* species, in *E. gibberosum* emasculation does not take place at the beginning of genitalia contact. At first there is a 'normal' insertion for about four minutes, during which both partners are completely motionless and the male apparently dies. Emasculation is not that efficient as in *T. argo*. It lasts more than three minutes. Breaking point of the palp usually is identical to *T. argo*. Holdfast structures of the palp are provided only by sclerites, not by haematodochal horns. The result is the same: the male palp is fastened to the epigynum for hours, while the female is occupied with mate consumption. The function of the coxal spurs on legs IV of the female remains uncertain, as they are not necessary for genitalia coupling.

Emasculation obviously allows long contact of genitalia and concurrent nutritional benefits to the female. The palp blocks the epigynum and therefore may serve as an ephemeral mating plug, reducing female receptivity for a while. Probably, the dismembered palp goes on with sperm transfer independently.

Could palp-amputation of the subadult male and emasculation behaviour have evolved convergently in both genera? It appears that such peculiar traits are unlikely to have evolved twice. However, some morphological structures differ considerably, in particular the copulatory ducts and coxal spurs of the female as well as holdfast structures of the male palp. For conclusions on relationships and the evolutionary constraints of such behaviour it will be necessary to analyse further representatives of these genera.

ACKNOWLEDGEMENTS

I am grateful to Dr. Konrad Thaler (Innsbruck) for help with collecting and for discussion, and to Dr. M. Grasshoff (Frankfurt/Main) for loan of comparative material. This work was supported by the University of Innsbruck.

REFERENCES

Branch, J.H. 1942. A spider which amputates one of its palpi. *Bulletin of the South California Academy of Science* 41, 139-140.

Knoflach, B. & van Harten, A. 2000. Palpal loss, single palp copulation and obligatory mate consumption in *Tidarren cuneolatum* (Tullgren, 1910) (Araneae, Theridiidae). *Journal of Natural History*, 34, 1639-1659.

Knoflach, B. & van Harten, A. 2001. *Tidarren argo* sp. nov. (Araneae : Theridiidae) and its exceptional copulatory behaviour: emasculation, male palpal organ as a mating plug and sexual cannibalism. *Journal of Zoology* 254, 449-459.

Levi, H.W. 1980. The male of *Echinotheridion* (Araneae: Theridiidae). *Psyche* 87, 177-179.

Ramirez, M.J. & Gonzalez, A. 1999. New or little-known species of the genus *Echinotheridion* Levi (Araneae, Theridiidae). *Bulletin of the British Arachnological Society* 11, 195-198.

Schmidt, G. 1980. Beobachtung einer Kopulation zwischen Spinnen zweier Gattungen, In: *Verhandlungen 8. Internationaler Arachnologen-Kongreß, Wien 1980* (J. Gruber ed.), pp. 229-232. Verlag H. Egermann, Wien.

Schmidt, G. 1981. Zur Spinnenfauna von La Gomera. *Zoologische Beiträge N.F.* 27, 85-107.

Wunderlich, J. 1992. *Die Spinnen-Fauna der makaronesischen Inseln*. Verlag J. Wunderlich, Straubenhardt.

European Arachnology 2000 (S. Toft & N. Scharff eds.), pp. 145-156.
© Aarhus University Press, Aarhus, 2002. ISBN 87 7934 001 6
(Proceedings of the 19th European Colloquium of Arachnology, Århus 17-22 July 2000)

Female genital morphology and sperm priority patterns in spiders (Araneae)

GABRIELE UHL

University of Bonn, Institute of Zoology, Department of Ethology, Kirschallee 1, D- 53115 Bonn, Germany
(g.uhl@uni-bonn.de)

Abstract

For spiders, gross female spermathecal morphology has been widely used as the major predictor of sperm priority pattern depending either upon taxonomic classification or on the number of ducts that connect with the spermathecae. In order to establish whether, or to what degree, the female reproductive tract follows a cul-de-sac (one duct connects to the spermatheca) or a conduit design (two ducts connect to the spermatheca at opposite ends) I present information on genital morphology of two haplogyne species (*Pholcus phalangioides*, Pholcidae; *Dysdera erythrina*, Dysderidae) and two entelegyne species (*Nephila clavipes*, Tetragnathidae; *Pityohyphantes phrygianus*, Linyphiidae). Predictions based on female anatomy and copulatory mechanisms are compared to available data on sperm utilization patterns.

Female genital anatomy deviates markedly from the expected pattern in all cases. There are more than the two predicted types of sperm storage sites: sperm can either be stored in the bursa, or in spermathecae connected by two ducts that lie close together, or in multiple sperm stores of different morphology. If males are able to insert their genital structures as far as to the lumen of the female sperm storage organ, male manipulation of sperm masses stored from previous males are possible and changes in sperm priority patterns can be expected. Combined information on detailed female anatomy and copulatory mechanism do not suffice to make reliable predictions on the pattern of sperm priority. Possible reasons for this discrepancy are briefly outlined.

Key words: Sexual selection, sperm storage, female choice.

INTRODUCTION

Female spiders show a propensity to mate with more than one male (Austad 1984; Elgar 1998), and maintain sperm for long periods in their sperm-storage organs (e.g. Uhl 1993a). Moreover, male spiders are unable to monopolize access to a female for the duration of her reproductive life, mainly because life expectancy for males is usually shorter than for females (Elgar 1998). Sperm from one of several males may be utilized randomly as sperm mix in the spermathecae leading to similar fertilization success of successive males or to fertilization success that depends on the relative number of sperm stored from each male. On the other hand, non-random utilization of sperm from a particular male is termed sperm precedence or priority, and is generally seen as a consequence of sperm stratification within the sperm storage organ of the female (Austad 1984; Elgar 1998; see Simmons & Siva-Jothy 1998 for definitions). Sperm from different males may remain stratified within the spermathecae either because sperm are non-motile, or are transferred or stored in distinctive portions, thus leading to a positional advantage. Whether the first or the last male to mate sires most of the offspring may thus purely depend on the position of the sperm mass within the female sperm storage organ.

The commonly accepted hypothesis on sperm precedence patterns in spiders relies on the fact that the sperm masses are stratified within the female genital tract (Austad 1984). Austad proposed that spider spermathecal morphology may represent a phylogenetic constraint which would lead to a non-adaptive pattern of sperm priority (Fig. 1). The term 'non-adaptive' in this context probably meant that adaptation was not recent but related back to an ancestral stage. Spiders (Araneoclada) are classified into two groups, the Entelegynae and the Haplogynae (Coddington & Levi 1991). Formerly, spiders were said to exhibit a fundamental dichotomy in the female spermathecal morphology that divides along phylogenetic lines with the Haplogynae possessing one duct and the Entelegynae possessing two ducts that connect with the spermathecae. If this were consistently so, the hypothesis that the two groups have distinctly different sperm priority patterns would be based on a firm morphological basis. However, in a number of genera within the entelegyne families Uloboridae, Tetragnathidae, Anapidae and the superfamily Palpimanoidea reversal to the haplogyne condition with only a single spermathecal duct occurred (see Coddington & Levi 1991, and even Austad himself 1984). Likewise, within the haplogyne Pholcidae, there are at least two species in which two ducts have evolved independently (Huber 1996). The idea that Haplogynae und Entelegynae each have uniform female genital morphology is thus not supported. As a consequence, inferring distinctly different sperm precedence patterns for taxa of the two groups is untenable.

To examine the hypothesis that sperm priority patterns depend on the morphology of female genitalia independent of phylogenetic position seems to be a more rewarding task. In Austad´s paper, this hypothesis is mixed with the previous one on phylogenetic constraints. If

phyletic limitation

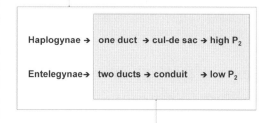

taxon independent view

Fig. 1. The original hypothesis by Austad (1984) proposed a phyletic limitation to sperm precedence patterns in spiders. The 'taxon independent' hypothesis only assumes a connection between the number of ducts that connect to the spermathecae and P_2 values.

female genital morphology determines sperm priority patterns, and sperm stratification occurred, species with one or two ducts that connect with the spermathecae should exhibit distinctly different sperm priority patterns (Fig. 1). I will use the terms haplogyne genitalia or haplogyne condition for species with one duct and entelegyne genitalia or entelegyne condition for species with two ducts, irrespective of taxonomic classification. Thus a spider, classified as belonging to the Haplogynae can exhibit female genital morphology of the entelegyne condition as in the case of the two Pholcid species mentioned earlier.

Sperm of haplogyne species passes along the single duct during copulation and again outward at oviposition (Fig. 2a). Austad (1984) termed this design, that is very similar to that in many insects (Walker 1980), 'cul-de-sac' (dead end). The entelegyne condition, on the other hand, was termed 'conduit' situation (one-way) and consists of a copulatory duct that leads to the spermatheca and a fertilization duct through which sperm reaches the eggs for fertilization (Fig. 2b). Predictions about sperm priority are that species with dead-end spermathecae should exhibit last male sperm priority as the last sperm to enter should lie closest to the single duct. This would represent a 'last in - first out' system. On the other hand, species

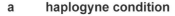

a   haplogyne condition

→ **last male sperm priority**

b   entelegyne condition

→ **first male sperm priority**

Fig. 2. Hypothetical sperm stratification inside the spermatheca of **(a)** spider with haplogyne genitalia and **(b)** spider with entelegyne genitalia and consequences of sperm priority patterns. S1: sperm from first male, S2: sperm from second male.

with one-way spermathecae are predicted to exhibit first male sperm priority, because the first sperm to enter should be closest to the fertilization duct and be the first to exit (first in - first out). The priority pattern should have implications for the mating strategy: species of the haplogyne condition should tend to guard mates just before egg laying, whereas species of the entelegyne condition should tend to guard penultimate females (see Elgar 1998).

Sperm utilization by females after multiple copulations is expressed as the proportion of offspring fathered by the last male to mate. In a typical experimental double-mating trial this is the proportion of eggs sired by the second male to mate, P_2 (Boormann & Parker 1976). Sperm mixing is expected to lead to P_2 values around 50% whereas sperm precedence of the first or second male leads to low or high P_2 values.

In order to test whether female genital morphology allows predictions on sperm priority patterns, the anatomy of the female reproductive tract needs to be examined in detail, thereby establishing whether, or to what de-

gree, it follows a cul-de-sac or conduit design. I therefore present detailed genital morphology for two haplogyne (*Pholcus phalangioides* (Fuesslin 1775) Pholcidae; *Dysdera erythrina* (Walckenaer, 1802), Dysderidae) and two entelegyne species (*Nephila edulis* (Labillardière, 1799), Tetragnathidae; *Pityohyphantes phrygianus* (C.L. Koch, 1836), Linyphiidae). For each species, I further present information on copulatory mechanisms. Possible access of male genitalia to the sperm storage site inside the female is crucial to the possibility that males physically manipulate sperm priority patterns. I also summarize published studies to check the predictions based on genital morphology and copulatory mechanism against paternity patterns.

Apart from physical male manipulation of stored sperm masses there are numerous other, more cryptic possible mechanisms, e.g. chemical manipulation. Products of the male reproductive organs that are transferred during copulation often induce female resistance to further mating, earlier oviposition, and even sperm transport (Eberhard 1997). On the other hand, female behavioural, morphological or physiological mechanisms that occur during or after copulation were shown to impose a bias on male reproductive success (Eberhard 1996). Although cryptic processes can be expected to play an important role, I will restrict this paper mainly to the question of whether female genital morphology and copulatory mechanism allow predictions on the pattern of sperm priority.

Pholcus phalangioides

The cellar spider *P. phalangioides* has only one genital opening through which copulation and egg laying are achieved as in typical haplogyne spiders. However, there are no spatially separated sperm storage organs of the cul-de-sac type (Fig. 3a). Sperm is stored in the bursa itself, and is embedded in a secretion produced by the female before copulation (Uhl 1994a). Complicated glands produce this matrix, and possibly also cause sperm activation (Uhl 1994b). What was formerly described as sper-

Fig. 3. Schematic presentation of female genital anatomy and place of sperm storage in **(a)** *Pholcus phalangioides* (Pholcidae), **(b)** *Dysdera erythrina* (Dysderidae) **(c)** *Nephila edulis* (Tetragnathidae) and **(d)** *Pityohyphantes phrygianus* (Linyphiidae). CD: copulatory duct, CF: copulatory fold, FD: fertilization duct, FF: fertilization fold, GO: genital opening, PD: posterior diverticulum, SP: spermatheca, UE: uterus eternus (= bursa). Dotted lines indicate the existence of folds instead of ducts. Shaded areas indicate places of sperm storage.

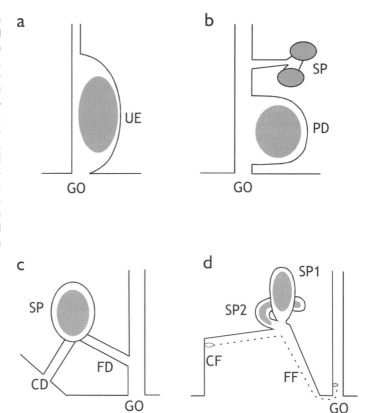

mathecae by Wiehle (1933) only rarely contains a few sperm. This structure was shown to have a different function: it is a fold connected to muscles that serves to open the genital valve before egg laying (Uhl 1994a). Overall, genital morphology in *P. phalangioides* deviates markedly from the expected pattern.

During copulation, several parts of the male pedipalp are inserted into the female: the procursus, the embolus, the appendix (a coupling structure) and the uncus (Uhl et al. 1995). The embolus is inserted directly into the female secretion where the sperm are stored and semithin sections from twice mated females showed no obvious stratification of sperm masses (Uhl 1994a). The male performs rhythmic twisting movements with the simultaneously inserted pedipalps during copulation, which result in extrusion of sperm near the centre of the female genital opening (Uhl et al. 1995).

Two different predictions arise from these findings: if female genital morphology played the most important role, sperm mixing should occur and P_2 should vary around 50%, whereas sperm displacement via pedipalp movement would predict a last male advantage. Investigations on sperm priority in the cellar spider showed a mean P_2 value of 66% for first broods of seven females (Yoward 1998, Tab. 1), despite a much shorter copulation duration in second matings compared to first ones (Uhl 1993b; Yoward 1998). The analysis of a larger sample size of 47 first broods showed high mean paternity value of 78% for second males (Schäfer & Uhl 2002). However, both investigations demonstrate highly variable paternity values (Tab. 1). The number of pedipalp movements a male performs during copulation is a good predictor of his fertilization success (Schäfer & Uhl 2002). Thus, the number of movements very likely influences the degree of displacement of previous male´s sperm in second matings.

Table 1. Compilation of data available both on sperm priority and genital mechanisms in spiders.
[a] : calculated from Yoward´s data for first broods to allow comparison between species
[b] : note that mean P2 values did not meet assumptions of normality.

Species (n)	Mean	P₂ in % Range (sd)(se)	median	Genital mechanism known?	Male manipulation possible?	Predictions possible ?	Source
Haplogyne genital structure							
Bursal storage							
Pholcus phalangioides (7)	0.66 [a]	0.39-1.00 (0.22)	0.66	yes	yes	yes	Yoward 1996; Uhl *et al.* 1995
(47)	0.78	0.00-1.00 (0.25)	0.89				Schäfer & Uhl 2002.
Holocnemus pluchei (40)	0.74 [b]	0.09-1.00 *(0.04)*	?	yes	yes	yes	Kaster & Jakob 1997; Huber 1995
Physocyclus globosus (12)	0.38	0.00-1.00 (0.30)	0.44	yes	yes	no	Eberhard et al. 1993; Huber & Eberhard 1997
Cul-de-sac							
Tetragnatha extensa (7)	~ 0.70	(0.36)	?	~ yes	no access	yes	West & Toft 1999
Entelegyne genital structure							
„Cul-de-sac" type							
Nephila clavipes (63)	0.18	0.00-1.00 (0.32)	0.02	~ yes	yes	no	Christenson & Cohn 1988
Nephila edulis (8)	0.66	(0.56)	0.83	yes	yes	yes	Schneider et al. 2000; Uhl & Vollrath 1998
Nephila plumipes (33)	0.46	*(0.05)*	0.42	~ yes	yes	no	Schneider & Elgar 2001
Latrodectus hasselti (11)	0.56	0.00-1.00		~ yes	yes	~yes	Andrade 1996; see text for morphological data on various species

Sperm extrusion was also found for another pholcid spider, *Physocyclus globosus* (Tacza-nowski, 1874) (Huber & Eberhard 1997). Sperm masses only emerged in copulations with non-virgin females, they appeared gradually rather than in step with male pedipalp movements and continued to emerge after copulation ended. Double mating experiments revealed a mean P2 of 38% (Eberhard et al. 1993; Tab.1). Copulation durations in virgin versus mated females are not nearly as different as in *P. pha-langioides* (Huber & Eberhard 1997; Uhl 1993b). In *P. phalangioides,* sperm extrusion can also occur in copulations with virgin females, extru-sion seems to appear in step with male move-ments and does not continue after copulation. Apparently, closely related species with similar genital morphology have evolved very differ-

ent male and female mechanisms that may bias paternity. Further investigations are needed that examine whether and to what extend a male removes his own sperm relative to his rival´s sperm. Also, sperm transfer may not be a time dependent process as is often assumed: it may either relate to pedipalp movements or to copulation duration. For *P. phalangioides* pre-liminary results suggest the former mechanism (Uhl, unpublished). In fact, time independent sperm transfer was found in *Frontinella commu-nis* (Hentz, 1850) (Austad 1982, syn. *F. pyramitela*) and *Micrathena gracilis* (Walckenaer, 1805) (Bukowski & Christenson 1997). Both studies investigated sperm release from the male palp by interrupting copulations and counting the remaining number of sperm in the palp. Other studies that suggest that sperm transfer were time related actually investigated sperm uptake and storage by the female, not sperm release. However, these may be distinctly separate processes that are often confounded (Bukowski & Christenson 1997).

Data on sperm priority and copulatory mechanism exists for yet another Pholcid spider, *Holocnemus pluchei* (Scopoli, 1763). High last male sperm precedence in this species (Kaster & Jakob 1997) may also be a result of male manipulation, as male genitalia reach the place of sperm storage (Huber 1995).

Dysdera erythrina

Semi-thin sections show that two structures function as sperm storage organs in *D. erythrina*: the so-called posterior diverticulum (PD), a large dilatation of the genital cavity similar to the structure found in *Pholcus phalangioides*, and a bilobed anterior spermatheca (Fig. 3b; Cooke 1966; Schult 1980; Uhl 2000). Thus, genital morphology deviates strongly from the presumed haplogyne pattern. Both structures are equipped with glandular tissue but the glandular tissue of the two is markedly different. The glandular tissue of the spermatheca is composed of complicated glandular units around cuticular ductules, whereas the glandular tissue of the posterior diverticulum is composed of rather simple gland cells (Uhl 2000). The products presumably differ, leading to possibly different storage conditions for the spermatozoa. Encapsulated spermatozoa are found in each lobe of the spermatheca. Sperm seems to be packed much tighter in the spermathecae than in the posterior diverticulum.

The male pedipalp in *D. erythrina* is a simple, blunt tipped structure, and thus probably reaches only as far as the posterior diverticulum (PD). Due to its size, it is highly unlikely that it enters the duct to the bilobed spermatheca. Thus, male manipulation is only possible for the PD, not for the spermatheca and it is tempting to suggest that the two types of sperm storage organs have evolved to allow (in the case of the PD) or prevent (in the case of the spermatheca) males from accessing previously stored sperm. Multiple organs may further facilitate specialization in short-term and long-term sperm storage as occurs in *Drosophila* (Pitnick et al. 1999). To date, there is no information on sperm priority pattern in *D. erythrina*.

Nephila edulis

Because *N. edulis* is an entelegyne spider, we would expect to find two ducts that connect on opposite sides of the spermathecae to form a one-way system. Female genital morphology in *N. edulis* does possess two ducts (in and out) connected to each of the two spermatheca, but, the details of the connections to the spermathecae differ from the assumptions made by the Austad hypothesis. The ducts are close together and the reproductive tract looks more like a cul-de-sac than a conduit (Fig. 3c). The spermatheca lacks a septum that would create the equivalent of a one-way system inside of the spermathecal lumen. Based on female morphology alone, the sperm from the last male should be closer to the fertilisation duct, and last male sperm priority should prevail. The male pedipalp in *N. edulis* consists of a compact genital bulb provided with a long conductor that supports the sperm transferring structure, the embolus (Uhl & Vollrath 1998). The embolus is rolled up inside the bulb and can be pushed out of the tip of the conductor. During copulation, the embolus reaches the lumen of the spermathecae (Uhl & Vollrath 1998) which makes stratification unlikely and speaks in favour of sperm mixing. Male manipulation of previous males' sperm, also seems possible.

An investigation on sperm priority in *N. edulis* shows a mean P_2 value of 66% (median 83%) based on 8 matings (Schneider et al. 2000) (Tab. 1), which tentatively suggests that sperm manipulation by subsequent males is possible. On the other hand, in *N. edulis* duration and frequency of copulation is a very good predictor of paternity independent of mating order, which suggests that sperm are utilized according to relative numbers. As mentioned above, caution has to be applied when assuming gradual sperm transfer on the basis of a correlation between copulation duration and sperm utilization. In *N. clavipes* (Linnaeus, 1767) this correlation was found although sperm transfer occurs during the first of many insertion bouts (T. E. Christenson pers. comm.). In *N. clavipes* clear first male advantage was demonstrated (mean P_2 value: 18%) (Christenson & Cohn 1988). Besides possible male manipulation an alternative explanation for the priority pattern would be that significantly different sperm numbers are taken up by the female depending on the mating order (T.E. Christenson pers. comm.). Investigations on sperm release and uptake by Cohn (1988) point in this direction. In *N. plumipes* (Latreille, 1804) equal sperm numbers seem to mix in the female spermatheca (mean P_2: 46%, Schneider & Elgar 2001, Tab. 1). Although female and male morphology appears to be very similar in the three *Nephila* species, transfer, storage and ultization mechanisms seem to be quite different. Clearly, further investigations are needed.

Due to extreme intraspecific male size variability in *Nephila* (Vollrath 1980), one might expect small males to be at a disadvantage if it comes to copulatory mechanisms and sperm transfer. Size variability could lead to different degrees of mating efficiency and fertilization success for males of different sizes. However, in *N. edulis* genital characters show negative allometric values when plotted against somatic characters, which means that small males have relatively large genitalia and large males relatively small genitalia (Uhl & Vollrath 2000). Although male somatic size has a coefficient of variation of about 45%, genital variability is only 20% ($p < 0.01$, Lewontin's method 1966). Mean values of male and female genitalic characters match surprisingly well: embolus length minus conductor length in the male has a mean of 1.3 mm while mean copulatory duct length in the female is 1.28 mm (Uhl & Vollrath 2000). There seem to be strong selective advantages leading towards intermediate, standardized sizes of male genitalia as in many other species of insects and spiders (Eberhard et al. 1998). It seems that males adapt their genital size to that appropriate to the most common female size.

As a consequence of this finding, it might be expected that copulatory mechanisms for males of different sizes should be similar. However, the study by Schneider et al. (2000) showed that small males had a mating advantage, they mated for longer and fertilized more eggs than large males.

Pityohyphantes phrygianus

We might expect linyphiids to represent 'proper' conduit type spiders, as the linyphiid *Frontinella communis* exhibits clear first male sperm priority with little variation in P_2 (Austad 1982).

The epigynum of *P. phrygianus* has a scape, with an atrium on both sides. There are two spermathecae on each side, one is straight and thumb-like (spI) and the other is twisted (spII) and surrounds the straight one half way (Fig. 3d). Both spermathecae extend from a massive U-shaped structure. As in *D. erythrina*, the composition of the associated glands differs between spermathecae, as only the twisted spermathecae exhibit a strip of glandular ductules. The internal characteristics are quite complicated: a fold rather than a tube functions as a copulatory 'duct', leading to the spermathecae (Uhl & Gunnarsson 2001). This copulatory fold is 'sealed' after copulation with a homogeneous secretion, probably to impair copulatory success of subsequent males. At the base of the spermathecae there is a valve-like structure that makes intromission of male genital structure unlikely. Surprisingly, a fertilisation duct does not exist in *P. phrygianus* (Uhl & Gunnarsson 2001). A fold extends from the base of the spermathecae laterally along the ventral wall towards the opening of the atrium. From there, it turns into the epigastric fold leading towards the gonoduct. This fold is the only connection between the spermathecae and the oviduct. Such folds have also been found in other Linyphiids, including *Lepthyphantes* (Saaristo & Tanasevitch 1996) and *Batyphantes gracilis* (Blackwall, 1841) (M. Saaristo pers. comm.).

Histological sections depicted by Engelhardt (1910) on *Linyphia triangularis* (Clerck, 1757) also point in this direction (but see van Helsdingen 1969).

In *P. phrygianus*, various apical parts of the male palp (the 'embolic division' sensu Merrett 1963) probably couple to the knob of the u-shaped base, an interpretation derived from the position of the mating plug. If this is the case, males are not expected to be able to physically manipulate stored sperm masses directly. The predicted sperm priority pattern based on female genital morphology is last male priority against which the production of a mating plug evolved.

Other species

Investigations on sperm priority patterns are generally rare for spiders, and information on genital morphology and copulatory mechanism for these species is often unavailable. However, there is some information on both aspects for Tetragnathidae and Theridiidae. Published data on sperm priority are summarized in Elgar (1998) and Uhl & Vollrath (1998).

The Tetragnathinae is an entelegyne spider subfamily, in which female genitalia of the haplogyne condition occur. West & Toft (1999) found that the last male fertilized about 70% of the eggs in first egg sacs of *Tetragnatha extensa* (Linnaeus, 1758) (Tab.1). From what is known about male and female genital morphology of *Tetragnatha* species it is questionable whether the male pedipalp has access to the spermathecae (e.g. Wiehle 1963; Uhl et al. 1992), which would lead to the prediction of last male sperm priority as was found for *T. extensa*. However, in a drawing of genitalia of *T. montana* Simon, 1900 in functional contact, it seems as if the emboli were inserted into the spermathecae (Huber & Senglet 1997). Unfortunately, female genitalia are not fully depicted which leaves copulatory mechanism obscure. The tetragnathid *Leucauge mariana* (Keyserling, 1881), on the other hand, possesses a conduit design with three successive sperm storage chambers (Eberhard & Huber 1998). Moreover, the male embolus may reach as far as to the lumen of the

first chamber and may thus be able to manipulate stored sperm.

In the entelegyne genus *Latrodectus* (Theridiidae), genital morphology of both sexes does not vary considerably between species. The two spermathecae are heavily sclerotized, dumb-bell shaped structures with anterior and posterior lobes. These lobes are connected by an intermediate, more slender part. In some species like in *L. hesperus* Chamberlin & Ivie, 1935, the spermathecae are of the functional cul-de-sac type with copulatory and fertilization duct close together (Bhatnagar & Rempel 1962, misidentified as *L. curacaviensis*), whereas in other species (e.g. *L. hystrix* Simon, 1890; *L. geometricus* C.L. Koch, 1841; *L. cinctus* Blackwall, 1865; *L. renivulvatus* Dahl, 1902: Knoflach & van Harten 2002; *L. revivensis*: Berendonck & Greven 2002) the ducts are further apart. At the end of copulation, the tip of the embolus typically breaks off and remains either in the narrow entrance of the spermatheca, as for example in *L. dahli* Levi, 1959, or is deeply inserted in the lumen of the spermatheca (*L. geometricus*) (Knoflach & Van Harten 2002). The broken embolus tips do not necessarily prevent the female from remating, but in the case of *L. renivulvatus*, in which whole emboli may break and plug the copulatory ducts (Knoflach & van Harten 2002) successful remating seems impossible. In *L. dahli* and *L. geometricus* Knoflach & van Harten report cases in which two tips were found in one spermathecal entrance. For the latter species Müller (1985) detected four tips in the spermatheca and one in the bursa. In *L. hesperus* and *L. revivensis*, however, only a single tip was found in each spermathecal entrance, whereas several more could be found in the bursa (Bhatnagar & Rempel 1962; Berendonck & Greven 2002). The data from *L. hesperus*, *L. reviviensis* and *L. renivulvatus* suggest that successful intromission of palps is only possible once for each spermatheca. If a male is allowed to inseminate only one spermatheca but not the other, an additional copulation with another male may occur, which will lead to sperm of different males being stored in different storage organs. The only information on sperm priority we have to date is from a study on yet another species, *L. hasselti* Thorell, 1870, in which sperm mixing with considerable variation occurs (Andrade 1996). It remains to be clarified whether mixed paternity in *L. hasselti* results from sperm mixing in a given spermathecae or from activation of sperm stored in different spermathecae. The latter situation may explain why some males have no paternity success whereas others fertilize 100% of the eggs.

Is sperm stratification a plausible assumption?

In the testis, male spiders produce encapsulated spermatozoa. These remain encapsulated during sperm induction into the male pedipalp and are transferred in an encapsulated state to the female (Alberti 1990). Whereas in *Leucauge mariana* (Eberhard & Huber 1998) sperm activation (better: decapsulation) within the female occurs soon after insemination, sperm are stored in an inactive state in *Pholcus phalangioides*, probably until shortly before egg laying (Uhl 1994a). Both studies investigated the conditions of sperm in situ, in the female sperm storage organ. A study on *Nephila clavipes* (Brown 1985) demonstrated that decapsulation took 7 to 18 days from mating, depending on whether the female had moulted the same day or mated later in adulthood. Brown squeezed the spermathecal content onto a slide containing physiological saline, which may have influenced the results as sperm become active when transferred to physiological saline. It should be noted that decapsulated sperm do not necessarily move in the female genital tract as it may require additional stimuli for sperm to become mobile. These findings show that although sperm stratification may occur in some taxa it is not a general characteristic of spiders. Female glandular secretion seems to trigger the process of activation both in *L. mariana* and *P. phalangioides* (Eberhard & Huber 1998; Uhl 1994b), which suggests that females have the potential of biasing the fertilization success of rival males by selectively activating stored sperm.

CONCLUSION

Each of the four spiders investigated deviates considerably from the assumptions underlying the modified Austad's hypothesis. Obviously, female genital morphology is extremely variable even within families. Data on morphology and sperm usage show that sperm precedence patterns cannot be predicted by the number of ducts connected to a spermatheca. Even if species are classified according to their specific design of female genital morphology, independent of their phylogenetic position, and even if knowledge on copulatory mechanisms is included, predictions on sperm precedence patterns are difficult to make. Unfortunately, there is only little information on detailed genital morphology, copulatory mechanisms and P_2 values for single spider species which would help to clarify the matter. Beyond morphology, manifold processes of male and female manipulation, male and female age, remating intervals and body size may influence sperm transfer, storage and usage. Adaptation, not constraints seems to play the major role in shaping sperm priority patterns. Thus, researchers should refrain from assuming particular precedence patterns solely on the basis of either taxonomic classification or the number of ducts present.

ACKNOWLEDGEMENTS

I cordially thank Søren Toft and Nikolaj Scharff for inviting me to give this talk at the 19th European Colloquium of Arachnology in Århus. William Eberhard, Jutta Schneider, Tina Berendonck, Barbara Knoflach-Thaler, Michael Schmitt and an anonymous referee gave valuable comments on a previous draft of the manuscript. I am especially grateful to Bill Eberhard, whose thorough comments helped to substantially improve this paper.

REFERENCES

Alberti, G. 1990. Comparative spermatology of Araneae. *Acta Zoologica Fennica* 190, 17-34.

Andrade, M.C.B. 1996. Sexual selection for male sacrifice in the Australian redback spider. *Science* 271, 70-72.

Austad, S.N. 1982. First male sperm priority in the bowl and doily spider, *Frontinella pyramitela* (Walckenaer). *Evolution* 36, 777-785.

Austad, S.N. 1984. Evolution of sperm priority patterns in spiders. In: *Sperm competition and the evolution of mating systems.* (R.L. Smith ed.), pp. 223-249. Harvard University Press, Cambridge Mass.

Berendonck, B. & Greven, H. 2002. Morphology of female and male genitalia in *Latrodectus revivensis* Shulov, 1948 (Araneae, Theridiidae) with regards to sperm priority patterns. In: *European Arachnology 2000* (S. Toft & N. Scharff eds.), pp. 157-167. Aarhus University Press, Aarhus.

Bhatnagar, R.D.S. & Rempel, J.G. 1962. The structure, function, and postembryonic development of the male and female copulatory organs of the black widow spider *Latrodectus curacaviensis* (Müller). *Canadian Journal of Zoology* 40, 465-510.

Boorman, E. & Parker, G.A. 1976. Sperm (ejaculate) competition in *Drosophila melanogaster*, and the reproductive value of females to male in relation to female age and mating status. *Ecological Entomology* 1, 145-155.

Brown, S.G. 1985. Mating behavior of the golden-orb-weaving spider, *Nephila clavipes*: II. Sperm capacitation, sperm competition, and fecundity. *Journal of Comparative Psychology* 99, 167-175.

Bukowski, T.C. & Christenson, T.E. 1997. Determinants of sperm release and storage in a spiny orbweaving spider. *Animal Behaviour* 53, 381-395.

Christenson, T.E. & Cohn, J. 1988. Male advantage for egg fertilization in the golden orb-weaving spider (*Nephila clavipes*). *Journal of Comparative Psychology* 102, 312-318.

Coddington, J.A. & Levi, H.W. 1991. Systematics and evolution of spiders (Araneae*). Annual Review of Ecology and Systematics* 22, 565-92.

Cohn, J. 1990. Is it the size that counts? Palp morphology, sperm storage, and egg hatching frequency in *Nephila clavipes* (Araneae, Araneidae). *Journal of Arachnology* 18, 59-71.

Cooke, J.A.L. 1966. Synopsis of the structure and function of the genitalia in *Dysdera crocata. Senckenbergiana Biologica* 47, 35-43.

Eberhard, W.G. 1996. *Female Control: Sexual Selection by Cryptic Female Choice.* Monographs in Behavior and Ecology, Princeton University Press, New Jersey.

Eberhard, W.G. & Huber, B.A. 1998. Courtship, copulation, and sperm transfer in *Leucauge mariana* (Araneae, Tetragnathidae) with implications for higher classification. *Journal of Arachnology* 26, 342-368.

Eberhard, W.G., Guzman-Gomez, S. & Catley, K.M. 1993. Correlation between spermathecal morphology and mating systems in spiders. *Biological Journal of the Linnean Society* 50, 197-209.

Eberhard, W.G., Huber, B.A., Rodriguez, S.R.L., Briceño, R.D., Salas, I. & Rodriguez, V. 1998. One size fits all? Relationships between the size and degree of variation in genitalia and other body parts in twenty species of insects and spiders. *Evolution* 52, 415-431.

Elgar, M.A. 1998. Sperm competition and sexual selection in spiders and other arachnids. In: *Sperm competition and sexual selection.* (T. R. Birkhead & A.P. Møller eds.), pp. 307-339. Academic Press, San Diego.

Engelhardt, V. 1910. Beiträge zur Kenntnis der weiblichen Copulationsorgane einiger Spinnen. *Zeitschrift für Wissenschaftliche Zoologie* 96, 32-117.

Huber, B.A. 1995. Copulatory mechanism in *Holocnemus pluchei* and *Pholcus opilionoides*, with notes on male cheliceral apophyses and stridulatory organs in Pholcidae (Araneae*). Acta Zoologica* 76, 291-300.

Huber, B.A. 1996. On American 'Micromerys' and *Metagonia* (Araneae, Pholcidae), with notes on natural history and genital mechanics. *Zoologica Scripta* 25, 341-362.

Huber, B.A. & Eberhard, W.G. 1997. Courtship, copulation, and genital mechanics in *Physocyclus globosus* (Araneae, Pholcidae). *Canadian Journal of Zoology* 74, 905-918.

Kaster, J.L. & Jakob, E.M. 1997. Last-male sperm priority in a haplogyne spider (Araneae: Pholcidae): correlations between female morphology and patterns of sperm usage. *Annals of the Entomological Society of America* 90, 254-259.

Knoflach, B. & van Harten, A. 2002. The genus *Latrodectus* (Araneae: Theridiidae) from mainland Yemen, the Socotra Archipelago and adjacent countries. *Fauna of Arabia* 19, 321-361.

Lewontin, R.C. 1966. On the measurement of relative variability. *Systematic Zoology* 15, 141-142.

Merrett, P. 1963. The palps of male spiders of the family Linyphiidae. *Proceedings of the Zoological Society* 140, 347-467.

Müller, G.H. 1985. Abgebrochene Emboli in der Vulva der 'Schwarzen Witwe' *Latrodectus geometricus* C.L. Koch 1841 (Arachnida: Araneae: Theridiidae). *Entomologische Zeitschrift* 95, 27-30.

Pitnick, S. Markow, T. & Spicer G.S. 1999. Evolution of multiple sperm storage organs in Drosophila. *Evolution* 53, 1804-1822.

Saaristo, M.I. & Tanasevitch, A.V. 1996. Redelimitation of the subfamily Micronetinae Hull, 1920 and the genus *Lepthyphantes* Menge, 1866 with descriptions of some new genera (Aranei, Linyphiidae). *Berichte des nat.-med. Vereins Innsbruck* 83, 163-186.

Schäfer, M.A. & Uhl, G. 2002. Determinants of paternity success in the spider *Pholcus phalangioides* (Pholcidae: Araneae): the role of male and female mating behaviour. *Behavioural Ecology and Sociobiology* 51, 368-377.

Schneider, J.M. & Elgar, M.A. 2001. Sexual cannibalism and sperm competition in the golden orb-spider *Nephila plumipes* (Araneoida): female and male perspective. *Behavioural Ecology* 12, 547-552.

Schneider, J.M., Herberstein, M.E., de Crespigny, F.C., Ramamurthy, S. & Elgar, M.A. 2000. Sperm competition and small size advantage for males of the golden orb-web spider *Nephila edulis*. *Journal of Evolutionary Biology* 13, 939-946.

Schult, J. 1980. Die Genitalstrukturen haplogyner Araneae unter phylogenetishem Aspekt (Arachnida). PhD thesis University Hamburg.

Simmons, L.W. & Siva-Jothy, M.T. 1998. Sperm competition in insects: mechanisms and the potential for selection. In: *Sperm competition and sexual selection*. (T.E. Birkhead & A.P. Møller eds.), pp. 341-434. Academic Press, San Diego.

Uhl, G. 1993a. Sperm storage and repeated egg production in female *Pholcus phalangioides* Fuesslin (Araneae). *Bulletin de la Societé Neuchâteloise des Sciences Naturelles* 116 (1), 245-252.

Uhl, G. 1993b. Mating behaviour and female sperm storage in *Pholcus phalangioides* (Fuesslin) (Araneae). *Memoirs of the Queensland Museum* 33, 667-674.

Uhl, G. 1994a. Genital morphology and sperm storage in *Pholcus phalangioides* (Fuesslin, 1775) (Pholcidae; Araneae). *Acta Zoologica* 75, 1-12.

Uhl, G. 1994b. Ultrastructure of the accessory glands in female genitalia of *Pholcus phalangioides* (Fuesslin, 1775) (Pholcidae; Araneae). *Acta Zoologica* 75, 13-25.

Uhl, G. 2000. Two distinctly different sperm storage organs in female *Dysdera erythrina* (Araneae: Dysderidae). *Arthropod Structure and Development* 29, 163-169.

Uhl, G. & Gunnarsson B. 2001. Female genitalia in *Pityohyphantes phrygianus*: a spider with skewed sex ratio. *Journal of Zoology* 255, 367-376.

Uhl, G. & Vollrath, F. 1998. Genital morphology of *Nephila edulis*: implications for sperm competition in spiders. *Canadian Journal of Zoology* 76, 39-47.

Uhl, G. & Vollrath, F. 2000. Extreme body size variability in the golden silk spider (*Nephila edulis*) does not extend to genitalia. *Journal of Zoology* 251, 7-14.

Uhl, G.; Huber, B.A. & Rose, W. 1995. Male pedipalp morphology and copulatory mechanism in *Pholcus phalangioides* (Fuesslin, 1775) (Araneae, Pholcidae). *Bulletin of the British Arachnological Society* 10, 1-9.

Uhl, G., Sacher, P., Weiss, I. & Kraus, O. 1992. Europäische Vorkommen von *Tetragnatha shoshone* (Arachnida, Araneae, Tetragnathidae*). Verhandlungen des naturwissenschaftlichen Vereins Hamburg N.F.* 33, 247-261.

Van Helsdingen, P. J. 1969. A reclassification of the species of *Linyphia* based on the functioning of the genitalia (Araneae, Linyphiidae), I. *Zoologische Verhandelingen* 105, 1-303.

Vollrath, F. 1980. Male body size and fitness in the web-building spider *Nephila calvipes*. *Zeitschrift für Tierpsychologie* 53, 61-78.

Walker, W.F. 1980. Sperm utilization strategies in nonsocial insects. *American Naturalist* 115, 780-799.

West, H.P. & Toft, S. 1999. Last-male sperm priority and the mating system of the haplogyne spider *Tetragnatha extensa* (Araneae: Tetragnathidae). *Journal of Insect Behavior* 12, 433-450.

Wiehle, H. 1933. *Holocnemus hispanicus* sp.n. und die Gattungen *Holocnemus* Simon und *Crossoprisa* Simon. *Zoologischer Anzeiger* 104, 241-252.

Wiehle, H. 1963. Spinnentiere oder Arachnoidea (Araneae). XII. Tetragnathidae—Streckspinnen und Dickkiefer. In: *Die Tierwelt Deutschlands und der angrenzenden Meeresteile* (F. Dahl ed.), pp. 1-76. Gustav Fischer, Jena.

Yoward, P.J. 1998. Sperm competition in *Pholcus phalangioides* (Fuesslin, 1775) (Araneae, Pholcidae) - shorter second copulations gain a higher paternity reward than first copulations. In: *Proceedings of the 17th European Colloquium of Arachnology, Edinburgh 1997* (P.A. Selden ed.), pp. 167-170. British Arachnological Society, Burnham Beeches, Bucks.

European Arachnology 2000 (S. Toft & N. Scharff eds.), pp. 157-167.
© Aarhus University Press, Aarhus, 2002. ISBN 87 7934 001 6
(Proceedings of the 19th European Colloquium of Arachnology, Århus 17-22 July 2000)

Morphology of female and male genitalia of *Latrodectus revivensis* Shulov, 1948 (Araneae, Theridiidae) with regard to sperm priority patterns *

BETTINA BERENDONCK[1,2] & HARTMUT GREVEN[1]

[1]*Zoologie II, Heinrich-Heine-Universität, D-40225 Düsseldorf, Germany*
(bettina.berendonck@uni-duesseldorf.de).
[2]*Jacob Blaustein Institute for Desert Research, Ben Gurion University, Sede Boker, Israel*

Abstract

The shape of the female spermatheca is assumed to play a decisive role in determining the sperm priority patterns in spiders that probably are reflected in the mating behaviour of a given species. To estimate the significance of the shape of the female sperm storage organs with regard to sperm priority patterns and the possible influence of the broken embolus tip for subsequent males in *Latrodectus revivensis* we examined the morphology of virgin and mated females' epigyna and male pedipalps by means of scanning electron- and light microscopy of cleared specimens. The female epigynal plate bears long mechanoreceptors. Her copulatory duct forms a narrow tube that takes up the male embolus during insertion. The spermatheca resembles the conduit type and first male sperm priority would be the expected pattern. However, the male embolus clearly shows a defined breaking point and its position inside the female spermathecal opening after copulation suggests that it might act as a mating plug by establishing a physical barrier against subsequent males.

Key words: *Latrodectus*, genital morphology, sperm priority pattern, mating plug

INTRODUCTION

Austad (1984) suggested that in spiders there exists a fundamental dichotomy of spermathecal morphology that divides them roughly along phylogenetic lines and mainly determines sperm priority patterns. In a 'cul-de-sac' spermatheca a single duct connects the sperm storage organ to the vagina resulting in a 'last in-first out' situation of sperm of subsequent males. In 'conduit' spermathecae there are two separate ducts, one duct through which the male transfers its seminal fluids (copulatory

duct) and a second duct from which sperm is released into the uterus externus during oviposition (fertilisation duct). The resulting 'first in-first out' situation of sperm from two different males should result in a first male sperm precedence.

In the entelegyne spider *Latrodectus revivensis* Shulov, 1948 a conduit type spermatheca should be expected and, thus, a first-male sperm priority pattern. However, Uhl (2002) has shown that the female genital morphology cannot be deduced from the classification into entelegynes or haplogynes. Even the presence of two separate ducts does not necessarily mean that a spermatheca is also 'functionally' of the conduit type (Uhl and Vollrath 1998). For

---

* In memory of Merav Ziv, Jacob Blaustein Institute for Desert Research, Ben Gurion University, Sede Boker, Israel

L. hasselti Thorell, 1870, Uhl and Vollrath (1998) assume a conduit female spermatheca of the cul-de-sac type (entrance and exit lie close to each other) judging from the illustrations of the genitalia of the related *L. curacaviensis* Müller, 1776 that were shown by Bhatnagar and Rempel (1962). However, illustrations of female spermathecae of various *Latrodectus* species by other authors (Levi 1959, 1983; Wiehle 1961, 1967; Abalos & Baez 1963, 1967; Kaston 1970; Levy & Amitai 1983; Müller 1985) suggest a spatial separation of the entrance and exit.

Males of many *Latrodectus* species are known to frequently lose the apical part of their embolus (= embolus tip) inside the female genital tract during copulation (Dahl 1902; Smithers 1944; Levi 1959; Wiehle 1961; Bhatnagar & Rempel 1962; De Biasi 1962; Abalos & Baez 1963, 1967; Kaston 1970). It is generally taken for granted that this embolus tip does not keep subsequent males from copulating with an already inseminated female (Abalos & Baez 1963; Breene & Sweet 1985). However, so far there is no evidence that a male successfully transfers sperm into a spermatheca that already has one embolus tip in its opening (not only in the copulatory ducts). In the present paper we give a detailed overview of the male and female genitalia of *L. revivensis* to gain more information about the structural prerequisites for sperm priority patterns. We studied the gross morphology of the female epigynal plate, the spermatheca and the connected ducts, the structure of the male embolus and the position of broken embolus parts inside the female genitalia by using light and scanning electron microscopy. Our results are compared with observations on other *Latrodectus* species.

MATERIAL AND METHODS

Latrodectus revivensis occurs in the central Negev desert of Israel (Levy & Amitai 1983). In the 1998 season, 30 mated females which had already built at least one egg-sac in the field, ten subadult females, ten subadult males and three adult males of unknown mating history, were collected in the vicinity of Sede Boqer,

Israel. Immature females and males were reared to adulthood.

The genitalia of the 24 females which had mated in the field were isolated and kept in 5 % KOH solution for several days until the tissue was dissolved. Cleared specimens were finally transferred into glycerine and checked with a light microscope. The spermatheca of five adult virgin females and the pedipalps of five virgin males were treated similarly.

The epigyna of three mated females from the field and the isolated spermathecae of five virgin and three mated females were isolated and cleaned with 5% KOH and contact lens cleaner. The cleaned spermathecae of two of the mated females were transversally cut with a razor-blade in the region where the copulatory duct opens into the spermatheca. The female epigyna, the spermathecae and the pedipalps of the five virgins and the three males of unknown mating history were dehydrated in a graded series of ethanol, critical-point dried, sputter-coated with gold and examined with a Jeol JSM-5400 scanning electron microscope.

RESULTS

Female epigynum

The epigynum of *L. revivensis* is an arched and heavily sclerotised plate which is situated on the ventral side of the female's opisthosoma immediately in front of the epigastric furrow. The epigynal plate is transverse, suboval and densely covered with thin hairs that partly overhang the nearly triangular opening of the atrium. The posterior margin of the opening is slightly curved backwards forming very probably a guide for the male pedipalp during genital coupling (Fig. 1). From each side of the atrium an opening leads into one of the paired copulatory ducts (synonymous with the 'bursae copulatrices' in Bhatnagar & Rempel 1962). Each duct twines around the spermatheca on that side (Fig. 2, not seen in Fig. 4 due to preparation). After two coils in a posteriolateral direction each duct makes a sharp turn back towards the spermatheca, completing another two coils. Along its main course the

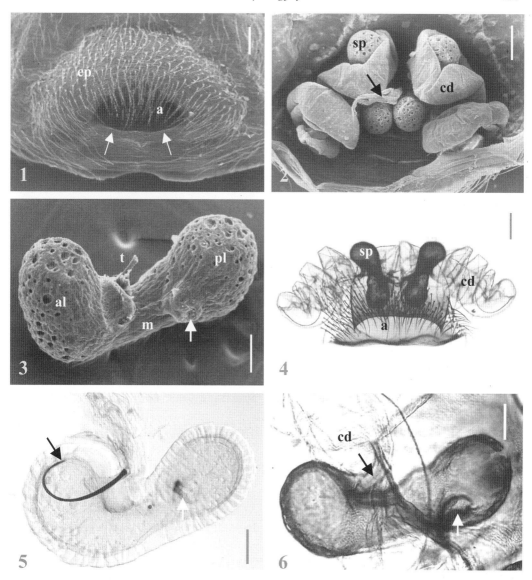

Fig. 1. Epigynal plate (ventral view). The posterior margin of the nearly triangular opening is slightly curved backwards (arrows). a: atrium, ep: epigynal plate. Scale bar: 200 μm. **Fig. 2.** Female egigynum (dorsal view) with copulatory ducts and cuticular lining of the duct that opens into the uterus externus (arrow). cd: copulatory duct, sp: spermatheca. Scale bar: 200 μm. **Fig. 3.** Left spermatheca (medial view). Note the embolus tip lodged inside the spermathecal opening. The outer cuticular papilla indicates the exit of the fertilisation duct (arrow). al: anterior lobe, m: narrow middle portion, pl: posterior lobe, t: embolus tip. Scale bar: 100 μm. **Fig. 4.** Cleared epigynum (ventral view) with slightly sclerotized copulatory ducts. a: atrium, cd: copulatory duct, sp: spermatheca. Scale bar: 200 μm. **Fig. 5.** Cleared spermatheca (medial view) with a male embolus tip in the opening. The opening of the pointed end of the embolus tip rests at the most anterior inner wall of the spermatheca (black arrow). Fertilisation duct (white arrow). Scale bar: 100 μm. **Fig. 6.** Cleared spermatheca (medial view). Close to the spermatheca the flat copulatory duct forms a heavily sclerotised tube on one side (black arrow). The funnel-shaped fertilisation duct originates from the apical, lateral wall of the posterior lobe, slightly bends anteriorly with its exit finally opening on the basal surface (white arrow). cd: copulatory duct. Scale bar: 100 μm.

copulatory duct is slightly sclerotised (Fig. 4). In its last part, close to the entrance of the spermatheca, the duct becomes very flat and one side forms a heavily sclerotised tube (Figs. 6, 16, 17). This organisation is retained in that section where the cuticle of the copulatory duct fuses with the cuticle of the spermatheca (Fig. 18). In transversely cut preparations one can see that the lumen of this part of the spermatheca is reduced to a very narrow slit that continues into the spermathecal lumen medially (Fig. 18). As in the copulatory duct, the slit is slightly widened laterally (not exceeding 17 µm) forming a tube (Figs. 18, 19) that takes up the male embolus tip during copulation (Figs. 5, 15, 16, 17).

The paired spermathecae are dumb-bell shaped and lined by very thick and heavily sclerotised walls (Figs. 3, 18). Each spermatheca consists of two rounded lobes that are connected by a narrow middle region (Figs. 3, 5, 6). The cuticle is penetrated by numerous pores of variable size (Figs. 2, 3). In the anterior and posterior lobes they have a diameter up to 40 µm. The fertilisation duct is formed by the spermathecal wall itself. It originates from the lateral apical wall almost in the middle of the posterior lobe (Figs. 5, 6). The lumen of the fertilisation duct is funnel-like; it is wide at the beginning and becomes narrower towards the end (Fig. 6). The whole duct bends slightly anteriorly; it finally opens on a cuticular papilla on the basal surface (Fig. 3) connecting the spermatheca to the uterus externus via a cuticula lined duct(Figs. 2, 6).

Male pedipalp

The two distal segments, the tarsus and pretarsus of the male pedipalp are modified and form the copulatory organ. The cymbium (tarsus) is spoon-shaped and covered with vari-ous sense organs such as tactile hairs, slit sensilla and trichobothria. The genital bulb (pretarsus) consists of various sclerites connected by membranous parts (haematodochae) (Figs. 7, 8). They surround the internal parts of the spermophore (synonymous to the male 'receptaculum seminis' in Wagner 1887, see Bhatnagar & Rempel 1962 for citation). The spermophore (Fig. 8) consists of the fundus, the reservoir and the ejaculatory duct. The fundus and the reservoir form a sclerotised spiral coil, with one end overlapping the other. From the reservoir originates the transparent and slender ejaculatory duct (Fig. 10) which has an outer diameter of approx. 4 µm and an inner diameter of about 3 µm (Fig. 14). The embolus is the only external part of the genital bulb that is introduced into the copulatory duct and the spermatheca of the female during copulation (Fig. 15). The long spirally coiled part of the embolus can be clearly divided into the dark brown and heavily sclerotised truncus and the membranous pars pendula that runs along its inner concave side (Figs. 8, 10, 11). In cross section the truncus forms a U-shaped channel closed by the pars pendula (Fig. 14). Inside the channel lies the ejaculatory duct originating from the reservoir (Figs. 10, 14). Apically, the embolus possesses a solid S-shaped sclerite, the embolus tip (Figs. 7, 8, 10, 12). SEM pictures show that the transition between the embolus tip and the rest of the embolus is marked by a saddle-like thickening (Fig. 12). The thickening is also visible in the embolus tips inside the female genital tract and indicates the site of fracture (Figs. 3, 5, 15, 16, 17). The total length of the tip varies among individuals (approx. 300-400 µm). The outer diameter of the tip is approximately 13 to 16 µm at the breaking point and 8 to 10 µm just beyond it (Figs. 12, 13). From here the diameter remains constant

Fig. 7. Right male pedipalp (lateral view) with its long spirally coiled embolus. em: embolus, t: embolus tip. Scale bar: 200 µm. **Fig. 8.** Cleared male pedipalp (lateral view). The genital bulb consists of various sclerites connected by membranous parts. The receptacle consists of the fundus, the reservoir and the ejaculatory duct inside the embolus. em: embolus, f: fundus, t: embolus tip, r: reservoir. Scale bar: 200 µm. **Fig. 9.** Opening of the embolus tip with a maximum diameter of approx. 6 µm. Scale bar: 10 µm. **Fig. 10.** Apical portion of the male embolus. The main part of the embolus, that ends at the breaking point (arrow)

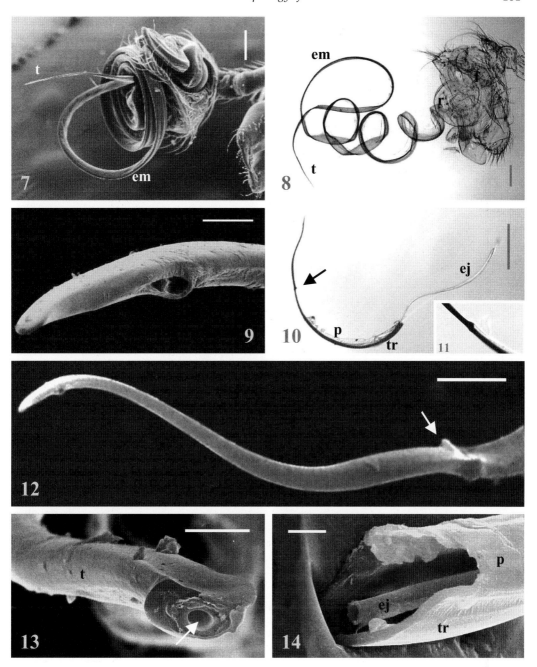

can be clearly divided into the dark truncus and the membranous pars pendula. They surround the ejaculatory duct. ej: ejaculatory duct, p: pars pendula, tr: truncus. Scale bar: 200µm. **Fig. 11.** Breaking point of the embolus, 4 x larger than Fig. 10. **Fig. 12.** The embolus tip is a S-shaped sclerite. The transition between the tip and the rest of the embolus is marked by a saddle-like thickening (arrow). Scale bar: 50 µm. **Fig. 13.** Broken off embolus tip. A canal with a diameter of approx. 3 µm runs through the tip (arrow). t: embolus tip. Scale bar: 10 µm. **Fig. 14.** Broken embolus. The truncus forms a U-shaped channel closed by the pars pendula. Inside lies the ejaculatory duct originating from the reservoir. ej: ejaculatory duct, p: pars pendula, tr: truncus. Scale bar: 10 µm.

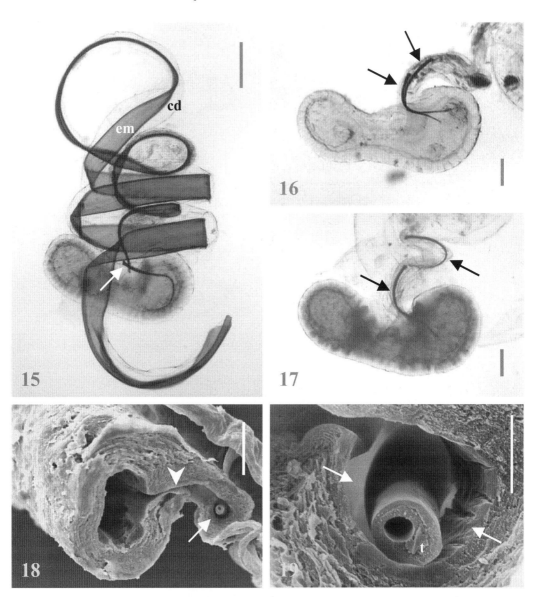

Fig. 15. The male lost its complete embolus inside the female copulatory duct. Clearly visible is the well-defined breaking point of the embolus (arrow). The tip is lodged deep inside the spermathecal opening. cd: copulatory duct, em: embolus. Scale bar: 200 µm. **Fig. 16.** Cleared spermatheca (lateral view) with two tips (arrows) reaching into the spermathecal lumen. Scale bar: 100 µm. **Fig. 17.** Cleared spermatheca (lateral view) with one tip in the opening (left arrow) and second tip (right arrow) lying in the copulatory duct behind the first one. Scale bar: 100 µm. **Fig. 18.** Spermatheca cut transversally in the region of the entrance. The narrow slit (arrow head) is slightly widened laterally forming a tube, that takes up the embolus tip (arrow). Scale bar: 200 µm. **Fig. 19.** The embolus tip lies in the tube-like opening of the spermatheca. Secretion fills the space between tip and cuticle (arrows). t: embolus tip. Scale bar: 10 µm.

for the proximal third of the tip. Distally, the diameter decreases steadily towards the opening and tapers to the slightly flattened apex (Fig. 12). A canal with a diameter of approx. 3 μm runs through the tip (Figs. 13, 19). This canal is continuous with the ejaculatory duct which ends at the breaking point (Fig. 14). The actual opening of the tip lies approx. 30 μm proximal from the apex. It is oval and has a diameter not exceeding 6 μm (Fig. 9).

Position of the male embolus tip inside the female spermathecal opening

In most mated females a part of the male embolus, in some cases even the whole embolus, was found inside the female genital tract (Fig. 15). It becomes obvious that the male screwed its coiled embolus completely into the similarly coiled copulatory duct of the female. The embolus tip was lodged deeply inside the lateral part (tube) of the slit-like opening of the female spermatheca (Fig. 18), where it finally snapped off at the breaking point. The space between the tip and the cuticle of the spermathecal opening was filled with a secretion (Fig. 19). In those cases where the tip broke off at the breaking point and the rest of the embolus was retreated after copulation only the wider part close to the saddle-like thickening of the tip remained lying inside the distal part of the copulatory duct (Figs. 3, 5, 15). About 2/3 of the embolus tip's length extended into the anterior lumen of the spermatheca. The actual site where the seminal fluid was expelled was found at the most anterior wall of the spermatheca (Fig. 5). Here the opening of the pointed end of the male embolus came to rest during insemination. In most females not more than one tip was found in each spermathecal opening. In two females two tips were seen in the opening, but neither of these tips was positioned deeply inside the opening as described before (Fig. 16). Only between 1/3 to 1/2 of the tips' lengths reached into the spermathecal lumen. In other specimens a second tip could be observed lying in the copulatory duct, behind a tip that had been already positioned in the spermathecal opening by another male (Fig. 17).

DISCUSSION

Our description of the female and male genitalia of *L. revivensis* revealed many similarities to other *Latrodectus* species described by various authors in greater or lesser detail throughout the last century. The most detailed studies were undertaken for *L. curacaviensis* by Bhatnagar and Rempel (1962) and *L. geometricus* C.L. Koch, 1841 by Abalos and Baez (1963, 1967). In the present study, the use especially of the scanning electron microscope revealed some new details, e.g. about the epigynal plate, the female copulatory duct, the spermatheca and the male embolus.

None of the authors who have characterised the outer aspects of any *Latrodectus* species mention the long and thin hairs that cover the epigynal plate and partly overhang the opening of the atrium (Shulov 1940, 1948; Levi 1959, 1966, 1983; Bhatnagar & Rempel 1962; Abalos & Baez 1963, 1967; Mackay 1972; Levy & Amitai 1983). These hairs are innervated and represent typical mechanoreceptors (unpubl.) that may enable the female to perceive stimuli provided by the male during courtship (Shulov 1940; Ross & Smith 1979; Anava & Lubin 1993) and, through this, information about his quality. In addition, the female might be able to obtain information about the success of the coupling process itself.

The copulatory ducts of *L. revivensis* are not simple tubes as described for other *Latrodectus* species (Bhatnagar & Rempel 1962; Kaston 1970; Breene & Sweet 1985). Close to the spermatheca the duct becomes flat. Moreover, a distinct tube is formed in this region, whose wall is more sclerotised than the rest of the copulatory duct. This tube may form a guidance system for the male embolus during copulation and might act as a bottle-neck which the male embolus has to pass, before the tip can enter the spermatheca to deposit the seminal fluids. In addition, musculature in this region appears more complex than suggested (see

Whitehead & Rempel 1959) and is now under investigation.

The paired spermathecae of *L. revivensis* are dumb-bell shaped organs. Each spermatheca is divided into distinct compartments: the anterior lobe, the posterior lobe and the narrow middle portion. The copulatory duct is connected to the anterior lobe; the fertilisation duct is part of the posterior lobe. This compartmentation should result in a stratification of sperm inside the lumen. A fair raffle of rival males' sperm proposed for many insects (Parker 1990), implying that the sperm of different males mix inside the spermatheca due to the active movement of the sperm cells, does not occur in *Latrodectus* species. Here, as in most araneomorphs, single coated sperm cells lie immobile inside the spermatheca until oviposition (Alberti 1990; for further references see Foelix 1997). The position of the sperm cells attained during insemination is most probably the decisive factor determining whether they will be used for fertilisation. In case a second male is able to insert his embolus tip in an already filled spermatheca (that seems to happen frequently at least in some *Latrodectus* species such as *L. geometricus* (see Müller 1985) and *L. pallidus* O.P. Cambridge, 1872 (Harari pers. comm.)), he has to develop considerable force to push his sperm from the opening of the embolus tip in the anterior lobe and through the narrow middle portion into the posterior lobe. Only in those sperm cells which lie in close proximity of the fertilisation duct the protective coat is dissolved (unpubl.). Thus, only those spermatozoa can be used for fertilisation. A second male should not only have to displace the rivals' sperm in the anterior and posterior lobes, but also has to push his sperm close to the lumen of the fertilisation duct. When both spermathecae are filled by the first male, a clear first male sperm priority would be expected. The main question is therefore, whether a male is able to fill only one or both spermathecae. Looking at the various combinations in which the spermathecae can be filled by two males (one male inserts his sperm into one spermatheca, a second male inserts his sperm into the second spermatheca; one is subsequently filled by both (e.g. in those females with two tips in one opening), the second spermatheca is just filled by one, etc.), we would expect highly variable P2 values. In case the male is able to deter the intromission of a rival's embolus tip into the spermathecal lumen with his own tip, the situation becomes even more complex, and P2 values ranging from 0.0 to 1.0 as found for *L. hasselti* (Andrade 1996) are not surprising.

The inner basal surface of the spermathecal cuticle showed pores of different sizes. As suggested by Bennet (1992) for theridiid species in general, the large pores in the cuticle of the anterior and posterior lobes of *L. revivensis* are primary pores, 'more or less simple canals which convey gland ducts through the walls of the spermathecal heads' (Bennet 1992, p. 3). Not much is known about the glands of the spermathecal epithelium and the secretions they pass into the spermathecal lumen (short review in Lopez 1987). Possible functions include pheromone release (Kovoor 1981), sperm nutrition and mechanical displacement of sperm. Also the activation of sperm in spiders is mediated by female secretion rather than being time dependent as proposed for *Nephila clavipes* (Brown 1985).

Theridiids including *L. revivensis* possess one of the most complex pedipalps known in spiders (Levi 1961). Although there are several parts of the genital bulb engaged in the coupling process between male and female genitalia, the male embolus is the only part that will be introduced into the female. This was demonstrated in a female of *L. revivensis* that retained long parts of the embolus in one of her copulatory ducts after mating. The loss of the whole embolus was reported for several *Latrodectus* species (Wiehle 1961; Müller 1985). It is assumed that a loss of the entire embolus happens more often in species, whose females possess more coils in their copulatory duct, e.g. *L. geometricus* (Abalos & Baez 1963; Abalos 1968). According to Comstock (1910) the embolus of *Latrodectus* species belongs to the free type and spiral subtype. Wiehle (1961) defined this structure as an 'Einführungsembolus' ('introducing embolus') of the subtype 'Querschleifenembolus mit besonderem Endstück' ('transversely coiled embolus with a special terminal piece') (Wiehle 1961, p. 480). In this type the plane of the coil is positioned vertical to the longitudinal axis of the cymbium. The embolus is more sclerotised at the convex side and accompanied by a membranous part on the inner, concave side. Towards the distal end the embolus becomes very slender and the terminal piece (= embolus tip) can be clearly distinguished from the rest (Wiehle 1961). The embolus of *L. revivensis* and other *Latrodectus* species perfectly combines rigidity and flexibility which enables the embolus to penetrate the variable copulatory duct of the female. The tip, however, does break off frequently inside the female genital tract (Dahl 1902; Smithers 1944; Levi 1959; Wiehle 1961, 1967; Bhatnagar & Rempel 1962; Abalos & Baez 1963, 1967; Kaston 1970; Breene & Sweet 1985; Andrade 1996) at a definite breaking point ('Sollbruchstelle' Wiehle 1961). As shown in the present study the embolus tip of *L. revivensis* is a solid sclerite with a canal inside. A translucent, circular tube inside the tip described for *L. geometricus, L. curacaviensis* and *L. mactans* by Abalos and Baez (1963) could not be found in *L. revivensis*. In the latter species the ejaculatory duct extends only up to the breaking point. The embolus tip of *L. revivensis* possesses a saddle like thickening close to the breaking point which is similar to the backward-directed tooth of the tip of *L. curacaviensis* (Bhatnagar & Rempel 1962). These structures might act as barbs when the male tries to retract his embolus, thus facilitating the rupture of the tip. The ejaculatory duct and the canal of the tip of *L. revivensis* are very narrow (approx. 3 μm in diameter) and each encapsulated sperm cell has to pass it one by one during sperm induction and ejaculation. As long as the embolus of a male is not mutilated (e.g. ripped off or uncoiled) after the first insertion a male might be able to copulate with another female (see Breene & Sweet 1985), but for several reasons a successful second copulation using the same pedipalp appears rather doubtful. An embolus devoid the tip is too short to reach the lumen of the spermatheca. Furthermore, the diameter of the broken embolus is too wide to fit into the narrow spermathecal opening. This could be demonstrated in the female specimen of *L. revivensis* which had an entire embolus in the copulatory duct. Spermatozoa expelled out of an embolus (with or without the tip) which had only reached the copulatory duct will be lost for fertilisation. Through the pressure a male is able to create, the sperm mass will be pushed backwards into the wider parts of the copulatory duct instead of entering the spermathecal lumen. Abalos and Baez (1967) wrote that 'when the apical element is situated in the canals of the female, the seminal mass is found in the canal' (p. 200). Furthermore, the recharge of the pedipalp is not possible without the solid tip, hence, in a broken embolus the ejaculatory duct lies loosely between the truncus and pars pendula. The expected 'sterility' of mated males that have lost their tips could be shown by experiments conducted in *L. hasselti* by Andrade (in press). A successful second insemination with an already used palp without an embolus tip appears impossible. Thereore, a male should invest all his sperm in the first copulation

It becomes clear by our pictures that a broken embolus tip prevents other males from successfully entering a previously filled spermatheca. As shown in the SEM pictures, the male embolus tip closes the female spermathecal opening very tightly. Only in those cases in which the first male did not position his embolus tip deeply inside the opening a second male managed to enter the lumen with his own tip. More frequently the second male did not reach the spermathecal lumen but lost his tip, and most probably his sperm, inside the copulatory duct. The frequency of the loss of the embolus tip, the number and position of the tips inside the genital tract of mated females from the field and the comparison with other *Latrodectus* species allows us to characterise the embolus tip of

L. revivensis as an effective mating plug (Berendonck et al. in prep.).

Austad (1984) reduced the problem of sperm precedence patterns of entelegyne spiders to a theoretical shape of the female spermatheca. Our findings suggest that females and males of *L. revivensis* have a variety of possibilities to control the outcome of matings that include epigynal structures, spermathecal organisation and embolus tips that close perfectly the spermathecal opening.

ACKNOWLEDGEMENTS

We thank J. Dunlop, G. Uhl and an anonymous referee for valuable comments on the manuscript and the University of Aarhus, Denmark, for the financial support that made it possible to participate in the 19th European Colloquium of Arachnology in Århus.

This research was supported by a graduate fellowship from the Jacob Blaustein Institute for Desert Research, Ben-Gurion University of the Negev, Israel. This is publication No. 320 of the Mitrani Department of Desert Ecology.

REFERENCES

Abalos, J.W. 1968. La transferencia espermática en los arácnidos. *Revista de la Universidad Nacional de Cordoba (Argentina)*, 2d ser. 9, 251-278.

Abalos, J.W. & Baez, E.C. 1963. On spermatic transmission in spiders. *Psyche* 70, 197-207.

Abalos, J.W. & Baez, E.C. 1967. The spider genus *Latrodectus* in Santiago dell Estero, Argentina. In: *Animal toxins* (F.E. Russell & P. R. Saunders eds.), pp. 59-74. Pergamon Press, Oxford .

Alberti, G. 1990. Comparative spermatology of Araneae. *Acta Zoologica Fennica* 190, 17-34.

Anava, A. & Lubin, Y. 1993. Presence of gender cues in the web of a widow spider, *Latrodectus revivensis*, and a description of courtship behaviour. *Bulletin of the British Arachnological Society* 9, 119-122.

Andrade, M.C.B. 1996. Sexual selection for male sacrifice in the Australian redback spider. *Science* 271, 70-72.

Andrade, M.C.B & Banta, E.N. In press. Value of male remating and functional sterility in redback spiders. *Animal Behaviour*

Austad, S.N. 1984. Evolution of sperm priority patterns in spiders. In: *Sperm competition and the evolution of mating systems* (R.L. Smith ed.), pp. 223-249. Harvard University Press, Cambridge, Massachusetts.

Bennet, R.G. 1992. The spermathecal pores of spiders with special reference to dictynoids and amaurobioids (Araneae, Araneomorphae, Araneoclada). *Proceedings of the Entomological Society of Ontario* 123, 1- 21.

Bhatnagar, D.S. & Rempel, J.G. 1962. Structure, function and postembryonic development of the black widow spider *Latrodectus curacaviensis* (Mueller). *Canadian Journal of Zoology* 40, 465-510.

Breene, R.G. & Sweet, M.H. 1985. Evidence of insemination of multiple females by the male black widow spider *Latrodectus mactans* (Araneae, Theridiidae). *Journal of Arachnology* 13, 331-335.

Brown, S.G. 1985. Mating behaviour of the Golden-Orb-Weaving spider, *Nephila clavipes*: II. Sperm capacitation, sperm competition and fecundity. *Journal of Comparative Psychology* 99, 167-175.

Comstock J.H. 1910. The palpi of male spiders. *Annals of the Entomological Society of America* 3, 161-185.

Dahl, F. 1902. Über abgebrochene Kopulationsorgane männlicher Spinnen im Körper des Weibchens. *Sitzungsberichte der Gesellschaft für Naturforschung zu Berlin*, 185-203.

De Biasi, P. 1962. Estrutura interna e presença de segmentos do êmbolo no epígino de Latrodectus geometricus (Araneidae: Theridiidae). *Papéis Avulsos do Departamento de Zoologia* 15, 327-331.

Foelix, R.F. 1997. *Biology of Spiders*. Harvard University Press, Cambridge, Massachusetts.

Kaston, B.J. 1970. Comparative biology of American black widow spiders. *Transactions of the San Diego Society of Natural History* 16, 33-82.

Kovoor, J. 1981. Une source probable de pheromone sexuelles: les glandes tegumentaires de la region genitale de femelles d'araignees. *Atti della Societa Toscana di Scienze Naturali Memorie B* 88, 1-15.

Levi, H.W. 1959. The spider genus *Latrodectus* (Araneae, Theridiidae*). Transactions of the American Microscopical Society* 78, 7-43.

Levi, H.W. 1961. Evolutionary trends in the development of palpal sclerites in the spider family Theridiidae. *Journal of Morphology* 108, 1-9.

Levi, H.W. 1966. The three species of *Latrodectus* (Araneae) found in Israel. *Journal of Zoology* 150, 427-432.

Levi, H.W. 1983. On the value of genitalic structures and coloration in separating species of widow spiders (*Latrodectus* sp.) (Arachnida: Araneae: Theridiidae). *Verhandlungen des Naturwissenschaftlichen Vereins Hamburg* 26, 195-200.

Levy, G. & Amitai, P. 1983. Revision of the widow spider genus *Latrodectus* (Araneae: Theridiidae) in Israel. *Zoological Journal of the Linnean Society* 77, 39-63.

Lopez, A. 1987. Glandular aspects of sexual biology. In: *Ecophysiology of spiders* (W. Nentwig ed.), pp. 121-132. Springer, Berlin.

Mackay, I. R. 1972. A new species of widow spider (Genus *Latrodectus*) from Southern Africa (Araneae; Theridiidae). *Psyche* 79, 236-242.

Müller, G. H. 1985. Abgebrochene Emboli in der Vulva der 'Schwarzen Witwe' *Latrodectus geometricus* C.L. Koch 1841 (Arachnida: Araneae: Theridiidae). *Entomologische Zeitschrift* 95, 27-30

Parker, G.A. 1990. Sperm competition games: raffles and roles. *Proceedings of the Royal Society of London B* 242, 120-126.

Ross, K. & Smith, R.L. 1979. Aspects of the courtship of the black widow spider, *Latrodectus hesperus* (Araneae: Theridiidae), with evidence for the existence of a contact sex pheromone. *Journal of Arachnology* 7, 69-77.

Shulov, A. 1940. On the biology of two *Latrodectus* spiders in Palestine. *Proceedings of the Linnean Society of London* 152, 309-328.

Shulov, A. 1948. *Latrodectus revivensis* sp. nov. from Palestine. *Ecology* 29, 209-215.

Smithers, R. H. N. 1944. Contributions to our knowledge of the genus *Latrodectus* in South Africa. *Annals of the South African Museum* 36, 263-312.

Uhl, G. 2002. Female genital morphology and sperm priority patterns in spiders (Araneae). In: *European Arachnology 2000* (S. Toft & N. Scharff eds.), pp. 145-156. Aarhus University Press, Aarhus.

Uhl, G. & Vollrath, F. 1998. Genital morphology of *Nephila edulis*: implications for sperm competition in spiders. *Canadian Journal of Zoology* 76, 39-47.

Wagner, W. 1887. Kopulationsorgane des Männchens als Criterium für die Systematik der Spinnen. *Horae Societatis Entomologicae Rossicae* 22: 3-132.

Whitehead, W. F.; Rempel, J. G. 1959. A study of the musculature of the black widow spider, *Latrodectus mactans* (Fabr.). *Canadian Journal of Zoology* 37, 831-870.

Wiehle, H. 1961. Der Embolus des männlichen Spinnentasters. *Zoologischer Anzeiger* Suppl. 24, 457-480.

Wiehle, H. 1967. Steckengebliebene Emboli in den Vulven von Spinnen (Arach., Araneae). *Senckenbergiana Biologica* 48, 192-202.

SPIDER COMMUNITIES

CONSERVATION ECOLOGY

European Arachnology 2000 (S. Toft & N. Scharff eds.), pp. 171-176.
© Aarhus University Press, Aarhus, 2002. ISBN 87 7934 001 6
(Proceedings of the 19th European Colloquium of Arachnology, Århus 17-22 July 2000)

Practical use of a single index to estimate the global range of rarity of spider communities in Western France

ALAIN CANARD & FRÉDÉRIC YSNEL

Laboratoire de Zoologie et d'Ecophysiologie, UMR 6553, Université de Rennes 1, Avenue du Général Le-clerc, 35042 Rennes Cedex, France (Frédéric.Ysnel@univ-rennes1.fr)

Abstract

A patrimonial index (I_p) was calculated for different habitats of a nature reserve and the relative contribution of each habitat to the global patrimonial value of the reserve is presented. Strong variations can be observed between the different values of the patrimonial index. These variations have to be related to the management of the habitat, to the presence of local habitats, and also to the fragmentation of the shrubby layers. For example, set-asides and local *Juncus maritimus* beds exhibit the highest values, while areas exposed to animal trampling and fragmented areas exhibit the lowest ones; we can notice that the values gradually increase from the lowest to the highest. The variablity of the I_p is linked to the number of species analysed, to the collecting method and varies strongly in the course of the year.

Key words: spider communities, nature reserve, biodiagnostics, Western France

INTRODUCTION

The principle of evaluation of a habitat by analysis of the global range of rarity of spider communities is based on the following concept: an area which has been partly or totally depleted of its living species is immediately colonised by the species living in adjacent areas as soon as the conditions have improved. These early arriving species have a high dispersal potential; they are ubiquitous species and form a community of low conservation (or patrimonial) value. On the contrary, an area which has a high number of rare specialised species (stenoecious species), with a low capacity of recolonisation, is an area which has maintained its original biodiversity. Such a spider community has a high conservation value. Referring to this idea, an index (I_p) based on the relative rarity of the spider species to estimate the conservation value of different communities was

elaborated (Canard et al. 1998). To assess how species richness among the communities influence the I_p value, we first present a brief theoretical analysis on the variations of the I_p values. Secondly, we describe an example of the practical use of this index as a contribution to the elaboration of the managing plan of a natural reserve in the west of France.

MATERIALS AND METHODS
Calculation of the index

A reference base indicating the distribution of all the spider species of Western France allows us to evaluate the relative rarity of each species (Fig. 1). The number of stations in the reference base are divided into several groups (group of species known from 1, 2, 3 ....10 stations to group of species known from 200 to 250 stations).

Fig. 2 gives a theoretical example of the

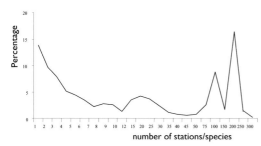

Fig. I. Reference base curve for the West of France (percentages from 23296 records).

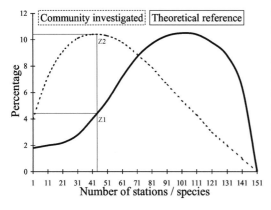

Fig. 2. Theoretical comparison of the investigated biotope curve and the reference base curve for a biotope containing few common species (after Canard et al. 1998).

principle of calculation of the index. The black curve indicates the percentage of species known from different groups in reference base; the doted line indicates the percentage of species of a hypothetical community investigated known from different groups. At point Z1 there are 4% of the total species of the theoretical reference base known from 41-50 stations. At point Z2 there are 10.4% of the species in the community investigated known from 41-50 stations. The patrimonial index (I_P) is calculated in a computer program which sums all the differences between Z1 and Z2 for each group of stations. Thus the global range of rarity of the community investigated is given by a single value: the patrimonial value (I_P). By referring to the database of the west of France, the patrimonial index may vary from -27 when

there are only very common species in the community, to +75 when there are only rare species in the community (known from only one station). These values are different from those in Canard & al. (1998); they agree with new data on species distribution. Another noticeable value is 'zero' which corresponds to a theoretical community composed of all the species of the reference base.

To examine how both the species richness and the overall rarity of the community affect the I_P values, we used simulated I_P values for several theoretical communities. These communities were composed of species known from 10, 40, 70, or 100 stations in the reference base. We then introduced one or two unknown species (not listed in the reference base) for each theoretical community to explore the range of variation of the I_P values.

Field analysis

The nature reserve investigated ('Réserve Ornithologique de Séné') is a coastal area of Western France (47°36'N, 2°42'W) consisting of a complex of salt-marshes, rush meadows, pastures, mowed and non-mowed meadows, fallow land, surrounded by hedges and thickets. The first aim of this nature reserve is to serve as a stop-over area for migrating birds on major flyways. For a few years, arthropods have been taken into account to develop a managing plan for the terrestrial areas of this nature reserve. Spiders were collected by nine series of pitfall traps (2 pitfall traps per plot) from 15 March 1998 to 30 October 1998. Results from a second period of pitfall trapping (15/02/99—30/03/99) were added. In both periods, spiders were removed from the traps every two weeks. Additional sampling was made by visual searches and beating of branches throughout this period. Eighteen plots have been investigated (Table 1) representing the different vegetation types found in the nature reserve. Special attention was paid to edge effects, and we considered the spider community of a field margin separately from that found in the center of the plot. We also considered the communities

Table 1. List of the 18 investigated plots ordered according to their patrimonial index (I_p). N: number of species; Es.: percentage of exclusive species found only in one habitat.

Areas investigated	N	Es. (%)	I_p
Fallow land (center)	46	11	-21.43
Mesophilous grassland	36	0	-22.94
Humid meadow (margin)	44	4.5	-23.37
Juncus maritimus beds	41	9.8	-23.64
Tree foliage	56	25.8	-24.24
Upper shore communities	49	4	-24.66
Salt-marshes	19	10.5	-24.85
Mowed humid meadow	41	14.5	-25.16
Blackthorn-bramble scrubs (ground-layer)	37	13.5	-25.12
Sub-Halophytic humid pasture	56	5.3	-25.28
Halophytic humid pasture	54	11	-25.60
Fallow land (margin)	73	22	-25.82
Mesophile grassland	40	22.5	-26.08
Non-mowed humid meadows	40	10	-26.53
Small deciduous woods (ground-layer)	43	23	-26.71
Mesophile pasture	43	0	-26.96
Gorse clumps	39	20.5	-27.43
Banked edge of salt basins	33	6	-27.55

found at the center and margin of the fallow land separately.

I_p values were also calculated both for all species collected by each method (pitfall traps, visual searches, beating of branches) and for the species that were caught exclusively by each method. For this purpose, data from all 18 habitats and all seasonal samples are combined.

RESULTS
Theoretical variations of I_p

If the I_p value does not vary for a community composed of species known from the same number of stations, whatever this number of species is, the integration of a new species (an unknown species in the reference base) induces a strong increase of the I_p value (from –26.1 to –9.95) for the community which has from 5 to 250 species (Fig. 3a). When the species diversity is high (from 50 to 250 species) the I_p value increases very slowly (from –25.82 to –22.36). The highest values of the I_p can be observed for the community with two new species what-ever the number of species of the community is (Fig. 3b). When the number of species in the investigated community is low, we can notice strong differences between the values of I_p. When the number of species in the commun-ities is high, the I_p values are very close. Thus the difference

between two I_p values depends on the number of species and the communities must be composed of at least 25-30 species for a reliable comparison.

I_p values for the nature reserve

247 spider species were identified in the whole nature reserve and the values of the species richness of the different plots allows the direct comparison of the I_p values. However, the salt-marshes exhibited the lowest diversity (19 species) while the maximal diversity (73 species) was found in the margin of a fallow field.

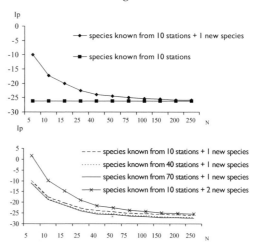

Fig. 3. Simulation of the patrimonial index (I_p) in relation to species richness (N)

Thus it can be underlined that, compared to other habitats, the I_P value of salt marshes is probably overestimated while the I_P value of the margin of a fallow field is probably underestimated. Table 1 lists these 18 plots according to their patrimonial index. The fallow land exhibited the highest patrimonial index whereas the banked edge of salt basins exibited the lowest, and the values gradually increased from the lowest one to the highest one. We can distinguish two groups among the different areas investigated. The low patrimonial value group consists of stations subjected to human trampling (small deciduous wood), animal trampling (banked edge of salt basins) or animal grazing (mesophile pasture). These areas have a negative effect on the global patrimonial value of the reserve. The low patrimonial value of the gorse clumps must also be related to the fragmentation of the investigated clumps. The high patrimonial values group have to be related to the specific orientation of biotopes (warm microclimate of the old fallow land) or to unusual and only locally occurring habitats in the region (e.g. salt-marshes or *Juncus maritimus* beds). It is also due to the presence of species which have seldom been recorded in France, or which are new to the French fauna (as for instance *Haplodrassus minor*). Concerning the spider communities, these areas have retained their own specificity without any disturbance. They have a positive effect on the overall patrimonial value of the reserve. However, whatever the patrimonial value of the different plots, their species richness have a positive effect on the richness of the spider fauna of the whole nature reserve.

The comparison between the patrimonial index of each habitat plot and the whole nature reserve with other values in the same biogeographic area (Fig. 4) clearly indicates that the spider community of fallow land is distinguish-able from other communities composed of 25 to 80 species because of its high patrimonial value. The I_P value of the whole nature reserve is higher than those of dry heathlands and atlantic heathlands but lower than those of

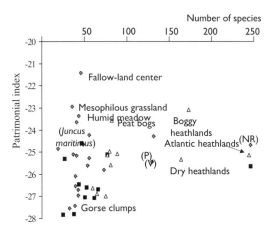

Fig. 4. Overview of the patrimonial index of different areas of the west of France including the nature reserve investigated. ◆ : habitats of the nature reserve; ■ : other habitats of Western France; △ : other habitats within the coastal area.
(Data from all 18 habitats and all seasonal samples: P : I_p value for species collected by pitfall traps; V : I_p value for species collected by visual-search; NR : I_p value of the whole nature reserve).

peat bogs and boggy heathlands. Fig. 5 shows the variation in the patrimonial index during the year (from March to October) for the samples from the 18 plots of the reserve. This variation is due to the staggered occurrence of adults of different species. Table 2 gives the patrimonial index values for the three different sampling methods used. More than 60% of the species were collected by only one method. Thus, a single sampling method cannot give a representative value of the patrimonial index. It is noticeable that the two values are very close, but considering the number of species collected, this small variation can be taken into account. Another remark concerns the high patrimonial value of the species exclusively caught by pitfall-traps. This may indicate that more rare species have been caught by pitfall traps. However, it may also indicate that there is a lack of data in the reference base concerning the distribution of species caught by pitfall traps.

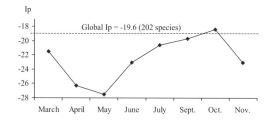

Fig. 5. Variation of the patrimonial index in the course of the sampling period, data from all plots combined.

Table 2. Patrimonial index (I_p) according to the sampling methods (data from the 18 habitats of the nature reserve surveyed through one year combined). N: total number of species collected; Es.: number of species ecxlusively caught by one sampling method.

Sampling method	N	I_p	Es.	Ip Es.
Pitfall trapping	135	-24.32	80	-19.90
Hand collecting	130	-25.44	71	-26.10
Branch beating	69	-24.24	40	-21.13

DISCUSSION

These results are consistent with previous studies demonstrating that the heterogeneity in mosaic landscape and the incidence of mowing or trampling typically affect the species richness (Duelli 1997; Dupont & Lumaret 1997); this short study provides a response, not only in terms of analysis of species richness but also in terms of evaluation of the conservation value of human management. Because the index is very sensitive to the presence/absence, and to the number of species collected, standardized methods have to be strictly followed when comparing several plots during the same periods. The index does not take into account the species density and only refers to the presence/absence criterion. One may object that the distribution of abundance among species can be important in the evaluation of the patrimonial value of localities. Unfortunately, the number of spiders collected strongly depends on the sampling effort (visual search), the sampling period, the design of the pitfall traps and the microenvironment around them. Thus we suggest that it is inadequate to integrate quantita-

tive data on spider densities to compare habitat conservation values at the community level from available data sets. Use of density is most appropriate to assess the different forms of rarity (Rabinowitz et al. 1986) when a study focuses on a particular species. However, in order to minimize the 'sampling effect' on the I_p values, the reference base can be split into several units according to the sampling method or to sampling seasons.

The originality of the patrimonial value of the different listed communities comes from the integration of the rarity degree on all the species of the community. This approach can complete other approaches based on the research of the so-called 'rare species' and whose status in Europe (Van Helsdingen 2000) still needs to be studied. Other kinds of single index have been proposed to evaluate the patrimonial value of the habitats. Ruzicka and Bohac (1994) have proposed a single index based on the percentage of representation of three species groups in the investigated community: spiders associated with protected territories (group I), with managed territories (group III) and others species (group II). Other indices based on the occurrence of species in geographic units (method of square mapping) have been pointed out to estimate the global range of rarity of insect communities (Eyre & Rushton 1989) or of spider communities (Gadjos & Sloboda 1996). In France the so-called 'indice biotique global de rareté' first proposed by Favet & Bigot (1993) (which is nowadays used to define the French status categories for invertebrate species) is the result of a rather subjective attribution for each species of a rarity index varying from 1 (endangered species) to 8 (very common species). Although the reference base still needs to be completed, the method we are working on presents a base which systematically integrates all the data on the species distribution, and the patrimonial index can be quickly up-dated with new data on the species distribution.

Whatever definition of rarity one uses, the results will be influenced by the spatial scale at which it is applied (Gaston 1994). In order to explore rarity at different range size, the reference base can also be split into several units from the geographic point of view (example presented here), to a smaller natural complex (as for instance a complex of littoral areas), to a particular type of biotope (heathlands, dunes ...), to a particular macroclimatic area within the geographic zone or the administrative district. The possibility of using different indices at different scales within a geographic area to assess the global range of rarity of spider communities is presently being analysed.

REFERENCES

Canard, A., Marc, P. & Ysnel, F. 1998. Comparative value of habitat biodiversity: an experimental system based on spider community analysis. In: *Proceedings of the 17th European Colloquium of Arachnology, Edinburgh 1997* (P.A. Selden ed.), pp. 319-323. British Arachnological Society, Burnham Beeches, Bucks.

Duelli, P. 1997. Biodiversity evaluation in agricultural landscapes: An approach at two different scales. *Agriculture, Ecosystems & Environment* 62, 81-91.

Dupont, P. & Lumaret, J.P. 1997. *Intégration des invertébrés continentaux dans la gestion et la conservation des espaces naturels. Analyse bibliographique et propositions.* (P. Dupont & J.P Lumaret eds.), pp. 35-53. Ministère de l'Environnement.

Eyre, M.D. & Rushton, S.P.1989. Quantification of conservation criteria using invertebrates. *Journal of Applied Ecology* 26, 159-171.

Gadjos, P. & Sloboda, K. 1996. Present knowledge of the arachnofauna of Slovakia and its utilization for biota quality evaluation and monitoring. *Revue Suisse de Zoologie* hors série 2, 235-244.

Gaston, K.J. 1994. *Rarity.* Chapman & Hall.

Favet, C. & Bigot, L. 1993. Expertise des milieux naturels: une méthode originale par cotation des populations d'insectes. *Insectes OPIE* 90, 25-28.

Ruzicka, V. & Bohac, J. 1994. The utilization of epigeic invertebrate communities as bioindicators of terrestrial environmental quality. In: *Biological monitoring of the environment: a manual of methods* (J. Salanki, D. Jeffrey & G.M. Hughes eds.), pp. 79-86. CAB International, Wallingford.

Rabinowitz, D.,Cairns, S. & Dillin,T. 1986. Seven forms of rarity and their frequency in the flora of the British Isles. In: *Conservation biology: The science of scarcity and diversity* (M.E. Soulé ed.), pp. 182-204. Sinauer, Associates, Sunderland, Massachusetts.

Van Helsdingen, P.J. 2000. Spider (Araneae) protection measures and the required level of knowledge. *Ekológia (Bratislava)* 19 Suppl. 4, 43-50.

European Arachnology 2000 (S. Toft & N. Scharff eds.), pp. 177-182.
© Aarhus University Press, Aarhus, 2002. ISBN 87 7934 001 6
(Proceedings of the 19th European Colloquium of Arachnology, Århus 17-22 July 2000)

Regional variation in spider diversity of Flemish forest stands

D. DE BAKKER[1], J.-P. MAELFAIT[2,3], K. DESENDER[1], F. HENDRICKX[2] & B. DE VOS[4]

[1]*Royal Belgian Institute of Natural Sciences, Department of Entomology, Vautierstraat 29, B-1000 Brussels, Belgium* (debakker@hotmail.com)

[2]*Ghent University, Laboratory for Animal Ecology, Zoogeography and Nature Conservation, K.L. Ledeganckstraat 35, B-9000 Ghent, Belgium* (JeanPierre.Maelfait@rug.ac.be)

[3]*Institute of Nature Conservation, Kliniekstraat 25, B-1070 Brussels, Belgium* (Jean-Pierre. Maelfait@instnat.be)

[4]*Institute for Forestry and Game Management, Gaverstraat 4, B-9500 Geraardsbergen, Belgium*

Abstract

In total, 55423 adult spiders (250 species) were collected in pitfall traps throughout a year of sampling in 56 forest stands in Flanders (northern part of Belgium). Three different regions were compared according to spider diversity in deciduous and pine forests. The total number of species and individuals of spiders did not differ regionally. However, the richness of endangered stenotopic forest species was significantly higher in the forest of the loam region than in the forests of the loamy sand and Campine region. This result confirms the high nature value of the forests of the Flemish loam region previously demonstrated for plants and carabid beetles. High priority should be given to their conservation.

Key words: spider diversity, forests, stenotopic forest species, Flanders, Belgium, geographical regions

INTRODUCTION

Human activities have caused severe changes in composition and diversity of most known ecosystems (including forests) and in their faunas (Tack et al. 1993; Barnes et al. 1998). The exploitation of forested areas in Western Europe began already some 5 to 7000 years ago (Tack et al. 1993). These anthropogenic influences lead to a reduction of total forest cover, the conversion of natural forests into simplified monocultures of mainly exotic tree species and to a severe fragmentation of remaining forests (Harris 1984; Warren & Key 1989; Saunders et al. 1991; Peterken 1996). Forest fragments are now isolated from each other by intensively managed agricultural land, industrial areas,

roads and urban settlements. Nevertheless, these 'artificial' woodlands can in due time evolve into semi-natural woodlands and even into more or less natural woodlands (Peterken 1996). Semi-natural woodlands in Flanders cover nowadays only about 3% of the region; forest cover of exotic tree species cover about 5% (Hermy 1989). Also, in Flanders most forests are small and fragmented. The average area of a forest unit is 19.2 ha, but almost 70% of the forests are less than 10 ha and 14% even less than 1 ha (Anonymous 1998). In spite of earlier efforts (Maelfait et al. 1990, 1991), the nature value of Flemish forest stands as concerns their invertebrate fauna was until recently only very fragmentarily known. To increase this knowledge, a project was initiated to evaluate the quality of forest stands by means of several invertebrate taxa, including spiders. All sampled forest stands were also characterised with respect to soil, litter, vegetation and tree cover parameters (De Vos 1998, 1999a,b). The first results concerning the spiders caught during this extensive sampling campaign in forests distributed over the whole region of Flanders, were given in De Bakker et al. (1998, 2000). This paper presents results concerning the observed diversity of spider assemblages and its regional variation. A future contribution will discuss the possibilities of spiders to be used as bio-indicators for evaluating forest site quality.

MATERIAL AND METHODS

Every forest stand was sampled using three pitfall traps emptied at approximately fortnightly intervals during a complete year cycle (April 1997 - April 1998). As can be seen in Fig. 1 and Table 1, the 56 sampled stands were localised in 40 forests more or less evenly distributed over the three main geographical regions of Flanders (the northern part of Belgium). The region indicated in Fig. 1 as the 'Loamy sand region' (LS) consists of sandy and sandy-loam soil. The forest complexes are small and often quite young. On the more sandy soils pine stands occur, otherwise most forests are deciduous. Only deciduous forest stands have been sampled in this region. The soils of the 'Loam region' (LO) consist of silt also deposited during the Pleistocene era. Mainly due to historical reasons (e.g. hunting areas for the nobillity) but also to relief (more slopes and hills), older and larger deciduous forests occur here. Pine forest stands are very exceptional; only deciduous stands have been sampled. The 'Campine region' (CA) occupies the northeast of Flanders. Niveo-eolian, pleistocene sands cover it. In general, soils are distinctly podsolised so that they have an endured horizon of illuvial humus and/or iron at about 30 to 40 centimetres below the surface. This gives the typical Campine soils the characteristics of be-

A.

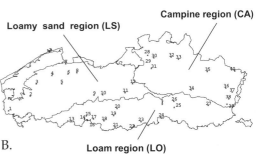

Loamy sand region (LS)

Campine region (CA)

B.   **Loam region (LO)**

Fig. 1. Present forest cover in Flanders **(A)**, and sampled forests per geographical region **(B)** (cf. Table 1).

ing both sandy and waterlogged, with water accumulating above the impermeable horizon of the soil profile. Hence, we find in the still existing natural areas a fine mosaic of very dry sandy areas and dunes intermixed with lakes, marshes and peat bogs. Until the nineteenth century the region was almost completely covered by heathlands, which had developed under a balanced agro-pastoral system. Although there are still important remnants of heathland, most of this semi-natural habitat type was turned into pine tree plantations (*Pinus sylvestris*) at the end of the 19th and the beginning of the 20th century, especially on the drier soils. This was done to fulfil the demands of the mining industry. Deciduous forest occurs or has been planted on the wetter soils.

Statistical analyses were done with the program STATISTICA (StatSoft 1997).

RESULTS

In total, 55423 adult spider individuals were caught representing 250 different species. This is a very large number of species considering that the Belgian and Flemish fauna contain 689 and 604 known species respectively (Maelfait et al. 1998). Of the 604 Flemish species, only 291

Table 1. Classification and characteristics of sampled forest stands. Sn./Fn.: stand/forest number, cf. Fig. 1. Tree %: tree cover. LS, LO, CA: regions cf. Fig. 1. DECI: deciduous forests; PINE: pine forests.

Sn.	Fn.	Type	Tree %	Main tree species
1	1	LS_DECI	90	*Quercus robur, Acer pseudoplatanus*
2	2	LS_DECI	75	*Quercus robur*
3	7	LS_DECI	80	*Quercus robur, Fraxinus excelsior*
4	3	LS_DECI	85	*Fagus sylvatica*
5	4	LS_DECI	90	*Quercus robur, Fagus sylvatica*
6	3	LS_DECI	100	*Quercus robur, Acer pseudoplatanus*
7	3	LS_DECI	50	*Populus x canadensis*
8	5	LS_DECI	70	*Quercus rubra, Fraxinus excelsior*
9	5	LS_DECI	70	*Fagus sylvatica, Quercus robur*
10	6	LS_DECI	80	*Fagus sylvatica*
11	8	LS_DECI	90	*Quercus robur*
12	9	LS_DECI	90	*Quercus robur, Fagus sylvatica*
13	9	LS_DECI	90	*Fraxinus excelsior*
14	10	LS_DECI	80	*Quercus robur*
15	11	LS_DECI	90	*Fagus sylvatica*
16	12	LS_DECI	80	*Alnus glutinosa*
17	13	LO_DECI	100	*Fagus sylvatica*
18	14	LO_DECI	70	*Fraxinus excelsior, Alnus glutinosa*
19	14	LO_DECI	80	*Fagus sylvatica*
20	15	LO_DECI	80	*Quercus robur*
21	16	LO_DECI	50	*Fagus sylvatica*
22	16	LO_DECI	80	*Populus x canadensis, Salix* sp.
23	17	LO_DECI	80	*Populus x canadensis*
24	18	LO_DECI	90	*Quercus rubra, Castanea sativa*
25	19	LO_DECI	80	*Fagus sylvatica*
26	19	LO_DECI	90	*Betula* sp.
27	20	LO_DECI	90	*Betula* sp., *Quercus robur*
28	21	LO_DECI	70	*Quercus robur*
29	21	LO_DECI	80	*Quercus robur*
30	22	LO_DECI	90	*Fagus sylvatica*
31	23	LO_DECI	90	*Fagus sylvatica*
32	23	LO_DECI	50	*Quercus robur, Carpinus betulus*
33	23	LO_DECI	100	*Quercus robur, Carpinus betulus*
34	23	LO_DECI	80	*Fagus sylvatica*
35	24	LO_DECI	85	*Quercus robur*
36	24	LO_DECI	90	*Betula* sp.
37	24	LO_DECI	50	*Fagus sylvatica*
38	25	LO_DECI	90	*Quercus robur*
39	26	LO_DECI	85	*Quercus robur, Q. petraea*
40	27	LO_DECI	60	*Populus x canadensis*
41	28	CA_PINE	75	*Pinus sylvestris*
42	28	CA_PINE	30	*Pinus sylvestris*
43	29	CA_PINE	70	*Pinus sylvestris*
44	30	CA_PINE	70	*Pinus sylvestris*
45	31	CA_PINE	80	*Pinus sylvestris*
46	32	CA_PINE	80	*Pinus sylvestris*
47	33	CA_PINE	75	*Pinus nigra* ssp. *laricio*
48	34	CA_PINE	70	*Pinus nigra* ssp. *laricio*
49	35	CA_PINE	70	*Pinus sylvestris*
50	35	CA_PINE	60	*Pinus sylvestris*
51	36	CA_PINE	60	*Pinus sylvestris*
52	37	CA_DECI	60	*Fraxinus excelsior*
53	38	CA_DECI	80	*Fagus sylvatica*
54	39	CA_DECI	1	*Betula* sp.
55	40	CA_DECI	70	*Betula* sp., *Quercus robur*
56	40	CA_DECI	70	*Betula* sp., *Alnus glutinosa*

are safe or at low risk. Of the remaining 313 species, 36 are insufficiently known and 8 are presumably extinct. The remaining 269 species are rare in Flanders because they are either stenotopic species occurring in few types of threatened habitats (206 species) or of which Flanders is at the southern, northern or western limit of their geographical range (species that are restricted geographically: 63 species). Of the 206 stenotopic species, 41 are bound to particular forest types (stenotopic forest species); the other 165 stenotopics normally thrive in a few types of other threatened habitats, like heathland, oligotrophic grasslands and others (stenotopic non-forest species). In our sampling campaign we found 24 of the 41 stenotopic forest species and 38 stenotopic non-forest species, e.g. species bound to riverbanks, heathlands and marshes. In Fig. 2 the number of individuals and species caught per forest stand is shown. The number of individuals caught during a complete year cycle has as range (minimum-maximum per stand) per region: loamy sand: 408-1486, loam: 387-1601, Campine: 606-1628. There is also no obvious difference between the discerned regions as concerns the number of species caught, with respective ranges: loamy sand: 36-61, loam: 28-75 and Campine: 40-75. The number of individuals of special species caught (sum of stenotopic forest and non-forest species and geographically restricted species) is on average higher in the loam region. In Fig. 3 it can be seen that the number of stenotopic forest species is markedly higher in the loam forests than in the other two regions. This difference is even more pronounced if the numbers of individuals of these stenotopic forest species are considered (Fig. 3, below). The difference between the regions was analysed in more detail by means of an analysis of variance (ANOVA) on log-transformed numbers, in which the regions were treated as independent variables. For the Campine region a distinction was made between deciduous (CA_DECI) and pine forest stands (CA_PINE). In the other two regions only deciduous forest stands were sampled

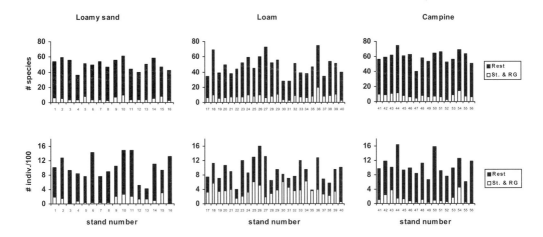

Fig. 2. Total number of spider species, number of stenotopic (St.) and species restricted geographically (RG) caught per forest stand (above); the numbers caught in hundreds of individuals (below).

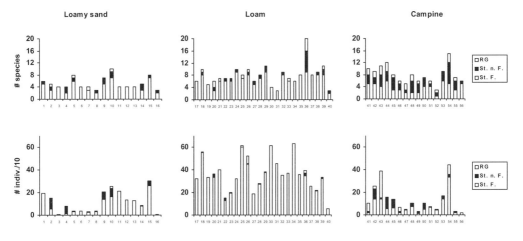

Fig. 3. Numbers of species and tens of individuals of stenotopic forest species (St. F.), stenotopic non-forest species (ST. n. F.) and of species restricted geographically (RG) per forest stand.

(LS_DECI and LO_DECI). For these four forest types it was tested if there were significant differences in the numbers of stenotopic forest species and individuals (St.F.spec., St.F.ind.), in stenotopic non-forest species and individuals (St.n.F.spec., St.n.F. ind.), in species and individuals that are restricted geographically (RG-species, RG-ind.) and in the remaining species and individuals (Rest spec., Rest ind.). All of these eight ANOVAs pointed out significant differences between the considered forest types (P < 0.05). The homogenous groups resulting

from Scheffé's post-hoc test are shown between brackets after the type of forest in Fig. 4.

From Fig. 4 we conclude that:
-the number of stenotopic forest species is significantly higher in the loam region than in the pine forests of the Campine region; the numbers of individuals of stenotopic forest species caught in the loam region is significantly higher than the numbers of individuals of these species caught in the forest stands of the three other forest types;

-the number of stenotopic non-forest species

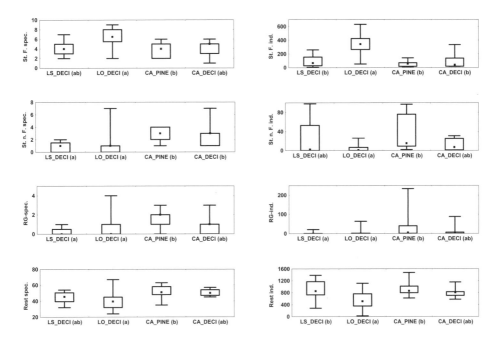

Fig. 4. Median (point), 25 and 75-percentile (box) and minimum and maximum (whisker) of number of species (left) and number of individuals (right) of the discerned species categories. St.F.spec.: stenotopic forest species; St.n.F.spec.: stenotopic non-forest species; RG-spec.: geographically restricted species; Rest spec.: species not covered by other categories (generalists); LS, LO, CA: regions cf. Fig. I; DECI: deciduous forests; PINE: pine forests. Letters in parentheses: same letter indicates no significant difference.

was highest in the two forest types of the Campine region; the number of individuals of these species is especially high in the Campine pine forests and low in the forests of the loam region;

-a similar result as above was obtained for the geographically restricted species;

-the number of the remaining, more eurytopic species and the number of individuals caught of these species was lower in the forests of the loam region than in the other types.

DISCUSSION

When we consider all spiders together, without making any distinction of degrees of stenotopy or rarity, the richness of the spider faunas of the forest stands was not different from region to region. This global result is caused by two essentially different phenomena compensating each other. In the loam region, the forests are richer in stenotopic forest spiders, but poorer in

eurytopic species, geographically restricted species and stenotopic species from non-forest habitats, such as wet or dry heathlands. The reverse is true for the forests of the loamy sand region and the Campine region. This can be understood for the Campine region because of the often quite open nature of the forests. The low tree cover of these forests (Table 1) offers possibilities for the intrusion of species bound to more open habitats, like for instance heathland spiders and geographically restricted species at the northern limit of their geographical range. The extreme fragmentation and small size of the forests in the loamy sand region seems to cause pronounced edge effects with intrusion of open habitat and eurytopic species. The richness of stenotopic forest spiders in the forests of the loam region indicates the high nature value of these forests, implying that a high priority should be given to their conservation. All stenotopic forest spider species are indeed threatened and therefore red listed. The high nature value of the forests of the loam region, as indicated by spiders, complements and strengthens the analogous results obtained for higher plants (Tack et al. 1993) and carabid beetles (Desender et al. 1999).

ACKNOWLEDGMENTS

We thank S. Toft and two anonymous reviewers for their suggestions, which permitted this report of our work to be improved.

REFERENCES

Anonymous 1998. Follow-up reports on the ministerial conferences on the protection of forests in Europe. Vol. II: Sustainable forest management in Europe. Special Report on the follow-up of the implementation of the resolutions H1 and H2 of the Helsinki Ministerial Conference. *Third Ministerial Conference on the protection of forests in Europe, Lisbon 1998*, 274 pp.

Barnes, B.V., Zak, D.R., Denton, S.R. & Spurr, S. H. 1998. *Forest ecology*. 4th ed. John Wiley & Sons, New York.

De Bakker, D., Maelfait, J.-P., Hendrickx, F., Van Waesberghe, D., De Vos, B., Thys, S. & De Bruyn, L. (1998). Relatie tussen bodemkwaliteit en spinnenfauna van Vlaamse Bossen: een eerste analyse. *Nieuwsbrief van de Belgische Arachnologische Vereniging* 13(3), 58-78.

De Bakker, D., Maelfait, J.-P., Hendrickx, F. & De Vos, B. (2000). A first analysis on the relationship between forest soil quality and spider (Araneae) communities of Flemish forest stands. *Ekológia (Bratislava)* 19, 45-58.

De Vos, B. 1998. Chemical element analysis of the forest floor in the macro-invertebrate soil fauna plots. *Report IBW Bb R: 98.005, Institute for Forestry and Game Management*.

De Vos, B. 1999a. Geselecteerde set standplaatsvariabelen ten behoeve van het onderzoek naar bodemfauna-indicatoren. *Ministerie van de Vlaamse Gemeenschap. Rapport IBW Bb R.99.006, Instituut voor Bosbouw en Wildbeheer*, 31 pp.

De Vos, B. 1999b. Positionele variabelen van de bodemfaunaproefvakken. *Rapport IBW Bb R: 99.008, Instituut voor Bosbouw en Wildbeheer*, 13 pp.

Desender, K., Ervynck, A. & Tack, G. 1999. Beetle diversity and historical ecology of woodlands in Flanders. *Belgian Journal of Zoology* 129, 139-156.

Harris, L.D. 1984. *The fragmented forest*. University of Chicago Press, Chicago.

Hermy, M. 1989. Bosgebieden. In: *Natuurbeheer* (M. Hermy ed.), pp. 145-168. Van de Wiele, Stichting Leefmilieu, Natuurreservaten & Instituut voor Natuurbehoud.

Maelfait, J.-P., Segers, H. & Baert, L. 1990. A preliminary analysis of the forest floor spiders of Flanders (Belgium). *Bulletin de la Société europeéne Arachnologique* 1, 242-248.

Maelfait, J.-P., Desender, K., Pollet, M., Segers, H. & Baert, L. 1991. Carabid beetles and spider communities of Belgian forest stands. *Proceedings of the 4th European Congress of Entomology/XIII. SIEEC, Gödöllö, Hungary*, 187-194.

Maelfait, J.-P., Baert, L., Janssen, M. & Alderweireldt, M. 1998. A Red List for the spiders of Flanders. *Bulletin van het Koninklijk Belgisch Instituut voor Natuurwetenschappen, Entomologie* 68, 131-142.

Peterken, G.F. 1996. *Natural woodland: ecology and conservation in Northern temperate regions*. Cambridge University Press, Cambridge.

Saunders, D.A., Hobbs, R.J. & Margules, C.R. 1991. Biological consequences of ecosystem fragmentation: a review. *Conservation Biology* 5, 18-32.

StatSoft, 1997. *STATISTCA for Windows*. Tulsa, USA.

Tack, G., Van Den Bremt, P. & Hermy, M. 1993. *Bossen van Vlaanderen: een historische ecologie*. Davidsfonds, Leuven.

Warren, M.S. & Key, R.S. 1989. Woodlands: past, present and potential for insects. In: *The conservation of insects and their habitats* (N.M. Collins & J.A. Thomas eds.), pp. 155-211. Academic Press, London.

European Arachnology 2000 (S. Toft & N. Scharff eds.), pp. 183-190.
© Aarhus University Press, Aarhus, 2002. ISBN 87 7934 001 6
(Proceedings of the 19th European Colloquium of Arachnology, Århus 17-22 July 2000)

Effects of winter fire on spiders

MARCO MORETTI

*Swiss Federal Research Institute WSL, Sottostazione Sud delle Alpi, PO Box 57, CH-6504
Bellinzona, Switzerland* (marco.moretti@wsl.ch)

Abstract

"Fire and biodiversity" is a much-discussed topic based on research in different countries under varying environmental conditions and fire regimes. Nevertheless, until now, no research has been carried out on faunal post-fire biodiversity in winter fire-prone ecosystems such as deciduous forests on the south-facing slopes of the Alps. The basic objectives of this study were to analyse the effects of single and recurring fires on spider communities and their subsequent diversity. Epigeic spiders were used as bioindicators to describe the ecological response of chestnut forest-floor habitats to winter fires in Southern Switzerland. Overall, 143 spider species were sampled from April to September using pitfall traps at both post-fire and intact chestnut coppice sites. Approximately 30% of the species found were present exclusively in burnt sites (compared with only 7% found exclusively in intact areas). These results suggest that post-fire succession was initiated by surviving individuals. No dominant pioneer species were observed in the burnt sites. Changes in communities were observable only within the first two years after a single fire, whereas changes persisted in areas with recurring fires. In the latter case, communities were characterised by the eudominance of a single species *(Pardosa saltans)* and by an increase in species richness. The presence of species strongly associated with particular fire regimes indicates a wide-mosaic structure of environmental and microclimatic conditions at the epigeic level, with a predominance of xeric conditions. Evidence was found that major silvicultural projects implemented before and during the early 1950's, combined with an intense fire history, have played an important role in the development of the spider communities in the chestnut forest of the Southern Alps.

Key words: forest fires, fire ecology, post-fire succession, fire regime, fire frequency, time elapsed since last fire, fauna, biodiversity, Southern Switzerland

INTRODUCTION

Fire is an important environmental factor in many ecosystems throughout the world. However, few studies have examined the impact of fires over periods exceeding 10-15 years (Huhta 1971; Merrett 1976), and little is known about longer-term effects of fire on spider communities. A further limitation of existing work is that most have been carried out in fire-prone regions (e.g. Mediterranean shrub-land, steppe, savanna, boreal forest, etc.) or in specific fire-climax ecosystems; much less is known about the role of fire in other ecosystems, such as temperate forests.

Our study concentrates on the impact of fire on spider communities in chestnut forests of the Southern Alps. Fires are frequent in these forests, but occur mainly during the dormant period between December and April (Conedera et al. 1996). The following objectives were pursued:
(i) analysis of the effects of single and recurring fires on spider diversity over a 25 year period,
(ii) identification of the species most affected by fires, both positively and negatively, and

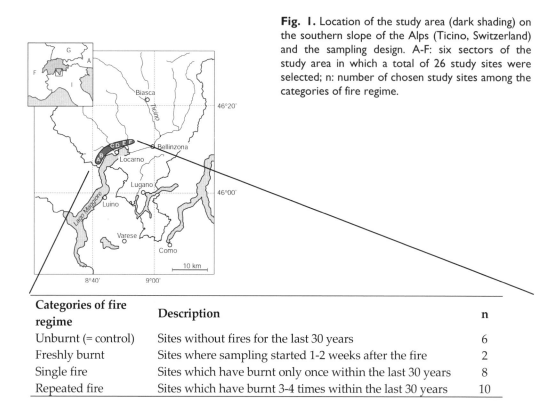

Fig. 1. Location of the study area (dark shading) on the southern slope of the Alps (Ticino, Switzerland) and the sampling design. A-F: six sectors of the study area in which a total of 26 study sites were selected; n: number of chosen study sites among the categories of fire regime.

Categories of fire regime	Description	n
Unburnt (= control)	Sites without fires for the last 30 years	6
Freshly burnt	Sites where sampling started 1-2 weeks after the fire	2
Single fire	Sites which have burnt only once within the last 30 years	8
Repeated fire	Sites which have burnt 3-4 times within the last 30 years	10

(iii) analysis of the ecological response of forest-floor habitats to wildfires using epigeic spiders as bioindicators.

The current project is part of a wider research program on the effects of wildfire on invertebrates and is supported by the WSL Swiss Federal Research Institute (Switzerland).

METHODS

Study area

The study region is located along a uniform, south-facing slope (450-850 meters a.s.l.) in Canton Ticino (46°09' N, 08°44' E) (Fig. 1). The forest is dominated by chestnut (*Castanea sativa* Mill.) coppices, situated on acidic substratum, corresponding to the *Phyteumo betonicifoliae-Quercetum castanosum* association (Ellenberg & Klötzli 1972; Keller et al. 1998). The climate is mild with humid summers and relatively dry winters. Average annual rainfall is 1800 mm. The mean annual temperature is 11°C (1°C in January and 22°C in July).

Sampling was based on a 'space-for-time substitution' (Pickett 1989), and 26 individual study sites were chosen. The sites were similar to each other in aspect, slope and soil, but differed in terms of fire frequency and time elapsed since last fire. The Wildfire Database of Southern Switzerland (Conedera et al. 1996) was the principal source of information on the regional fire history over the past 30 years (1968-1997). The sites were classified according to fire regime (Fig. 1).

Sampling methods and data analysis

Epigeic spiders were sampled using pitfall traps (plastic beakers, Ø 15 cm) with an overhead roof for rain protection (Uetz & Unzicker 1976; Obrist & Duelli 1996). The traps were filled with 2% formaldehyde solution and soap was added to reduce liquid surface tension. Three pitfall traps were installed on each of the 26 study sites (totaling 78 traps), and were emptied on a weekly basis from March through

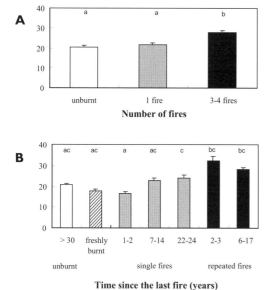

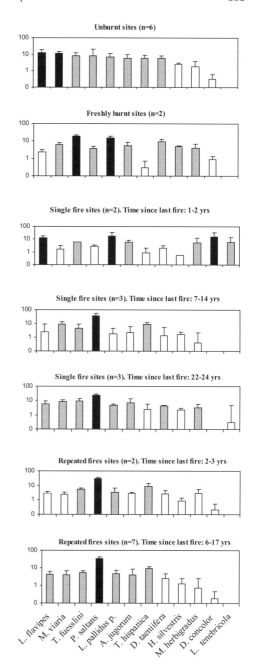

Fig. 2. Spider species richness (mean ± SE) grouped in classes of **(A)** 'fire frequency' (ANOVA F = 30.69, P < 0.001) and of **(B)** 'time elapsed since the last fire' (ANOVA, F = 19.14, P < 0.001). Bars with different letters are significantly different (P < 0.05; ANOVA with subsequent Scheffé post-hoc test).

September 1997 (28 sampling periods in all). The distance between traps was never less than 10 m. All adult spiders were classified into species using standard keys (i.e. Heimer & Nentwig 1991) and the nomenclature followed Platnick (1989). Specimens of each species were deposited in the Natural History Museum of Lugano.

In order to describe the response of forest habitats to fire, the study focused on the abundance of species, post-fire succession of dominant and subdominant species, and analysis of post-fire successional species strongly associated with particular fire regimes (differential species).

RESULTS

Species richness

A total of 143 spider species and 10196 individuals were collected. Of these, 64 species (46% of the total) were represented by less than 5 individuals, and 39 species (28%) were observed at only one study site.

Fig. 3. Relative dominance (%) of spider species at study sites belonging to different classes of fire-regimes. Dominant species (>10%; black bars), subdominant species (3.2-9.9%; grey bars) and recedent species (<3.2%; white bars). The species follow the dominance gradient of the unburnt sites.

Table 1. Number and percentage of species associated with particular fire regimes and of those which were found in all types of study sites. For each category of fire frequency six study sites were selected (except for the two only freshly burnt sites available).

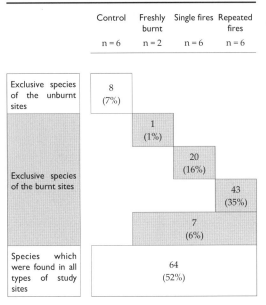

	Control	Freshly burnt	Single fires	Repeated fires
	n = 6	n = 2	n = 6	n = 6
Exclusive species of the unburnt sites	8 (7%)			
Exclusive species of the burnt sites		1 (1%)	20 (16%)	43 (35%)
			7 (6%)	
Species which were found in all types of study sites			64 (52%)	

Species richness at single fire sites was similar to values encountered at unburnt sites, while at repeated fire sites, species richness was found to be significantly higher (Fig. 2A). Elapsed time since the last fire had a considerable effect on the number of species found. At freshly burnt sites the number of species decreased but changes were not significant (Fig. 2B). On the other hand, in instances of single fires, this initial decrease was followed by a gradual increase, leading to a significantly higher number of species after 22-24 years. At repeated fire sites, there were no significant trends in species richness with respect to the time elapsed since the last fire.

Species structure of the community

Fig. 3 summarises some important differences in dominant and subdominant species composition in relation to fire regime. On unburnt sites, *Lepthyphantes flavipes* and *Microneta viaria* were the dominant species (dominance > 10%),

and *Tegenaria fuesslini, Pardosa saltans, Lepthyphantes pallidus, Amaurobius jugorum, Trochosa hispanica* and *Dasumia taeniifera* were subdominant (dominance 3.2-9.9%). On freshly burnt sites, *Lepthyphantes flavipes* and *Microneta viaria* decreased considerably, while *Tegenaria fuesslini* and *Lepthyphantes pallidus* became the dominant species. *Lepthyphantes flavipes* continued to be dominant 1-2 years after a single fire, and *Lepthyphantes pallidus* and *Dyplostyla concolor* increased considerably in abundance, while the number of *Microneta viaria* declined sharply during this period. After 7-14 years, *Pardosa saltans* became strongly dominant at many single fire sites (dominance > 32%). In contrast, on recurring fire sites, *Pardosa saltans* became the dominant species after only 2-3 years, and remained dominant for 6-17 years.

Differential species

Table 1 shows that approximately 30% of all recorded species were found exclusively on burnt sites. The analysis yielded a mean number of 43 species (35%) confined exclusively to repeated fire sites, compared with only 8 species (7%) found exclusively on unburnt sites; 64 species (52%) showed no differences in distribution. Some of the species which were strongly associated with particular fire regimes are listed in Table 2.

DISCUSSION

There have been numerous studies on the effects of fire on epigeic and soil fauna. They vary greatly in sampling methods and other factors such as habitat conditions, time of study, fire regime and size of burnt areas. Consequently, caution is necessary when making direct comparisons between the present and past studies.

As far as the post-fire succession of spiders is concerned, in contrast to Huhta (1971), Merrett (1976), Schaefer (1980), and Koponen (1993, 1995), no evidence was found of pioneer species colonising study sites for limited periods after a fire. Although the abundance of dominant and subdominant species of unburnt sites decreased after a fire, most of them maintained an impor-

Table 2. Species and number of sampled individuals associated with particular fire regimes. For each category of fire frequency, six study sites were selected (except for the two only freshly burnt sites available).

	Species	Control	Freshly burnt	Single fires	Repeated fires
		(n = 6)	(n = 2)	(n = 6)	(n = 6)
Selected exclusive species of the burnt sites	*Walckenaeria furcillata*				5
	Haplodrassus signifer				6
	Zora silvestris				7
	Trochosa terricola				35
	Zelotes erebeus				14
	Episinus truncatus			2	4
	Cercidia prominens			1	5
	Trichoncus cf. sordidus			1	32
	Hahnia ononidum		4		10
	Meioneta gulosa		1	1	28
Selected species which 'profit' from the fire	*Poecilochroa conspicua*	1		2	4
	Alopecosa pulverulenta	2		1	7
	Centromerus sellarius	1		3	8
	Walckenaeria mitrata	1			11
	Zelotes apricorum	1		4	19
	Zodarion gallicum	14		6	25
	Micaria fulgens	6		5	79
	Callilepis nocturna	1	1	1	11
	Gnaphosa bicolor	9	2	7	35
	Hahnia helveola	14	14	31	29
	Xerolycosa nemoralis	7	10		43
	Zora spinimana	4	2	2	43
	Pardosa saltans	262	13	260	996
Selected species 'affected' by fire	*Ballus chalybeius*	24		2	9
	Coelotes mediocris	75		11	50
	Tapinocyba maureri	105	1	55	4
	Lepthyphantes flavipes	214	8	80	132
	Microneta viaria	217	22	105	138

tant position in the community. One possible explanation for the in situ survival of many species is that fires do not usually burn the fuel uniformly (Marxer & Conedera 2000), leaving some pockets under stones, logs, or surface tree roots where individuals of many species can survive (Huhta 1971; Merrett 1976; Riechert & Reeder 1972). It seems possible that post-fire succession is mainly determined by altered habitat conditions and through post-fire environmental variations at the epigeic level (Pyne et al. 1996).

On the whole, spider communities of single fire sites tended to return to the pre-fire species composition and structure over a relatively short period (within approximately 7 years). This is especially true for the Linyphiidae, a family characteristic of intact forests (Uetz 1975, 1979; Bultman & Uetz 1982; Leclerc & Blandin 1990). The rapid recovery is probably due to resprouting of chestnut stools (Delarze et al. 1992; Hofmann et al. 1998) resulting in accelerated restoration of the litter layer.

In fire ecosystems resident communities are sometimes significantly different from those in surrounding unburnt areas, enhancing biodiversity in the region as a whole. In ecosystems where fires are frequent, spider and other arthropod communities seem to be adapted to recurrent events of low to medium intensity (e. g. Riechert & Reeder 1972; Force 1981; Weaver 1985; Siemann et al. 1997; York 1998). Nevertheless, Springett (1976) and York (1998) pointed out that in frequently burnt dry sclerophyll forests of Australia, many native forest species are lost and may be replaced by species associated with more open habitats.

In our study, the species strongly associated with particular fire regimes (differential species), especially those found almost exclusively in frequently burnt sites, are characterized by being able to live under a wide diversity of environmental conditions, ranging from humid to extremely xeric habitats. On the other hand, no 'forest species' was lost. This diversity could be explained in terms of an increase in habitat structural heterogeneity, where characteristic elements of both sparse and dense vegetation occur in close proximity, providing a rich mosaic of microclimatic conditions. This heterogeneity provides a wide range of microhabitats capable of supporting a large number of species. In fact, most of the species found exclusively on burnt sites are associated with open, dry conditions such as grasslands and exposed rocky habitats (Maurer & Hänggi 1990; Hänggi et al. 1995).

In conclusion, our results show that spider communites of unburnt sites in Southern Swit-

zerland are mainly dominated by litter-dwellers. Species richness of sites that have experienced only a single fire is generally similar to that of unburnt sites. On the other hand, sites with recurring fires tend to have larger numbers of species and a characteristic species composition which differs from that of unburnt and single fire sites. However, the relatively high abundance of *Pardosa saltans* in intact forests shows that communities present in fire-free forest sites, which have not burnt for at least 30 years, are pre-adapted to disturbance. This confirms the hypothesis that spider communites in chestnut forests of the Southern Alps have been strongly conditioned by fire history and human activites such as coppicing. From the paleo history of fire we know that in Southern Switzerland this kind of disturbance has existed since the Neolithic period (Tinner et al. 1998).

ACKNOWLEDGEMENTS

We would like to thank Dr. Ambros Hänggi, Marco Conedera, Nicola Patocchi, Dr. Peter Duelli and Dr. Martin Obrist for their comments on the results. Our thanks are also due to the various people who helped with field work (Peter Wirz, Franco Fibbioli, Karl Sigrist) and in checking spider identifications (Dr. Ambros Hänggi, Xaver Heer and Nicola Patocchi).

REFERENCES

Bultman, T.L. & Uetz, G.W. 1982. Abundance and community structure of forest floor spiders following litter manipulation. *Oecologia* 55, 34-41.

Conedera, M., Marcozzi, M., Jud, B., Mandallaz, D., Chatelain, F., Frank, C., Kienast, F., Ambrosetti, P. & Corti, G. 1996. *Incendi boschivi al Sud delle Alpi: passato, presente e possibili sviluppi futuri*. Rapporto di lavoro PNR 31. Hochschulverlag AG ETH, Zürich.

Delarze, R., Caldelari, D. & Hainnard, P. 1992. Effects of fire on forest dynamics in southern Switzerland. *Journal of Vegetation Science* 3, 55-60.

Ellenberg, H. & Klötzli, F. 1972. Waldgesellschaften und Waldstandorten der Schweiz. *Mitteilungen der Schweizerischen Anstalt für das Forstliche Versuchswesen* 48, 587-930.

Heimer, S. & Nentwig, W. 1991. *Spinnen Mittel europas*. 1st edn. Paul Parey, Berlin.

Hofmann, C., Conedera, M., Delarze, R., Carraro, G. & Giorgetti, P. 1998. Effets des incendies de forêt sur la végétation au Sud des Alpes Suisses. *Mitteilungen der Eidgenössischen Forschungsanstalt für Wald, Schnee und Landschaft* 73, 1-90.

Huhta, V. 1971. Succession in the spider communities of the forest floor after clearcutting and prescribed burning. *Annales Zoologici Fennici* 8, 483-542.

Keller, W., Wohlgemuth ,T., Kuhn, N., Schütz M. & Wildi O. 1998. Waldgesellschaften der Schweiz auf floristischer Grundlage. *Mitteilungen der Eigenössischen Forschungsinstitut für Wald, Schnee und Landschaft* 73 (2),1-357.

Koponen, S. 1995. Postfire succession of soil arthropod groups in a subarctic birch forest. *Acta Zoologica Fennica* 196, 243-245.

Koponen, S. 1988. Effect of fire on spider fauna in subarctic birch forest, Northern Finland. In: *XI. Europäisches Arachnologisches Colloquium* (J. Haupt ed.), Vol. 38, pp. 148-153. TUB-Dokumentation, Kongresse und Tagungen, Berlin.

Koponen, S. 1993. Ground-living spiders (Araneae) one year after fire in three subarctic forest types, Québec (Canada). *Memoirs of the Queensland Museum* 33, 575-578.

Leclerc, J. & Blandin, P. 1990. Patch size, fine-scale co-occurrence and competition in forest litter linyphiids. *Acta Zoologica Fennica* 190, 239-242.

Marxer, P. & Conedera, M. 2000. *Experimental fire in Switzerland. Final report of the PROMETHEUS project*. Swiss Federal Research Institute WSL - Sottostazione Sud delle Alpi, Bellinzona.

Maurer, R. & Hänggi, A. 1990. *Katalog der schweizerischen Spinnen*, 1st edn. CSCF, Neuchâtel.

Merrett, P. 1976. Changes in the ground-living spider fauna after heathland fires in Dorset. *Bulletin of the British Arachnological Society* 3, 214-221.

Obrist, M. & Duelli, P. 1996. Trapping efficiency of funnel- and cup-traps for epigeal arthropods. *Mitteilungen der Schweizerischen Entomologischen Gesellschaft* 69, 361-369.

Pickett, S.T.A. 1989. Space-for-time substitution as an alternative to long-term studies. In: *Long-term studies in ecology: approaches and alternatives* (G.E. Likens ed.). Springer-Verlag, New York.

Platnick, N.I. 1989. *Advances in spider taxonomy 1981-1987. A supplement to Brignolis Catalogue of the Araneae*. Manchester.

Pyne, S.J., Andrews, P.L., & Laven, R.D. (1996) *Introduction to Wildland Fire*, 2nd edn. John Wiley & Sons, New York.

Riechert, S.E. & Reeder, W.G. 1972. Effects of fire on spiders distribution in southwestern Wisconsin prairies. In: *Proceedings of the Second Midwest Prairie Conference* (J. Zimmerman ed.), pp. 73-90. Madison, Wisconsin.

Schaefer, M. 1980. Effects of an extensive fire on the fauna of spiders and harvestmen (Araneida and Opilionida) in Pine forests. In: *VIII Internationaler Arachnologen-Kongress* (J. Gruber ed.), pp. 103-108. Verlag H. Egermann, Wien.

Siemann, E., Haarstad, J., & Tilman, D. 1997. Short-term and long-term effects of burning on oak savanna arthropods. *American Midland Naturalist* 137, 349-361.

Springett, J.A. 1976. The effect of prescribed burning on the soil fauna and on litter decomposition in Western Australian forests. *Australian Journal of Ecology* 1, 77-82.

Tinner, W., Conedera, M., Ammann, B., Gäggeler, H.W., Gedey, S., Jones, S.R. & Sägesser, B. 1998. Pollen and charcoal in lake sediments compared with historically documented wildfires in southern Switzerland since 1920. *Holocene* 8, 31-42.

Uetz, G.W. 1975. Temporal and spatial variation in species diversity of wandering spiders (Araneae) in deciduous forest litter. *Environmental Entomology* 4, 719-724.

Uetz, G.W. 1979. The influence of variation in litter habitats on spider communities. *Oecologia* 40, 29-42.

Uetz, G.W. & Unzicker, J.D. 1976. Pitfall trapping in ecological studies. *Journal of Arachnology* 3, 101-111.

Weaver, J.C. 1985. Spider response to recurrent fire: reclaiming a hypothesis from field data. In: *Research and creative activities forum*, pp. 1-10, University of Missouri, Columbia.

York, A. 1998. Managing for biodiversity: what are the long-term implications of frequent fuel-reduction burning for the conservation of forest invertebrates? In: *III International Conference On Forest Fire Research* (D.X. Wiegas ed.), Vol. 2, pp. 1435-1445. ADAI University of Coimbra, Portugal, Luso.

European Arachnology 2000 (S. Toft & N. Scharff eds.), pp. 191-197.
© Aarhus University Press, Aarhus, 2002. ISBN 87 7934 001 6
(Proceedings of the 19th European Colloquium of Arachnology, Århus 17-22 July 2000)

Epigeic spiders of alder swamp forests in Eastern Poland

MARZENA STAŃSKA, IZABELA HAJDAMOWICZ & MAREK ŻABKA

University of Podlasie, Department of Zoology, Prusa 12, 08-110 Siedlce, Poland. (stanska@ap.siedlce.pl)

Abstract

Studies of epigeic spiders collected by pitfall traps in selected alder swamp forests in eastern Poland are presented. A list of 14 spider species shared by all analysed plots is given. The fauna of alder swamp forests is described as typical for wet European lowland forests, with no habitat indicator species being recognised. Despite similarities in species composition and dominance structure, species diversity of spiders in primeval alder swamp forests was higher than in managed forests.

Key words: spider assemblages, alder swamp forests, primeval forests, Białowieża Forest, Poleski National Park, Poland

INTRODUCTION

Due to drainage and forestry management only remnants of alder swamp forests (*Carici elongatae-Alnetum*) have been preserved in Europe, mostly in river valleys or/and in dips with limited water run-off (Kwiatkowski 1994; Matuszkiewicz 1984). Human impact has caused both the destruction of alder swamp forest communities and the decrease of biodiversity (Kwiatkowski 1994).

Because of their rarity, and for logistic reasons, the alder swamp forests are not popular areas for arachnological studies. In Poland the spider fauna of alder forests was investigated by Dąbrowska-Prot & Łuczak (1968), Dąbrowska-Prot et al. (1973), Staręga (1988), Sielicki & Staręga (1996) and Stankiewicz (1999). There is also some data from Germany (Baehr 1983; Löser et al. 1982), Belgium (Maelfait et al. 1995) and Austria (Komposch 2000).

The aims of our study were: 1) To investigate whether it is possible to distinguish a specific spider species assemblage characteristic for alder swamp forests. 2) To determine whether there are differences in species composition and dominance structure between spider assemblages of primeval and managed alder swamp forests.

STUDY AREA

The spiders of alder swamp forests were sampled from three collecting plots in Białowieża Forest and Poleski National Park (Fig. 1):

-Plot 1: primeval forest in Białowieża National Park (52°43'N, 23°50'E, 140 m above sea level), strictly protected since 1921. There is no management, and human access is restricted. Most trees are about 130 years old.

-Plot 2: managed forest in Białowieża Forest (52°40'N, 23°51'E, 145 m a.s.l.) with no dead trees; the average age of trees is 70 years. The plot is located near a village and 200 m from a camping site.

-Plot 3: forest in Poleski National Park (51°4' N, 23°06' E, 169 m a.s.l.) in the initial succession stage on a mire, most trees are about 30 years old.

Hummocks and dips are very distinctive structural features of alder swamp forests (*Carici elongatae-Alnetum*). The hummocks are covered with alder (*Alnus glutinosa*), birch

Fig. 1. Location of analysed alder swamp forests. 1: Białowieża - primeval forest, 2: Białowieża - managed forest, 3: Poleski National Park, 4: Wigierski National Park (Stankiewicz 1999), 5: Solnicki Forest (Sielicki & Staręga 1996).

(*Betula pubescens*) and in Białowieża Forest also with spruce (*Picea abies*). The scrub layer is mostly made up of *Frangula alnus* and *Alnus glutinosa*. The moss layer is poor and dominated by *Climacium dendroides*. The dips are filled with stagnant water for most of the year and covered by marshy vegetation of *Carex elongata*, *Dryopteris thelypteris*, *Solanum dulcamara*, *Iris pseudacorus*, *Menyanthes trifoliata* (Kwiatkowski 1994; Matuszkiewicz 1984; Sokołowski 1993).

MATERIAL AND METHODS

Epigeic and litter-dwelling spiders were sampled by pitfall traps in Poleski National Park from 1995 to 1998, and in Białowieża Forest in 1998-1999. In each plot ten traps were placed on the hummocks and they were operating throughout the year, being emptied fortnightly, from March to November.

Pitfall traps (plastic cups of 7 cm diameter and 10 cm depth) contained a solution of propylene glycol as preservative, with the addition of detergent to prevent the spiders from escaping (Topping & Luff 1995). The study period differed between particular plots. However, as

stated by Merrett (1967) this does not significantly influence the dominance structure although increasing the length of the study period enriches the species list.

The comparison of spider assemblages in primeval and managed alder swamp forests was only made for Białowieża Forest plots.

To compare the dominance structure and species composition of the spider communities, Renkonen's similarity index (Re) and Sørensen's similarity index (So) were applied, respectively (Trojan 1978). Sørensen's similarity index (So) was calculated according to the following formula:

So = 2c/a+b x 100,

where a, b are the numbers of species in the two compared communities, c the number of shared species. Renkonen's similarity index (Re) was calculated according to the formula:

$Re = \sum D_{min}$,

where D_{min} is the lower value of dominance coefficient (D) of species shared by the two communities.

The dominance groups were proposed after Woźny (1992): E = eudominants (>10%), D = dominants (5.1-10%), I = influents (2.1-5%), R = recedents (1.1-2%), + = subrecedents (≤1%). The nomenclature follows Platnick (1993, 1997) and Żabka (1997).

RESULTS

Altogether 115 species and 8587 specimens were collected; 7172 of them were identified to species level, 1982 from plot 1; 886 from plot 2 and 4304 from plot 3.

Most species (77) were found in the Poleski National Park (plot 3); 71 species were found in plot 1 and 50 in plot 2, both in the Białowieża Forest; 27 species were shared by all studied plots (Table 1).

In the primeval forest (plot 1) the number of exclusive species (34) was higher than in the managed forest in plot 2 (13), and the percentage of specimens of exclusive species was considerably higher there (7.9% and 1.7%, respectively). In the primeval forest (plot 1) eudominants, dominants, influents and recedents were

Tab. 1. List of spider species in particular study plots. 1: Białowieża - primeval forest, 2: Białowieża - managed forest, 3: Poleski National Park. E: eudominants (>10%), D: dominants (5.1-10%), I: influents (2.1-5%), R: recedents (1.1-2%), +: subrecedents (≤1%), *: subrecedents represented by one specimen in the plots 1 and 2.

Study plots	1	2	3	Study plots	1	2	3
Pirata hygrophilus Thorell	E	E	E	Hilaira excisa (O. P.-Cambridge)	*		
Pachygnatha listeri Sundevall	E	D	E	Lepthyphantes alacris (Blackwall)	+		
Centromerus sylvaticus (Blackwall)	R	E	D	Lepthyphantes angulatus (O. P.-C.)	+		
Pachygnatha clercki Sundevall	D	D	R	Linyphia hortensis Sundevall	*		
Trochosa spinipalpis (F.O.P.-Cambridge)	+	+	D	Meioneta innotabilis (O. P.-Cambridge)	*		
Bathyphantes nigrinus (Westring)	I	I	I	Neriene emphana (Walckenaer)	+		
Oedothorax retusus (Westring)	I	I	+	Pirata piscatorius (Clerck)	*		
Diplostyla concolor (Wider)	I	I	+	Pityohyphantes phrygianus (C.L. Koch)	*		
Diplocephalus picinus (Blackwall)	R	I	R	Pseudeuophrys erratica (Walckenaer)	*		
Dicymbium nigrum (Blackwall)	I	R	+	Theridion tinctum (Walckenaer)	*		
Allomengea vidua (L. Koch)	+	I		Tiso vagans (Blackwall)	*		
Walckenaeria nudipalpis (Westring)	R	R	+	Walckenaeria kochi (O. P.-Cambridge)	*		
Antistea elegans (Blackwall)	R	+	+	Xysticus cristatus (Clerck)	*		
Walckenaeria vigilax (Blackwall)	+	R	+	Ceratinella brevis (Wider)		+	+
Ozyptila trux (Blackwall)	+	+	R	Erigone atra Blackwall		*	+
Tallusia experta (O. P.-Cambridge)	+	+	R	Pachygnatha degeeri Sundevall		*	+
Walckenaeria alticeps (Denis)	+	+	R	Pirata latitans (Blackwall)		*	+
Walckenaeria atrotibialis O. P.-Cambridge	+	+	+	Agroeca proxima (O. P.-Cambridge)		*	
Clubiona lutescens Westring	+	+	+	Lepthyphantes flavipes (Blackwall)		*	
Robertus arundineti (O. P.-Cambridge)	+	+	+	Lepthyphantes minutus (Blackwall)		*	
Gongylidium rufipes (Linnaeus)	+	+	+	Neriene furtiva (O. P.-Cambridge)		*	
Lepthyphantes tenebricola (Wider)	+	+	+	Pardosa amentata (Clerck)		+	
Tapinocyba insecta (L. Koch)	*	+	+	Pardosa pullata (Clerck)		*	
Pirata piraticus (Clerck)	*	+	+	Pirata tenuitarsis Simon		+	
Ozyptila praticola (C. L. Koch)	+	+	+	Porrhomma pygmaeum (Blackwall)		+	
Drapetisca socialis (Sundevall)	+	+	+	Walckenaeria antica (Wider)		*	
Centromerus arcanus (O. P.-Cambridge)	+	*	+	Hygolycosa rubrofasciata (Ohlert)			I
Oedothorax gibbosus (Blackwall)	D	I		Bathyphantes parvulus (Westring)			+
Agraecina striata (Kulczyński)	I		D	Centromerita bicolor (Blackwall)			+
Robertus lividus (Blackwall)	R	I		Centromerus semiater (L. Koch)			+
Cicurina cicur (Fabricius)	R	+		Clubiona pallidula (Clerck)			+
Helophora insignis (Blackwall)	+	I		Clubiona phragmitis C.L. Koch			+
Lepthyphantes cristatus (Menge)	+	R		Clubiona rosserae Locket			+
Bathyphantes approximatus (O. P.-C.)	+	+		Dicymbium tibiale (Blackwall)			+
Coelotes atropos (Walckenaer)	+	*		Drassyllus pusillus (C.L. Koch)			+
Gonatium rubellum (Blackwall)	+	+		Ero cambridgei Kulczyński			+
Savignia frontata Blackwall	*	+		Euophrys frontalis (Walckenaer)			+
Walckenaeria obtusa Blackwall	+	+		Gongylidiellum murcidum Simon			+
Pirata uliginosus (Thorell)	+		D	Lepthyphantes alutacius Simon			+
Agyneta subtilis (O. P.-Cambridge)	+		+	Lepthyphantes angulipalpis (Westring)			+
Araneus diadematus Clerck	*		+	Maro minutus O. P.-Cambridge			+
Centromerus levitarsis (Simon)	+		+	Maso sundevalli (Westring)			+
Clubiona germanica Thorell	*		+	Meioneta affinis (Kulczyński)			+
Clubiona terrestris Westring	*		+	Metellina mengei (Blackwall)			+
Linyphia triangularis (Clerck)	+		+	Metellina segmentata (Clerck)			+
Lophomma punctatum (Blackwall)	+		+	Micrargus herbigradus (Blackwall)			+
Macrargus rufus (Wider)	+		+	Neon reticulatus (Blackwall)			+
Microneta viaria (Blackwall)	+		+	Neriene montana (Clerck)			+
Neriene clathrata (Sundevall)	+		+	Pardosa paludicola (Clerck)			+
Pardosa lugubris (Walckenaer)	+		+	Pelecopsis mengei (Simon)			+
Walckenaeria cuspidata Blackwall	*		+	Pocadicnemis juncea Locket et Millidge			+
Zora spinimana (Sundevall)	*		+	Pocadicnemis pumila (Blackwall)			+
Agyneta conigera (O. P.-Cambridge)	+			Porrhomma oblitum (O. P.-Cambridge)			+
Amaurobius fenestralis (Stroem)	+			Taranucnus setosus (O. P.-Cambridge)			+
Ceratinella brevipes (Westring)	*			Trochosa ruricola (De Geer)			+
Erigonella hiemalis (Blackwall)	*			Zelotes latreillei (Simon)			+
Haplodrassus cognatus (Westring)	*			Zelotes subterraneus (C.L. Koch)			+
Haplodrassus silvestris (Blackwall)	*						

represented by 15 species (86.5% of all specimens). In managed forest (plot 2) their number reached 17 species (90.5%). The number of subrecedents was considerably higher in plot 1 than in plot 2 (56/33 species and 13.5%/9.5%, respectively). The number of species represented by one specimen was higher in plot 1 than in plot 2 (22 and 12, respectively).

Dominance structure was analysed only for adult spiders. *Pirata hygrophilus* was the most numerous eudominant in all plots (>30% of specimens). *Pachygnatha listeri* was eudominant in plots 1 and 3 (10.1% and 20.7%, respectively) and dominant in plot 2 (6.0%). *Centromerus sylvaticus* and *Pachygnatha clercki* were shared by all sites, but they were numerous only in two plots, being eudominant in plot 2 and dominant in plot 3. *Oedothorax gibbosus* occurred abundantly and exclusively in plots 1 (dominant) and 2 (influent), whereas in plot 3 other dominants (*Agraecina striata, Trochosa spinipalpis* and *Pirata uliginosus*) were present. *Agraecina striata*, a rather rare species, was numerous (influent) in plot 1. *Bathyphantes nigrinus* was the only influent shared by all study plots.

The dominance structure and species composition of epigeic spiders communities showed considerable similarities (Table 2), the highest similarity was observed for plots 1 and 2, both in Białowieża Forest. There was less similarity between spider assemblages from plots 1 and 3 than from plots 2 and 3.

The similarity of dominance structure (Re) of epigeic spider assemblages was higher than similarity of species composition (So). The differences in species composition were probably the result of study site locations (Fig. 1). Plots 1 and 2 were placed in NE Poland in one large forest complex, whereas plot 3 was situated in SE Poland and surrounded by mires, fields and meadows.

In all investigated plots, several rare European spider species were found (sensu Hänggi et al. 1995). *Centromerus levitarsis* and *Meioneta innotabilis* were recorded in plot 1 and *Pirata tenuitarsis* in plot 2, both plots in Białowieża Forest. *C. levitarsis* was also found in Poleski

Tab. 2. Similarity of species composition according to Sørensen's index (%) and dominance structure according to Renkonen's index (%). 1: Białowieża - primeval forest, 2: Białowieża - managed forest, 3: Poleski National Park.

Study sites	Sørensen's index		Renkonen's index	
	2	3	2	3
1	61.2	56.8	70.9	59.1
2	-	48.8	-	55.6

National Park (plot 3). Other rare species in plot 3 include *P. tenuitarsis, Centromerus semiater, Clubiona rosserae, Ero cambridgei* and *Taranucnus setosus*. All of them prefer peat bogs and fens.

DISCUSSION

The analysis of species composition of epigeic spiders in Białowieża Forest and in Poleski National Park allows us to construct a list of 27 species common to the three studied plots (Table 1). The comparison of our data with the results for other alder swamp forests in NE Poland (Stankiewicz 1999; Sielicki & Staręga 1996; Fig. 1) reduced the list of species shared by all five sites to 14 (Table 3).

Although the species composition of the five areas differs (Table 4), eudominant and dominant species are similar, though not identical (Table 3). For instance, *Oedothorax gibbosus* was dominant or influent in Białowieża Forest, while in Wigierski National Park (Stankiewicz 1999) it belonged to dominants and was missing in Poleski National Park. *Trochosa terricola* and *Ozyptila praticola* were two additional eudominants found by Sielicki & Staręga (1996) in Solnicki Forest. *Trochosa terricola* was missing in our study and belonged to the subrecedents in Stankiewicz (1999). *Ozyptila praticola* belonged to the subrecedents in our plots and was missing in the study by Stankiewicz (1999). Both species are known to occur in drier habitats, e.g. woodlands and open areas (Hänggi et al. 1995; Prószyński & Staręga 1971).

Table 3. Spider species from Polish alder swamp forests shared by all studied plots. I: Białowieża - primeval forest, 2: Białowieża - managed forest, 3: Poleski National Park, 4: Wigierski National Park (Stankiewicz 1999), 5: Solnicki Forest (Sielicki & Staręga 1996). E: eudominants (>10%), D: dominants (5.1-10%), I: influents (2.1-5%), R: recedents (1.1-2%), +: subrecedents (=1%).

Study plots	I	2	3	4	5
Pirata hygrophilus Thorell, 1872	E	E	E	E	E
Pachygnatha listeri Sundevall, 1830	E	D	E	I	E
Pachygnatha clercki Sundevall, 1823	D	D	R	E	+
Centromerus sylvaticus (Blackwall, 1841)	R	E	D	I	+
Trochosa spinipalpis (F.P.-Cambridge, 1895)	+	+	D	R	D
Bathyphantes nigrinus (Westring, 1851)	I	I	I	R	R
Diplocephalus picinus (Blackwall, 1841)	R	I	R	+	I
Diplostyla concolor (Wider, 1834)	I	I	+	+	I
Gongylidium rufipes (Linnaeus, 1758)	+	+	+	+	I
Lepthyphantes tenebricola (Wider, 1834)	+	+	+	+	R
Ozyptila trux (Blackwall, 1846)	+	+	R	+	+
Walckenaeria alticeps (Denis, 1952)	+	+	R	+	+
Clubiona lutescens Westring, 1851	+	+	+	+	+
Walckenaeria atrotibialis (O. P.-Cambridge, 1878)	+	+	+	+	+

Table 4. Numbers of shared spider species and similarity according to Sørensen's index (%) for Polish alder swamp forests.

Study sites	Wigierski National Park		Solnicki Forest	
	No. of shared spp.	Sørensen index	No. of shared spp.	Sørensen index
Białowieża primeval forest	46	52.0	32	45.7
Białowieża managed forest	34	43.6	25	39.4
Poleski National Park	44	48.1	31	42.5

In comparison to the list of 14 species shared by all Polish sites, the list for similar forests in Belgium (Maelfait et al. 1995) did not include *Trochosa spinipalpis* and *Walckenaeria alticeps*, and the dominance structure was different. *Pirata hygrophilus* and *Diplocephalus picinus* were eudominants, *Diplostyla concolor* was dominant, *Centromerus sylvaticus*, *Pachygnatha listeri*, *Bathyphantes nigrinus* were influents, *Pachygnatha clercki* was recedent, and the remaining 14 species found in Polish forests were subrecedents in Belgium.

The list of 14 common species was compared with lists from similar alder forests in Upper Bavaria (Löser et al. 1982) and Central Alps (Komposch 2000). *Centromerus sylvaticus*, *Diplostyla concolor*, *Clubiona lutescens*, *Pirata hygrophilus* and *Bathyphantes nigrinus* were also found in the forest of Central Alps but only the latter two were numerous. Only *Diplostyla concolor* and *Walckenaeria atrotibialis* were shared with the species list from Bavarian *Alnus incana*-forest and only *Pirata hygrophilus* with the species list of Bavarian *Alnus glutinosa*-forests. The distinctiveness of the spider communities in Austrian and German alder forests is probably due to their location in the mountain zone.

It would be tempting to consider the 14 species shared by Polish sites as a characteristic spider assemblage of lowland alder swamp forests. However, the same species occur in other types of forests, e.g. ash-alder flood plain forest (Stankiewicz 1999), humid oak-lime-hornbeam forest (Stańska unpubl. data) and inundation forests (Gajdoš 1995). In humid oak-lime-hornbeam forests in Białowieża Forest all 14 shared species were found and the dominance structure of epigeic spider communities in one of these forests and in plot 1 showed considerable similarity (Re = 58%). In the ash-alder flood plain forest in Wigierski National Park 13 of the 14 (without *Bathyphantes nigrinus*) species were found and the dominants were almost the same as in our plots: *Pirata hygrophilus, Pachygnatha listeri, Centromerus sylvaticus* and *Lepthyphantes tenebricola* (Stankiewicz 1999). Also in the inundation forests (*Saliceto–Populetum*) the majority of the 14 species were found and the dominance structure was similar (Gajdoš 1995).

The proposed list of shared species seems characteristic for lowland wet/humid deciduous forests, including lowland alder forests. However all of those species are eurytopic, i.e. widespread in other habitats across Europe (Hänggi et al. 1995), and taken separately they should not be considered as indicator species for particular plant associations (sensu Neet 1995).

In spite of considerable similarity of spider dominance structure in Białowieża Forest plots (Re = 70.9), the 'tail' of subrecedent species and the list of species represented by one specimen were considerably longer in the primeval forest. The percentage of specimens of exclusive species was also distinctly higher in plot 1. The above data suggest that the spider assemblage in managed forest is less diverse than in primeval forest, probably because of the human impact, as other conditions (e.g. humidity) are very similar.

ACKNOWLEDGEMENTS

The study was supported by Komitet Badań Naukowych (Grants 6PO4G01417 and 6PO4G01011) and by Akademia Podlaska in Siedlce (Grants 18/91/S and 512/93/W). The authorities of Białowieża National Park and Poleski National Park are acknowledged for logistic help. Miss Jaynia Tarnawski (Sydney, Australia) and Mr. Graham Wishart (Gerringong, Australia) corrected the English.

REFERENCES

Baehr, B. 1983. Vergleichende Untersuchungen zur Struktur der Spinnengemeinschaften (Araneae) im Bereich stehender Kleingewässer und der angrenzenden Waldhabitate im Schönbuch bei Tübingen. Dissertation, Fakultät für Biologie der Eberhard-Karls-Universität Tübingen, Tübingen.

Dąbrowska-Prot, E. & Łuczak, J. 1968. Spiders and mosquitoes of the ecotone of alder forest (*Carici elongatae-Alnetum*) and oak-pine forest (*Pino-Quercetum*). *Ekologia Polska* Ser. A 16, 461-483.

Dąbrowska-Prot, E., Łuczak, J. & Wójcik, Z. 1973. Ecological analysis of two invertebrate groups in the wet alder wood and meadow ecotone. *Ekologia Polska* Ser. A 21, 753-812.

Gajdoš, P. 1995. The epigeic spider communities of lowland forests in the surroundings of the Danube River on the territory of Slovakia and their usage for biota monitoring. In: *Proceedings of the 15th European Colloquium of Arachnology* (V. Růžička ed.), pp. 73-83. Institute of Entomology, České Budějovice.

Hänggi, A., Stöckli, E. & Nentwig, W. 1995. *Habitats of Central European spiders.* Centre Suisse de Cartographie de la Faune, Neuchâtel. [*Miscellanea Faunistica Helvetiae* 4, 1-451]

Komposch, C. 2000. Harvestmen and spiders in the Austrian wetland 'Hörfeld-Moor' (Arachnida: Opiliones, Araneae). *Ekológia (Bratislava)* 19 Suppl. 4, 65-77.

Kwiatkowski, W. 1994. Krajobrazy roślinne Puszczy Białowieskiej [Vegetation landscapes of Białowieża Forest.] *Phytocoenosis* 6 (N.S.), 35-87.

Löser, S., Meyer, E. & Thaler, K. 1982. Laufkäfer, Kurzflügelkäfer, Asseln, Webespinnen, Weberknechte und Tausendfüsser des Naturschutzgebietes 'Murnauer Moos' und der angrenzenden westlichen Talhänge (Coleoptera: Carabidae, Staphylinidae; Crustacea: Isopoda; Aranei; Opiliones; Diplopoda). *Entomofauna* Suppl. 1, 369-446.

Matuszkiewicz, W. 1984. *Przewodnik do oznaczania zbiorowisk roślinnych Polski.* PWN Warszawa.

Maelfait, J.-P., de Knijf, G., de Becker, P. & Huybrechts, W. 1995. Analysis of the spider fauna of the riverine forest nature reserve 'Walenbos' (Flanders, Belgium) in relation to hydrology and vegetation. In: *Proceedings of the 15th European Colloquium of Arachnology* (V. Růžička ed.), pp. 125-135. Institute of Entomology, České Budějovice.

Merrett, P. 1967. The phenology of spiders on heathland in Dorset. Families Lycosidae, Pisauridae, Agelenidae, Mimetidae, Theridiidae, Tetragnathidae, Argiopidae. *Journal of Zoology* 156, 239-256.

Neet, C.R. 1996. Spiders as indicator species: lesson from two case studies. *Revue Suisse de Zoologie* Hors série 2, 501-510.

Platnick, N.I. 1993. *Advances in spider taxonomy 1988-1991, with the synonymies and the transfers 1940-1980.* New York Entomological Society and American Museum of Natural History Publ., New York.

Platnick, N.I. 1997. *Advances in spider taxonomy 1992-1995, with redescriptions 1940-1980.* New York Entomological Society and American Museum of Natural History Publ., New York.

Prószyński, J. & Staręga, W. 1971. Pająki - Aranei. In: *Katalog fauny Polski.* [*Catalogus faunae Poloniae.*] (A. Riedel ed.), pp. 1-382. PWN, Warszawa.

Sielicki, M. & Staręga, W. 1996. Pająki (*Araneae*) ekotonu ols - łąka w okolicach Białegostoku. *Fragmenta Faunistica* 39, 169-177.

Sokołowski, A.W. 1993. Fitosocjologiczna charakterystyka zbiorowisk leśnych Białowieskiego Parku Narodowego. [Phytosociological characterisics of forest communities in the Białowieża National Park.]. *Parki Narodowe i Rezerwaty Przy rody* 12, 5-189.

Stankiewicz, A. 1999. Pająki (*Araneae*) zbiorowisk leśnych Wigierskiego Parku Narodowego. PhD thesis, University of Białystok, Białystok.

Staręga, W. 1988. Pająki (*Aranei*) Gór Świętokrzyskich. *Fragmenta Faunistica* 31, 185-359.

Topping, C.J. & Luff, M.L. 1995. Three factors affecting the pitfall trap catch of linyphiid spiders (Aranaea: Linyphiidae). *Bulletin of the British Arachnological Society* 10, 35-38.

Trojan, P. 1978. *Ekologia ogólna.* PWN, Warszawa.

Woźny, M. 1992. Wpływ wilgotności podłoża na zgrupowania pająków oraz dynamika liczebności gatunków dominujących borów sosnowych Wzgórz Ostrzeszowskich. *Acta Universitatis Wratislaviensis* 1124, 25-82.

Żabka, M. 1997. Salticidae, pająki skaczące (Arachnida: Araneae). In: *Fauna Poloniae.* (A. Riedel ed.). Muzeum i Instytut Zoologii PAN, Warszawa.

European Arachnology 2000 (S. Toft & N. Scharff eds.), pp. 199-205.
© Aarhus University Press, Aarhus, 2002. ISBN 87 7934 001 6
(Proceedings of the 19th European Colloquium of Arachnology, Århus 17-22 July 2000)

Spider species and communities in bog and forest habitats in Geitaknottane Nature Reserve, Western Norway

REIDUN POMMERESCHE

The Norwegian Centre for Ecological Agriculture (NORSØK), N-6630 Tingvoll, Norway
(Reidun.Pommeresche@norsok.no)

Abstract
Species and communities of epigeic spiders in different bogs and forests in Western Norway were investigated. Fifty different sites were sampled using pitfall traps in 1997. Five groups of spider communities could be classified: spider communities of wet, open areas; communities of open forests; communities of shady pine forests; spider communities of humid deciduous forests, and spider communities of dry deciduous forests. There was a good correlation between spider communities and vegetation on the site demonstrated by DCA and DCCA. Environmental variables like productivity of wood, soil humidity, tree cover, bush cover as well as heat index turned out to be significant, explaining the gradient of communities and species of spiders found in the investigated vegetation types. The number of species per site varied from 21 to 51. *Gonatium rubens* was typical for the open areas (bogs and *Calluna*-pine forests), and *Gonatium rubellum* was typically found in the bilberry-pine and deciduous forests. *Pirata hygrophilus* and *Notioscopus sarcinatus* were typically found on the bogs. New species to Norway were *Euophrys frontalis*, *Maro lepidus* and *Porrhomma oblitum*.

Key words: Spiders, diversity, community, bogs, forests, DCA, DCCA, Norway

INTRODUCTION

Most of the research on spider diversity and faunistics in Norway is from the southern part of the country (Hauge & Wiger 1980; Hauge & Kvamme 1983; Tveit & Hauge 1983; Ellefsen & Hauge 1986; Hauge et al. 1991). Diversity and habitat preferences of spiders may differ in different geographical regions due to different climate, altitude or due to other abiotic factors. Hänggi et al. (1995) have worked out a synopsis on habitat preferences of spiders of Central Europe. Many of the Norwegian species are represented in this work, but it does not include many large Norwegian surveys. In the present paper, information on spider species and communities in bogs and forests typical for Western Norway is presented. The most abun-dant and frequent spider species and species typical of the main spider community assemblages and vegetation types are discussed, and the relationship between spider communities and environmental factors are pointed out.

MATERIAL AND METHODS
Description of the sites

The study area was in the Geitaknottane Nature Reserve, in Kvam south-east of Bergen, Western Norway (60°07′ N, 5°52′ E). It is situated 120-300 m above sea level and consists of a mosaic of exposed bedrock, bogs, pine- and deciduous forests. Fifty sites representing the main vegetation types in the area were chosen for study:

- Bogs (7 sites): Treeless bogs dominated by

Calluna vulgaris, Molinia caerulea and different *Sphagnum* spp.

- Open *Calluna*-pine forests (15 sites): The field layer was dominated by *Calluna vulgaris* and *Molinia caerulea*. A few pines (*Pinus sylvestris*) were scattered over the area.

- Bilberry-pine forests (12 sites): Tree cover of *Pinus sylvestris* mixed with birches (*Betula* spp.). The field layer dominated by *Vaccinium myrtillus*, some *Vaccinium vitis-idaea* and *Deschampsia flexuosa*.

- The deciduous forests (15 sites) consisted of five different vegetation types: Two bilberry-oak forests dominated by oak (*Quercus robur, Q. petraea*), with some pines (*Pinus sylvestris*). The field layer was similar to that of the bilberry-pine forests. Three humid grey alder-bird cherry forests consisted of grey alder (*Alnus incana*), some birches (*Betula* spp.) and bird cherry (*Prunus padus*). These forests had a dense field layer of grasses and herbs. Three humid grey alder-ash forests composed of birches (*Betula* spp.), grey alder (*Alnus incana*) and hazel (*Corylus avellana*) were sampled. The field layer consisted of *Ranunculus ficaria* and *Anemone nemorosa* in spring, later in the season various herbs and grasses. Six localities of dry Elm-lime forests had elm (*Ulmus glabra*) and ash (*Fraxinus excelsior*) in the canopy. A rich field layer consisted of *Matteuccia struthiopteris, Carex sylvatica* and various herbs. One locality was a hazel (*Corylus avellana*) shrub in a scree, with some bilberry and herbs.

- Spruce plantation (1 site): Dark, dense spruce plantation (*Picea abies*) with sparse vegetation and a thick layer of spruce needles. The vegetation on the different localities was classified according to Fremstad (1997).

Environmental factors

The following environmental factors were measured at each locality: Slope of the locality (degrees), exposure (0-360°) of the site, stratification (vegetation layers), tree cover (canopy cover (%)), forest productivity class (Fitje 1984), stand basal area (estimated using a relascope (m²/ha)), bush cover (%), soil humidity score

(defined by the vegetation type (Fremstad 1997)), heat index (Parkers index (Parker 1988, Økland & Eilertsen 1993)).

Sampling procedures

Eight pitfall traps (glassjars 11 cm high and 6.5 cm in diam., one-third filled with a 4% formalin solution) were used at each site to collect the spiders. The traps were emptied four times during the study period (April - November 1997).

Statistical analysis

Based on the number of specimens per spider species per locality, a two-way indicator species analysis, TWINSPAN (Hill 1979), was used to group the 50 spider communities. Ordination of communities and species was carried out using detrended correspondence analysis (DCA), while the ordination of the spiders in relation to environmental variables was carried out using detrended canonical correspondence analysis (DCCA) (Jongman et al. 1995). DCA and DCCA were run in CANOCO 3.12 (ter Braak 1991), based on number of specimens per spider species per locality. Rare species (low frequency) were downweighted. Forward selection of environmental variables and a Monte Carlo permutation test (999 permutations) were used to find the statistically significant (p < 0.05) environmental variables in DCCA.

RESULTS

Groups of species and communities

TWINSPAN split the spider species into five distinct community groups (Fig. 1). TWINSPAN-group 1 (TWIN 1) consisted of spider species and communities in wet, open areas (the bogs), TWIN 2 of 13 open *Calluna*-pine forests. TWIN 3 grouped spider species and communities of 12 bilberry pine forests, two *Calluna*-pine forests, two billberry-oak forests and one hazel shrub. TWIN 4 comprised spider species and communities of humid deciduous forests (three grey alder-bird cherry forests, three grey alder-ash forests, one elm-lime forest and one spruce plantation). TWIN 5 comprised spider species and communities of dry deciduous forests (five elm-lime forests).

Fig. 1 indicates that spider communities on the bogs (TWIN 1) had more species in common with the open *Calluna*-pine forests (TWIN 2) than with the other forests. Indicator species of the community groups, separated by means of TWINSPAN, were *Trochosa terricola*, *Gonatium rubens* and *Pardosa pullata*, all typical for open areas (TWIN 1 and 2). Species typical for the forests (TWIN 3-5) were *Lepthyphantes alacris* and *Gonatium rubellum*, for the bogs (TWIN 1) *Pirata hygrophilus* and for the deciduous forests (TWIN 4, 5) *Helophora insignis*.

Some species were most frequently found in particular TWINSPAN groups, although not separated as indicator species. *Notioscopus sarcinatus*, *Pirata piraticus* and *Pocadicnemis pumila* were most frequent in the bogs (TWIN 1); *Gnaphosa bicolor*, *Zelotes clivicola* and *Hahnia ononidum* were most frequent in the open *Calluna*-pine forests (TWIN 2). The dominance of the different forest species *Lepthyphantes tenebricola* and *L. alacris* vary within the forest communi-

ties (TWIN 3, 4 and 5). *Diplostyla concolor* was found mainly in the deciduous forests (TWIN 4 and 5).

Ordination of species and communities
As a second step in the analysis of the spider data, DCA-analysis was used to detect major gradients of faunistic variation (Fig. 2). DCA-axis 1 is interpreted to represents the gradient of spider species and communities from bogs, via species and communities found in open *Calluna*-pine forests, bilberry pine forests to spider species and communities of different deciduous forests. DCA-axis 1 has a gradient length of 4.837 SD, an eigenvalue of 0.796 and explains 23.6% of the variation in the spider species/community data. The eigenvalue of DCA-axis 2 is 0.214, and it explains only 6.4% of the variation. Spider communities of the bogs and the *Calluna*-pine forests were more clearly separated as distinct groups in the ordinations than were the communities found in the bilberry-pine forests and the deciduous forests (Fig. 2).

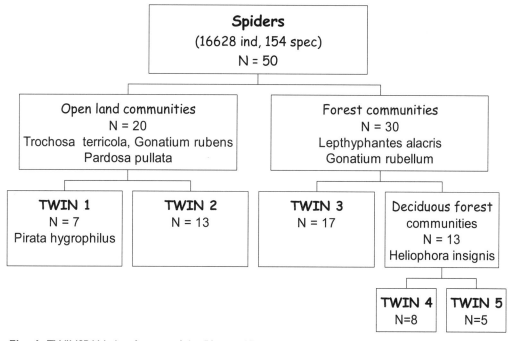

Fig. 1. TWINSPAN classification of the 50 sites (Geitaknottane, Western Norway) into five groups based on the spider data (1997), along with indicator species. N is the number of sites.

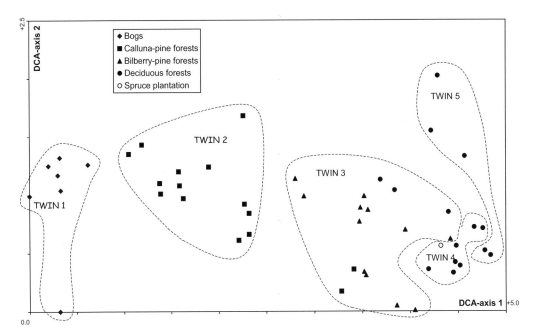

Fig. 2. DCA-ordination of the spider communities found in 50 different sites (Geitaknottane, Western Norway), based on activity abundance of spider species in 1997. The dotted lines indicate the different TWINSPAN-groups (cf. Fig. 1).

These interpretations of the DCA were largely confirmed by the results of the DCCA (Fig 3). Environmental variables like increasing forest productivity, decreasing soil humidity score, increasing bush cover, increasing tree cover and increasing slope of the site, turned out to be the significant ones, explaining the gradient of communities of spiders along DCCA-axis 1. A strong species-environment correlation (0.945) of DCCA-axis 1 support the importance of these factors. Increasing heat index was the only significant explanatory variable for DCCA-axis 2. The eigenvalue of DCCA-axis 1 is 0.697 and this axis explains 20.7% of the variation in the spider data. The eigenvalue of DCCA-axis 2 (0.139) explains considerably less (4.1%) and should be used with caution.

Diversity and habitat preference

In all, 16628 adult spiders were found, representing 154 species from 19 families. The number of species ranged from 21 to 51. On aver-

age, 27 (min 21; max 37) species were recorded in the deciduous forests (TWIN 4 and 5), 33 (30-38) in the bogs (TWIN 1), 32 (23-42) species in the bilberry-pine forests (TWIN 3) and 43 (32-50) species from the *Calluna*-pine forests (TWIN 2). Species new to Norway were *Euophrys frontalis*, *Maro lepidus* and *Porrhomma oblitum*.

Considering the ten most abundant species in the different TWINSPAN-groups, a complete change in species composition is indicated along the gradient from bogs (TWIN 1) to deciduous forests (TWIN 4 and 5). A change in species composition is also indicated by the gradient length of DCA-axis 1 (4.837 SD). The cursorial spiders (Lycosidae, Liocranidae, Tetragnathidae) dominated in the bogs and in the *Calluna* pine forests, while the linyphiids dominated in the bilberry-pine and deciduous forests. In the bogs (TWIN 1), *Pirata hygrophilus* showed the highest activity abundance (49.2%) followed by *Pardosa pullata* (17.2%), *Notioscopus sarcinatus* (3.9%), *Pardosa amentata* (3.3%) and *Trochosa terricola* (3.3%). In the *Calluna*-pine forests (TWIN 2) *P. pullata* (15.8%) dominated followed by *Alopecosa taeniata* (7.4%), *T. terricola* (6.4%), *Lepthyphantes mengei* (5.8%) and *Agyneta cauta* (4.2%). In the bilberry-pine forests (TWIN 3) *Lepthyphantes alacris* (24.2%) was the most abundant species, followed by *L. tenebricola* (10.2%), *Centromerus arcanus* (8.0%), *Agyneta cauta* (6.1%), *A. conigera* (5.8%) and *Dicymbium tibiale* (5.5%). The humid grey alder forests (TWIN 4) and the drier elm-lime forests (TWIN 5) were both dominated by *Lepthyphantes tenebricola* (30.3% and 20.6% respectively). In TWIN 4 the second most abundant species was *L. alacris* (13.4%), followed by *D. tibiale* (12.4%), *C. arcanus* (7.7%) and *Helophora insignis* (5.8%). For the elm-lime forests (TWIN 5) the second most abundant species was *Pardosa lugubris* (17.2%), followed by *Diplocephalus latifrons* (11.7%), *L. alacris* (7.5%) and *L. zimmermanni* (7.3%).

The correlation between the vegetation types and the spider fauna of the sites was quite high at the community level, but at the species level the pattern was more complex. Species found in many different vegetation types are *Centromerus arcanus* (found in 49 out of 50 sites), *Agyneta conigera* (42/50), *Lepthyphantes mengei* (40/50) and *Walckenaeria cuspidata* (39/50). Species found only in particular vegetation types in the study area are *Pirata hygrophilus* and *Notioscopus sarcinatus*, both found only in the bogs.

DISCUSSION

The DCA gradient of spider communities, from communities found in bogs, *Calluna*-pine forests, bilberry-pine forests to the communities of deciduous forests, was confirmed and partly explained by the environmental factors in DCCA. There was only a minor reduction in the biological variation being explained by the two first axes, from 30.0% (23.6 + 6.4) in DCA to 24.8% (20.7 + 4.1) in DCCA. The strong species-environment correlation (0.945) of DCCA-axis 1 also support this statement (ter Braak 1990).

According to the results of TWINSPAN and DCA/DCCA, the correlation between vegetation types and the spider fauna on the site is quite high at the community level. Exceptions are the spider communities in three deciduous forests and two *Calluna*-pine forests, all

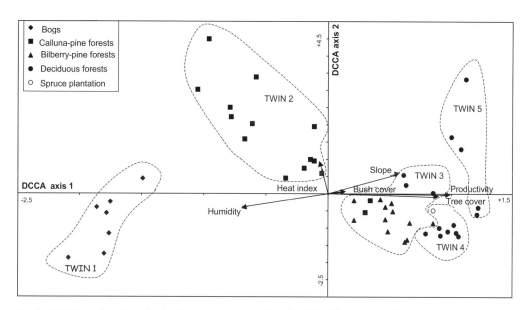

Fig. 3. DCCA-ordination of the spider communities found in 50 different sites (Geitaknottane, Western Norway), based on activity abundance of spider species in relation to 6 recorded environmental factors (1997). The dotted lines indicate the different TWINSPAN-groups (cf. Fig. 1).

grouped together with the spider communities found in bilberry-pine forests (TWIN 3). This is most likely explained by similar structures of the vegetation in the field layer. The presence of a woody field vegetation (*Vaccinium myrtillus* and *Calluna vulgaris*), with similar structure, density, and shade conditions, may affect the spiders living there. The DCA-ordination of one bog spider community separated from the other bog spider communities (lower left part of Fig. 2) is primarily due to high abundance of *Pardosa amentata*, not found at the other bog localities.

TWIN 3, 4 and 5 were not separated clearly as distinct community groups along the DCA/DCCA-axis (Fig. 2 and 3). The fact that these three TWINSPAN groups have some dominant species in common and few abundant species typical for one TWINSPAN-group, can result in different grouping in the TWINSPAN, but do not separate them clearly in the ordination. Similar microclimate or shade conditions in the bilberry-pine forests and the deciduous forests may result in similarity in the spider composition of these small habitats. However, neither plant composition or structure of the field layer were similar on these sites.

Along the first DCCA-axis, the spider species and communities in the bogs were ordinated with high scores for humidity and low scores for tree cover and productivity. The spider communities found in the different deciduous forests were ordinated with higher scores for tree cover and productivity, and lower scores for humidity. Some spider communities from *Calluna*-pine forests have relatively high scores for the heat index. These patterns partly confirm the habitat preference of the spider species presented by Reinke & Irmler (1994) and Hänggi et al. (1995). One exception is *Pirata hygrophilus*, which in the present study was most abundant and frequent in the bogs (treeless, open areas), while in Europe it is most frequently reported from humid forests (Maelfait et al. 1995; Thaler 1997).

ACKNOWLEDGEMENTS
I express my gratitude to Erling Hauge, Torstein Solhøy, Magne Sætersdal, Kjetil Aakra and John Skartveit for discussions and help. This paper is part of a graduate thesis (Pommeresche 1999), completed in connection with the forest biodiversity project 'Miljøregistrering i Skog' at the Norwegian Forest Research Institute (NISK), Bergen, Norway, 1997-2000.

REFERENCES
Ellefsen, G. & Hauge, E. 1986. Spiders (Araneae) from the Eidanger peninsula, Grenland, Telemark, SE Norway. *Fauna Norvegica* Ser. B 33, 33-39.
Fitje, A. 1984. *Tremåling*. Landbruksforlaget, Oslo.
Fremstad, E. 1997. *Vegetasjonstyper i Norge*, Norsk Institutt for Naturforskning, Trondheim.
Hänggi, A., Stöckli, E., & Nentwig, W. 1995. *Lebensräume mitteleuropäischer Spinnen*. Centre Suisse de Cartographie de la Faune, Neuchâtel. [*Miscellanea Faunistica Helvetiae* 4, 1-451]
Hauge, E., Bruvoll, A., & Solhøy, T. 1991. Spiders (Araneae) from islands of Øygarden, West Norway. Species associations, with ecological and zoogeographical remarks. *Fauna Norvegica* Ser. B 38, 11-26.
Hauge, E. & Kvamme, T. 1983. Spider from forest-fire areas in southeast Norway. *Fauna Norvegica* Ser. B 30, 39-45.
Hauge, E. & Wiger, R. 1980. The spider fauna (Araneae) from 12 habitats in the Vassfaret region, south-eastern Norway. *Fauna Norvegica* Ser. B 27, 60-67.
Hill, M.O. 1979. *TWINSPAN - a FORTRAN program for arranging multivariate data in an ordered two-way table by classification of the individuals and attributes.* Cornell University, New York.
Jongman, R.H.G., ter Braak, C.J.F., & Tongeren, O.F.R. 1995, *Data analysis in community and landscape ecology.* Cambridge University Press, Cambridge.
Maelfait, J.-P., de Knijf, G. , de Becker, P. &

Huybrechts, W. 1995. Analysis of spider fauna of the riverine forest nature reserve 'Walenbos' (Flanders, Belgium) in relation to hydrology and vegetation. In: *Proceedings of the 15th European Colloquium of Arachnology* (V. Růžička ed.), pp. 125-135. Institute of Entomology, České Budějovice.

Parker, K.C. 1988. Environmental relationships and vegetation associations of columnar cacti in the northern Sonora desert. *Vegetatio* 78, 125-140.

Pommeresche, R. 1999. Diversitet, samfunnsstrukturer og habitatpreferanser hos epigeiske edderkopper (Araneae) i ulike vegetasjonstyper innen Geitaknottane naturreservat, indre Hordaland. Cand. scient. Thesis. University of Bergen, Norway.

Reinke, H.D. & Irmler, U. 1994. Die Spinnenfauna (Araneae) Schleswig-Holsteins an Boden und in der bodennahen Vegetation. *Faunistisch-Ökologische Mitteilungen* Suppl. 17, 1-148.

ter Braak, C.J.F. 1990. *Update notes: CANOCO 3.10.* Agricultural Mathematic Group, Wageningen.

ter Braak, C.J.F. 1991. *CANOCO-a FORTRAN program for CANOnical Community Ordination by (partial) (detrended) (canonical) correspondence analysis, principal components analysis and redundancy analysis. 3.12.* Microcomputer Power, New York.

Thaler, K. 1997. Beiträge zur Spinnenfauna von Nordtirol 3. *Veröffentlichungen des Museum Ferdinandeum* 75/76, 97-146.

Tveit, L. & Hauge, E. 1983. The spider fauna of Kristansand and Setesdalen, S. Norway. *Fauna Norvegica* Ser. B 31, 23-45.

Økland, R.H. & Eilertsen, O. 1993. Vegetation-environmental relationships of boreal coniferous forests in the Solhomfjell area, Gjerstad, S Norway. *Sommerfeltia* 16, 19-24.

European Arachnology 2000 (S. Toft & N. Scharff eds.), pp. 207-214.
© Aarhus University Press, Aarhus, 2002. ISBN 87 7934 001 6
(Proceedings of the 19th European Colloquium of Arachnology, Århus 17-22 July 2000)

Similarities between epigeic spider communities in a peatbog and surrounding pine forest: a study from southern Lithuania

VYGANDAS RĖLYS[1] & DALIUS DAPKUS[2]

[1]*Department of Zoology, Vilnius University, Čiurlionio 21/27, LT-2009, Vilnius, Lithuania* (vygandas.
relys@gf.vu.lt)
[2]*Department of Zoology, Vilnius Pedagogical University, Studentų 39, LT-2034, Vilnius, Lithuania*

Abstract
Epigeic spider communities of an open raised bog, pine bog and surrounding dry pine forest were studied in southern Lithuania during 1999. The research was carried out by means of pitfall traps. A total of 108 spider species were registered. Large differences in species composition were revealed in spite of very similar numbers of species. Only nine species occurred in all three habitats. The highest similarity was found between the open raised bog and pine bog communities. The similarity between the open bog and dry pine forest communities was low. Only *Haplodrassus signifer* and *Agroeca brunnea* can be stated as common for both habitats. Some rare spider species were found (*Centromerus unidentatus*, *Zornella cultrigera*, *Euophrys westringi*). Altogether 15 spider species were registered as new for the Lithuanian fauna.

Key words: Araneae, peatbog, pine forest, communities, Lithuania

INTRODUCTION

Peatbogs and other wetlands are very sensitive and endangered ecosystems in Central Europe (Raeymaekers 1999, Succow 2000). The same situation exists in Lithuania where 6685 peatlands have been recorded (Janukonis 1995). Some peatbogs have remained in a natural state, but most of them have been drained and become highly fragmented, isolated, or naturally overgrown by forest. Some small undisturbed peatbog fragments exist on the edges of large excavated peatlands.

Anthropogenic impact leads to changes in plant and animal communities of peatbogs (Succow & Jeschke 1990). A negative human impact on spider communities has been studied by Hiebsch (1973, 1985), Hänggi & Maurer (1982), Platen (1989), Schikora (1993) and Albrecht (1998). The fauna of spiders living in Central European peatbogs is well known (Hänggi et al. 1995). However, there is little information regarding the relationship between spider communities living in peatbogs and surrounding areas. Moreover, the increasing fragmentation and uniformity of peatbog fragments following anthropogenic influence on the landscape, make these questions urgent. Some studies concentrated on the relationship between spider and other arthropod communities in forests, agricultural fields, and the surrounding areas (Duelli et al. 1990; Kromp & Steinberger 1992; Kajak & Lukasiewicz 1994; Luczak 1995; Downie et al. 1996; Topping 1997, 1999; Hänggi & Baur 1998; Riecken 1998). Questions relating to the edge effect, dispersal, isolation and fragmentation have been studied. Some questions on species communities and distribution of species between peatbogs and surrounding habitats have been analysed too (Almquist 1984; Vilbaste 1980; Freudenthaler 1989). However, diverse wetland habitats have mostly been studied, and no special attention has been paid to surrounding non-wetland habitats. Information on the relationship between peatbogs and surrounding non-wetland habitats can be obtained from papers of Koponen (1979), Hiebsch (1980), Vilbaste (1981), Löser et al. (1982), Schikora (1997) and Rupp (1999).

At present, peatbogs and other wetland fragments are mainly surrounded by drier habitats such as forests or meadows. We started the research into the epigeic spider fauna in various types of peatbogs and surrounding habitats in Lithuania. The aim was to evaluate the diversity and community structure of spiders in relation to peatbog size, level of isolation, anthropogenic impact, etc. In order to evaluate the 'naturalness level' of spider communities in small peatbogs or their fragments, comparative investigations were also carried out in protected areas, as well as in large intact and strictly protected peatbogs. This paper deals with the relationship between epigeic spider communities in the peatbog and surrounding forest in the largest mire complex of Lithuania.

MATERIALS AND METHODS

The research was carried out in the northwestern part of the Čepkeliai State Strict Nature Reserve (54°01' N, 24°26' E). It is located in southern Lithuania on the border with Belarus. Čepkeliai (5858 ha) is the largest mire complex in Lithuania. More than 50% is covered by large open sphagnum bogs, while the rest of the territory consists of fens, transitional bogs, small lakes and forested islands. Mires occur on sandy fluvioglacial lowland. The bog is surrounded by large dry pine forests.

Three study sites were chosen: pine bog (*Pinus sylvestris-Ledum palustre-Sphagnum* spp. community), open sphagnum bog (*Calluna vulgaris-Eriophorum vaginatum-Sphagnum rubellum*), and dry pine forest dominated by *Pinus sylves-* *tris-Vaccinium myrtillus-Pleurozium schreberi* communities bordering the bog.

Pitfall traps were used for collecting the material. Six plastic jars (volume 300 ml, depth 10 cm, diameter 7 cm) filled with 100 - 120 ml of 5% formaldehyde solution mixed with detergent were used at each locality. The distance between traps, disposed in a line, was ca. 5 m. The distance between the dry pine forest and the pine bog was 150 m, and the sites in the open bog and the pine bog were located at a distance of 120 m. The traps were in operation from 14 April to 11 October 1999. They were emptied once every three weeks.

The Sørensen coefficient of similarity (QS) was calculated for the whole set of species, and for the sets of species making up more than 0.5% (>5 individuals) of specimens in each community. If more than 0.5% of specimens of common species was found in each community, they were assumed to belong to the same population spread over all compared habitats.

The nomenclature of spiders follows Platnick (1993).

RESULTS
General overview of the material
The material collected comprised 2577 specimens of spiders representing 108 species. Of these, 55 species (965 specimens) were registered in the open bog; 57 species (882 specimens) in the pine bog, and 54 species (730 specimens) in the dry pine forest (Table 1). Despite the similar numbers of captured species, very low species similarities between communities were found (Table 2). Nine species: *Alopecosa pulverulenta*, *Agroeca brunnea*, *Agroeca proxima*, *Agyneta cauta*, *Diplocentria bidentata*, *Walckenaeria alticeps*, *Haplodrassus signifer*, *Zora spinimana*, and *Zelotes latreillei* (8.3% of all species) occurred in all habitats. Five of these species clearly had their highest abundance in one community (Appendix 1). Only *Agroeca brunnea* was represented by more than 2 specimens in all three habitats. No marked differences in the abundance of *Agyneta cauta* were registered in the open bog and pine bog. The abundance of

Table 1. The main parameters of spider communities investigated in Čepkeliai Reserve (southern Lithuania) in 1999. All species found were used in the calculations.

Habitat	Number of species	Number of specimens	Species found only in this community		Species represented by 1-2 specimens		No. of species (>0.5%) with max. abundance in this habitat
			No.	%	No.	%	
Open bog	55	965	18	32.7	30	54.5	13
Pine bog	57	882	11	19.3	28	49.1	11
Dry pine forest	54	730	30	55.5	27	50.0	22

Table 2. Sørensen similarity coefficients (QS) of the peatbog and pine forest spider communities studied in southern Lithuania in 1999. QS_{all}: all species were included. $QS_{0.5\%}$: only species with > 0.5% relative abundance.

Habitats compared	QS_{all}	$QS_{0.5\%}$	No. species in common	No. species (>0.5% of all ind.) in common
Open bog - pine bog	60.7	26.8	34	15
Open bog - dry pine forest	22.0	3.61	12	2
Pine bog - dry pine forest	37.8	5.42	21	3

Zora spinimana was similar in the pine bog and dry pine forest. The highest relative abundance of any species registered in all communities was not more than 5%. More than 49% of all species were represented by only 1 or 2 specimens in each community (Table 1). Fifteen species found during the research were new to the Lithuanian fauna (Appendix 1).

Similarities between communities

High similarity of species composition in the communities of the open bog and pine bog (*Sphagnum* spp. dominating the ground layer in both), and also of the pine bog and dry pine forest (*Pinus silvestris* and shrub vegetation) was expected.

The open bog and pine bog had the most similar species compositions of spiders (Table 2). High numbers of shared species (34), and shared species making up more than 0.5% of all specimens in each community (15) showed very close similarity between these two communities. It can be inferred that the populations of these 15 species inhabited both of the stud-

ied bog habitats. Most of these species had their highest abundance in one community. *Scotina palliardi*, *Pirata insularis*, and *Gnaphosa nigerrima* were most abundant in the open bog, while *Pirata uliginosus* and *Gnaphosa microps* dominated the pine bog. Of *Notioscopus sarcinatus*, 24 specimens were trapped in the pine bog, while only one in the open bog. The abundance of some species (*Pardosa sphagnicola*, *Aulonia albimana*, *Gonatium rubens*, *Walckenaeria nodosa*, and *Centromerus arcanus*) was similar in both communities. *Pardosa sphagnicola*, *Aulonia albimana*, and *Scotina palliardi* made up more than 5% of all specimens in both bog communities. Three species (*Agroeca brunnea*, *Agroeca proxima*, and *Agyneta cauta*) occurring in all three communities had maximum abundance in the pine bog. *Trochosa spinipalpis* and *Pardosa hyperborea* occurred only in the open bog, the latter was a dominant species (12.8%).

The similarity of species composition in the pine bog and dry pine forest was lower (Table 2). These habitats had 21 species in common, but only three species (*Agroeca brunnea*, *Zora spinimana,* and *Agyneta subtilis*) made up more than 0.5% of all specimens in each community. Only *Zora spinimana* did not show differences in abundance between the pine bog and pine forest. *Zelotes clivicola, Centromerus sylvaticus,* and *Zelotes subterraneus* were clearly represented in the pine forest, while only singletons occurred in the pine bog. *Haplodrassus signifer, Diplocentria bidentata,* and *Walckenaeria alticeps* were registered in all studied communities, but their abundance was highest in the pine forest. The latter two species were rare in bog communities.

The lowest species similarity was registered between the open bog and dry pine forest communities. These habitats had 12 species in common. Three of them were registered only in these habitats (Appendix 1). Only two common species (*Agroeca brunnea* and *Haplodrassus signifer*) represented more than 0.5% of specimens in each community. Both species were captured in the pine bog too.

DISCUSSION

The data show that there were some similarities between spider communities of the peatbog habitats and dry pine forest, but they were minimal, manifested mostly as shared species occurring in the pine bog and dry pine forest. It can be assumed that populations of such species are spread over both habitats. Only *Agroeca brunnea* showed no great specificity in habitat selection. Typical pine forest species occurring in the pine bog (*Centromerus sylvaticus*), or in both bog habitats (*Diplocentria bidentata* and *Walckenaeria alticeps*), were represented here only by low numbers of specimens and could be considered accidental. It can be concluded that the dispersal of spiders from dry pine forests has no major influence on the peatbog spider communities. Hiebsch (1980) stressed low similarities between the communities of bogs and pine forests too. Our results show that the number of woodland species was low in the intact pine bog habitat. The presence of pine trees and a shrub layer in the pine bog seemed to be insufficient to make this habitat suitable

for most of the epigeic woodland species. Löser et al. (1982), Freudenthaler (1989) and Rupp (1999) found that a typical woodland species *Trochosa terricola* was an important element of peatbog spider communities. In the present case, all specimens (161) of this species occurred in the dry pine forest. It can be expected that dry pine forests separating or bordering small peatbog fragments may prevent dispersal of spiders between peatbog fragments and recolonisation following extinction of typical peatbog species.

Only a very few species preferring open habitats other than wetlands were found during the present investigation (3 specimens of *Pardosa prativaga,* 2 *Pardosa pullata,* 1 *Metopobactrus prominulus,* 1 *Xysticus ulmi,* 2 *Xysticus cristatus,* 7 *Pachygnatha degeeri,* and 1 *Meionta rurestris*). It can be supposed that large pine forests function as a barrier to some of these species, especially because open areas are sparse in this forest region. On the other hand, similar results revealing low occurrence of such species in peatbogs surrounded by various habitats have been presented by other authors (Almquist 1984; Freudenthaler 1989; Schikora 1993, 1997). This supports the assumption that peatbogs are not, in general, suitable habitats for this group of species.

Schikora (1997) found low similarity between spider communities of a peatbog and surrounding lime (*Tilia cordata*) forest. Rupp (1999) noticed high species similarity between a peatbog and surrounding wet meadows, but low similarity between a peatbog and adjacent *Alnus-Fraxinus* riverine forest. Löser et al. (1982) revealed very different spider communities in peatbogs and surrounding *Luzulo-Fagetum* habitat, where only *Trochosa terricola, Lepthyphantes pallidus, Lepthyphantes cristatus,* and *Micrargus herbigradus* were common in both habitats. All these studies, as well as the present one, show low similarities between spider communities in peatbogs and surrounding woodland habitats.

ACKNOWLEDGEMENTS
VR is greatly indebted to the Organizing Committee of 19th European Colloquium of Arachnology for granting financial support for his participation in the Colloquium.

REFERENCES

Albrecht, H. 1998. Untersuchungen zur Spinnenfauna (Arachnida: Araneida) dreier anthropogen beeinflußter Hochmoore im Türinger Wald: Ein Vergleich 1971/1972 - 1996. *Thüringische Faunistischen Abhandlungen* 5, 91-115.

Almquist, S. 1984. Samhällen av spindlar och lockespindlar på Knisa myr, Öland. *Entomologisk Tidsskrift* 105, 143-150.

Downie, I.S., Coulson, J.C. & Butterfield, J.E.L. 1996. Distribution and dynamics of surface-dwelling spiders across a pasture-plantation ecotone. *Ecography* 19, 29-40.

Duelli, P., Studer, M., Marchand, I. & Jakob, S. 1990. Population movements of arthropods between natural and cultivated areas. *Biological Conservation* 54, 193-207.

Freudenthaler, P. 1989. Ein Beitrag zur Kenntnis der Spinnenfauna Oberösterreichs: Epigäische Spinnen an Hochmoorstandorten bei St. Oswald im Österreichischen Granit- und Gneishochland (Arachnida: Aranei). *Linzer biologischen Beiträge* 21/2, 543-575.

Hänggi, A., Stökli, E. & Nentwig, W. 1995. *Habitats of Central European spiders.* Centre Suisse de Cartographie de la Faune, Neuchâtel. [*Miscellanea Faunistica Helvetiae* 4]

Hänggi, A. & Baur, B. 1998. The effect of forest edge on ground-living arthropods in a remnant of unfertilized calcareous grassland in the Swiss Jura mountains. *Mitteilungen der Schweizerischen Entomologischen Gesellschaft* 71, 343-354.

Hänggi, A. & Maurer, R. 1982. Die Spinnenfauna des Lörmooses bei Bern - ein Vergleich 1930/1980. *Mitteilungen der Naturforschenden Gesellschahft Bern* 39 (NF), 159-183.

Hiebsch, H. 1973. Beitrag zur Spinnenfauna des Naturschutzgebietes 'Saukopfmoor'. *Abhandlungen und Berichte des Museums der Natur Gotha* 1973, 35-56.

Hiebsch, H. 1980. Beitrag zur Spinnenfauna des Naturschutzgebietes Bergen-Weissacker Moor im Kreis Luckau. *Brandenburgische Naturschutzgebiete, Folge 37. Naturschutzarbeit Berlin u. Brandenburg* 16 (1), 20-28.

Hiebsch, H. 1985. Zur Spinnenfauna der geschützten Hochmoore des Thüringer Waldes. *Landschaftspflege und Naturschutz in Thüringen* 22 (3), 71-78.

Janukonis, A. 1995. Resources of peat. In: *The cadastre of Lithuanian peatlands* (R. Liužinas ed.), pp. 21-22. Ministry of Environmental Protection, Vilnius. [In Lithuanian]

Kajak, A. & Lukasiewicz, J. 1994. Do semi-natural patches enrich crop fields with predatory epigean arthropods. *Agriculture, Ecosystems and Environment* 49, 149-161.

Koponen, S. 1979. Differences of spider fauna in natural and man-made habitats in a raised bog. In: *The use of ecological variables in environmental monitoring* (H. Hytteborn ed.), pp. 104-108. The National Swedish Environment Protection Board, Report PM 1151, Uppsala.

Kromp, B. & Steinberger, K.H. 1992. Grassy field margins and arthropod diversity: a case study on ground beetles and spiders in eastern Austria (Coleoptera: Carabidae; Arachnida: Aranei, Opiliones). *Agriculture, Ecosystems and Environment* 40, 71-93.

Löser, S., Meyer, E. & Thaler, K. 1982. Laufkäfer, Kurzflügelkäfer, Asseln, Webespinnen, Weberknechte und Tausendfüßler des Naturschutzgebietes 'Murnauer Moos' und der angrenzenden westlichen Talhänge (Coleoptera: Carabidae, Staphylinidae; Crustacea: Isopoda; Aranei; Opiliones; Diplopoda). *Entomofauna* Suppl. 1, 369-446.

Luczak, J. 1995. Plant-dwelling spiders of the ecotone between forest islands and surrounding crop fields in agricultural landscape of the Masurian Lakeland. *Ekologia Polska* 43, 79-102.

Platnick, N.I. 1993. *Advances in spider taxonomy 1988-1991. With synonymies and transfers 1940-1980.* Entomological Society and

American Museum of Natural History Press, New York.

Raeymaekers, G. 1999. *Conserving mires in the European Union: actions co-financed by LIFE. Nature.* European Commission, DG XI, Environment, Nuclear Safety and Civil Protection. Office for Official Publications of the European Communities, Luxembourg.

Riecken, U. 1998. The importance of semi-natural landscape structures in an agricultural landscape as habitats for stenotopic spiders. In: *Proceedings of the 17th European Colloquium of Arachnology, Edinburgh 1997* (P.A. Selden eds.), pp. 301-310. British Arachnological Society, Buckinghamshire.

Rupp, B. 1999. Ökofaunistische Untersuchungen an der epigäischen Spinnenfauna (Arachnida: Araneae) des Wörschacher Moores (Steiermark, Bez. Liezen). *Mitteilungen der Naturwissenschaftlichen Vereins Steiermark* 129, 269-279.

Schikora, H.B. 1993. Die epigäiche Spinnenfauna (Arachnida: Araneae) eines Hochmorreliktes in Norddeutschland von dem Hintergrund anthropogener Lebensraumveränderungen. *Mitteilungen deutscher Gesselschaft für allgemeine und angewandte Entomologie* 8, 373-382.

Schikora, H.B. 1997. Wachsende Regenmoorflächen im Zehlaubruch (Kaliningrad-Region): Extremlebensraum für epigäische Spinnen (Arachnida: Araneae)? *Verhandlungen der Gesellschaft für Ökologie* 27, 447-452.

Succow, M. 2000. *Landschaftsökologische Moorkunde.* Fischer, Stuttgart.

Succow, M. & Jeschke, L. 1990. *Moore in der Landschaft. Entstehung, Haushalt, Lebewelt, Verbreitung, Nutzung und Erhaltung der Moore.* Urania Verlag Leipzig, Jena, Berlin.

Topping, C.J. 1997. Predicting the effect of landscape heterogeneity on the distribution of spiders in agroecosystems using a population dynamics driven landscape-scale simulation model. *Biological Agriculture & Horticulture* 15, 325-336.

Topping, C. J. 1999. An individual-based model for dispersive spiders in agroecosystems: simulations of the effects of landscape structure. *Journal of Arachnology* 27, 378-386.

Vilbaste, A. 1980. The spider fauna of Estonian mires. *Eesti NSV Teaduste Akadeemia Toimetised, 29. Köide Biologia* 4, 313-327.

Vilbaste, A. 1981. The spider fauna of Estonian mires. *Eesti NSV Teaduste Akadeemia Toimetised, 30. Köide Biologia* 1, 7-17.

Appendix 1. Composition of three peatbog and pine forest communities of spiders studied in Southern Lithuania in 1999. *: species new to Lithuania (also known from other localities, unpublished). **: species new to Lithuania found only in this study area.

	Open bog		Pine bog		Dry pine forest	
	No. ind.	%	No. ind.	%	No. ind.	%
Pardosa hyperborea (Thorell)	124	**12.8**				
Trochosa spinipalpis (F.O.P.-Cambr.)	19	2.0				
Taranucnus setosus (O.P.-Cambr.)*	3	0.3				
Ceratinella brevis (Wider)	3	0.3				
Pardosa prativaga (L. Koch)	3	0.3				
Hypsosinga sanguinea (C.L. Koch)	2	0.2				
Centromerus unidentatus Miller**	2	0.2				
Meioneta mossica Schikora**	2	0.2				
Walckenaeria dysderoides (Wider)	2	0.2				
Episinus angulatus (Blackwall)	2	0.2				
Drassyllus pusillus (C.L. Koch)	2	0.2				
Centromerus levitarsis (Simon)	1	0.1				
Metopobactrus prominulus (O.P.-Cambr.)	1	0.1				
Clubiona stagnatilis Kulczynski	1	0.1				
Dolomedes fimbriatus (Clerck)	1	0.1				
Drassodes pubescens (Thorell)	1	0.1				
Drassyllus lutetianus (C.L. Koch)	1	0.1				
Xysticus ulmi (Hahn)	1	0.1				
Pardosa sphagnicola (Dahl)	212	**22.0**	162	**18.4**		
Aulonia albimana (Walckenaer)	188	**19.5**	164	**18.6**		
Scotina palliardi (L. Koch)	107	**11.1**	47	5.3		
Gonatium rubens (Blackwall)	23	2.4	18	2.0		
Lepthyphantes angulatus (O.P.-Cambr.)	21	2.2	10	1.1		
Pachygnatha degeeri Sundevall	6	0.6	1	0.1		
Antistea elegans (Blackwall)	6	0.6	1	0.1		
Agyneta decora (O.P.- Cambr.)*	5	0.5	2	0.2		
Theonoe minutissima (O.P.-Cambr.)*	2	0.2	1	0.1		
Phrurolithus minimus C.L. Koch**	2	0.2	1	0.1		
Pirata insularis Emerton	34	3.5	19	2.2		
Gnaphosa nigerrima L. Koch	13	1.3	5	0.6		
Walckenaeria nodosa O.P.-Cambr.*	7	0.7	6	0.7		
Pirata uliginosus (Thorell)	20	2.1	159	**18.0**		
Gnaphosa microps Holm*	29	3.0	58	6.6		
Centromerus arcanus (O.P.-Cambr.)	41	4.2	43	4.9		
Notioscopus sarcinatus (O.P.-Cambr.)	1	0.1	24	2.7		
Pocadicnemis pumila (Blackwall)	2	0.2	6	0.7		
Walckenaeria atrotibialis (O.P.- Cambr.)	2	0.2	6	0.7		
Tallusia experta (O.P.- Cambr.)	2	0.2	3	0.3		
Walckenaeria cuspidata Blackwall	1	0.1	3	0.3		
Walckenaeria nudipalpis (Westring)	1	0.1	3	0.3		
Cnephalohotes obscurus (Blackwall)	4	0.4	4	0.5		
Pardosa pullata (Clerck)	1	0.1	1	0.1		
Neon reticulatus (Blackwall)	1	0.1	1	0.1		
Alopecosa pulverulenta (Clerck)	20	2.1	7	0.8	1	0.1
Agroeca brunnea (Blackwall)	11	1.1	40	4.5	13	1.8
Agyneta cauta (O.P.- Cambr.)	12	1.2	19	2.2	2	0.3
Agroeca proxima (O.P.- Cambr.)	3	0.3	11	1.2	1	0.1
Diplocentria bidentata (Emerton)	2	0.2	1	0.1	26	3.6
Walckenaeria alticeps (Denis)	2	0.2	2	0.2	19	2.6
Haplodrassus signifer (C.L. Koch)	6	0.6	1	0.1	14	1.9
Zora spinimana (Sundevall)	1	0.1	6	0.7	7	1.0
Zelotes latreillei (Simon)	2	0.2	1	0.1	1	0.1
Robertus lividus (Blackwall)			6	0.7		
Lepthyphantes cristatus (Menge)			3	0.3		

Appendix I, continued. Composition of three peatbog and pine forest communities of spiders studied in southern Lithuania in 1999. *: species new to Lithuania (also known from other localities, unpublished). **: species new to Lithuania found only in this study area.

	Open bog		Pine bog		Dry pine forest	
	No. ind.	%	No. ind.	%	No. ind.	%
Mangora acalypha (Walckenaer)			2	0.2		
Stemonyphantes lineatus (Linnaeus)			2	0.2		
Micrargus herbigradus (Blackwall)			2	0.2		
Pirata hygrophilus Thorell			2	0.2		
Floronia bucculenta (Clerck)			1	0.1		
Neriene radiata (Walckenaer)			1	0.1		
Hahnia pusilla C.L. Koch			1	0.1		
Dictyna arundinacea (Linnaeus)			1	0.1		
Neon valentulus Falconer			1	0.1		
Agyneta ramosa Jackson			5	0.6	1	0.1
Zora silvestris Kulczynski*			4	0.5	1	0.1
Cercidia prominens (Westring)			2	0.2	1	0.1
Zelotes clivicola (L. Koch)			1	0.1	39	5.3
Centromerus sylvaticus (Blackwall)			1	0.1	21	2.9
Agyneta subtilis (O.P.- Cambr.)			4	0.5	12	1.6
Zelotes subterraneus (C.L. Koch)			1	0.1	7	1.0
Cicurina cicur (Fabricius)			1	0.1	3	0.4
Micrargus apertus (O.P.- Cambr.)*			1	0.1	2	0.3
Saaristoa abnormis (Blackwall)*			1	0.1	2	0.3
Pachygnatha listeri Sundevall			2	0.2	2	0.3
Zelotes petrensis (C.L. Koch)			1	0.1	1	0.1
Trochosa terricola Thorell					161	**22.1**
Walckenaeria cucullata (C.L. Koch)					77	**10.5**
Zora nemoralis (Blackwall)					45	6.2
Pardosa lugubris (Walckenaer)					40	5.5
Tapinocyba pallens (O.P.- Cambr.)					38	5.2
Centromerus aequalis (Westring)					37	5.1
Haplodrassus soerenseni (Strand)					26	3.6
Alopecosa aculeata (Clerck)					22	3.0
Zornella cultrigera (L. Koch)**					20	2.7
Macrargus rufus (Wider)					16	2.2
Minyriolus pusillus (Wider)					15	2.1
Gnaphosa muscorum (L. Koch)					10	1.4
Macrargus carpenteri (O.P.- Cambr.)					8	1.1
Xysticus erraticus (Blackwall)					8	1.1
Araneus angulatus Clerck					4	0.5
Xysticus luctuosus (Blackwall)					3	0.4
Centromerita bicolor (Blackwall)					2	0.3
Porrhomma pallidum Jackson					2	0.3
Walckenaeria acuminata Blackwall					2	0.3
Philodromus cespitum (Walckenaer)					2	0.3
Euophrys westringi (Thorell)*					2	0.3
Segestria senoculata (Linnaeus)					1	0.1
Metellina mengei (Blackwall)					1	0.1
Araneus diadematus Clerck					1	0.1
Nuctenea umbratica (Clerck)					1	0.1
Lepthyphantes mansuetus (Thorell)**					1	0.1
Meioneta rurestris (C.L. Koch)					1	0.1
Pelecopsis elongata (Wider)					1	0.1
Agelena labyrinthica (Clerck)					1	0.1
Clubiona subsultans Thorell					1	0.1
Euryopis flavomaculata (C.L. Koch)	1	0.1			1	0.1
Xysticus cristatus (Clerck)	1	0.1			1	0.1
Euophrys petrensis C.L. Koch*	2	0.2			3	0.4

European Arachnology 2000 (S. Toft & N. Scharff eds.), pp. 215-222.
© Aarhus University Press, Aarhus, 2002. ISBN 87 7934 001 6
(Proceedings of the 19th European Colloquium of Arachnology, Århus 17-22 July 2000)

Epigeic spider communities in inland dunes in the lowlands of Northern Germany

SABINE MERKENS

University of Osnabrück, FB Biologie/Chemie, Ökologie, Barbarastr. 11, D-49069 Osnabrück, Germany
(sabine.merkens@surfeu.de)

Abstract

The spider communities of 13 open inland dunes which lie scattered in the lowlands of Northern Germany have been analysed by means of pitfall trapping. The spider communities of the sandy places clearly differ from the communities of the neighbouring habitats which are heathland, dry grassland or pine forest. But there is no uniform spider community in the *Spergulo-Corynephoretum* habitat which is unique to inland dunes. The concrete vegetation and environmental factors of the site and its surroundings determine the species composition. Vegetation cover (especially lichen, moss and herbs) and the kind of neighbouring habitat mainly influence the species composition of the inland dunes. The influence of the geographical location and of the climate (atlantic in the west, more subcontinental in the east of Northern Germany) on the spider communities is discussed.

Key words: Araneae, inland dunes, Northern Germany, ecology, environmental factors, vegetation cover, biogeography

INTRODUCTION

Inland dunes are rare habitats in the lowlands of Northern Germany. They are sparse in plants but contain a large number of specialised and rare spider species. The hithero available information about spiders in the inland dunes of Northern Germany is very poor and restricted to single locations (e.g. Lademann 1995; Finch 1997). Until now we have to rely on investigations from dune habitats in Southern Germany (e.g. Braun 1956; Leist 1994; Bauchhenß 1995). The comparability with dune habitats in the north is limited because the climate and soil conditions are different.

In this study, the spider communities of 13 open inland dunes in Northern Germany have been analysed and compared. The dunes differ in vegetation cover, neighbouring habitats, size and geographical location. The significance of these environmental features for the distribution of spider species is of special interest.

The following questions were asked in this study:
(i) - Is there a community of spider species which is characteristic of open inland dunes and unique to these habitats?
(ii) - In what way do the distinguishing features of the habitats influence the species composition?
(iii) - What role does the geographical location play?

MATERIAL AND METHODS

Investigation areas

The investigation areas lie scattered in the lowlands of Northern Germany: along the rivers Ems, Weser, Elbe and Oder and in the Lüneburg Heath (Fig. 1). The greatest distance between the western areas along the Ems and the eastern area along the Oder is about 500 km. The climate in the lowlands of Northern Germany changes from atlantic in the west to more subcontinental in the east: the annual precipitation sums are 700-750 mm along the Ems and 550-650 mm along the Elbe and Oder. The annual amplitudes of air temperature are greater in the east than in the west (Deutscher Wetterdienst 1965).

The inland dunes investigated differ in vegetation cover: some are sparsely covered with *Corynephorus canescens* and *Carex arenaria*, some are completely covered with lichen and some have more moss and herbs. They also differ in size: the smallest area is about 0.1 ha, the greatest about 13.6 ha. They are surrounded by different neighbouring habitats: by heathland, dry grassland or *Pinus sylvestris*-forest (more detailed descriptions of the sampling sites in Merkens 2000).

Spiders
The spiders were caught in pitfall traps. Four pitfall traps, each filled with 4% formalin solution, were installed in the sandy sites (*Spergulo-Corynephoretum*) in all investigation areas. Two pitfall traps were installed in the neighbouring habitats. In six investigation areas the pitfall traps were used for two years (October 1995 to November 1997) and in the other seven investigation areas they were used for 17 months (October 1995 to March 1997). To compare the areas the data were standardized (individual sums x 100/number of sampling days/number of pitfall traps).

Vegetation structure
The vegetation structure was documented by the measurements of the average cover of lichen, moss, grass, herbs, straw and total vegetation cover in an area of 1 m² around each pitfall trap. Five classes of vegetation cover were defined and separately assessed for the abovementioned vegetation types: 0 = 0% cover, 1 = 1-25% cover, 2 = 26-50% cover, 3 = 51-75% cover, 4 = 76-100% cover. The average cover was calculated on the basis of 100 estimates of 10 cm².

Statistical methods
The faunistic data were subjected to a detrended correspondence analysis (DCA). By means of this procedure locations with similar species composition, and species with similar distribution patterns come close in the ordination diagrams. The locations of pitfall traps are arranged in an order which represents the most effective ecological gradient of the species communities (Jongman et al. 1987).

The correlation between distribution patterns of species and the quality of environmental features was analysed by the Pearson correlation for continuously scaled features

Fig. 1. Geographical location of the investigation sites in Northern Germany.

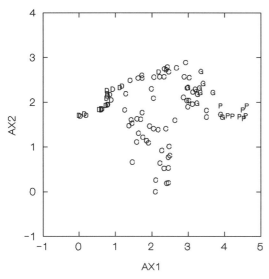

Fig. 2. Detrended correspondence analysis of the spider fauna of all sites: *Spergulo-Corynephoretum* (C), dry grassland (D), heathland (G), *Pinus sylvestris*-forests (P). Each symbol represents one trap.

(cover of the individual elements of the vegetation, habitat size) and by the Spearman rank correlation for ordinally scaled features (neighbouring habitats, geographical location; ordinal numbers were allocated to the different habitats and geographical locations: 1 = heathland, 2 = grassland, 3 = pine forest; 1 = Ems, 2 = Weser, 3 = Lüneburg Heath, 4 = Elbe, 5 = Oder). The correlation coefficients determine the degree of correlation between the abundance of species and the environmental features (Sachs 1978).

RESULTS

Altogether 35991 mature spiders from 286 species and 21 families were caught by pitfall traps. The Linyphiidae, Lycosidae, Gnaphosidae and Salticidae were the most frequent families in the sandy sites. In the investigation areas up to 46% of the species and up to 66% of the individuals were stenotopic species of sandy habitats. The stenotopy of the species has been estimated on the basis of the results of this study and is therefore regionally valid.

Is there a community of spider species which is characteristic of open inland dunes

and unique for these habitats? The habitat types can clearly be distinguished along the first axis of the DCA ordination diagram (Fig. 2): heathland and forests on the right, open sand in the middle and grassland on the left. The spider community of the open inland dunes was generally separated from the communities of the neighbouring habitats. Along the second axis the site scores of the dunes show great variation, while the site scores of the surrounding habitats vary little. This means that there is no uniform spider community of the inland dunes. The species composition differs between the individual sites.

In what way do the habitat variables and surroundings influence the species composition? The results of the correspondence analysis of the sandy sites are shown in Fig. 3. There is a gradient of vegetation cover along the first axis: the sparsely covered and open sites are positioned on the left, the more overgrown sites on the right. As for the neighbouring habitats there is a gradient on the second axis: the sites with neighbouring grassland are separate from the sites which are neighbour to a forest. As for the size of the area and the geographical location the distribution of the sites in the diagram is irregular (data not shown).

On the basis of the calculated correlation coefficients between the abundance of several species and the environmental features it is possible to distinguish three species groups (Table 1): (i) - The spider species of the first group correlate positively with the cover of lichen (and grass). This species group is composed heterogeneously. Some of the species are exclusively found at the sites which are covered with lichen (e.g. *Alopecosa schmidti*, *Drassyllus praeficus*), others clearly prefer this variant (e.g. *Typhochrestus digitatus*, *Aelurillus v-insignitus*). Some species are stenotopic of the sandy habitat, but show no preference for any of the variants of the plant community (e.g. *Zelotes longipes*, *Hypsosinga albovittata*). Some species are pioneers which are able to cope with a wide ecological amplitude and have an efficient spreading strategy (e.g. *Erigone atra*, *Araeoncus*

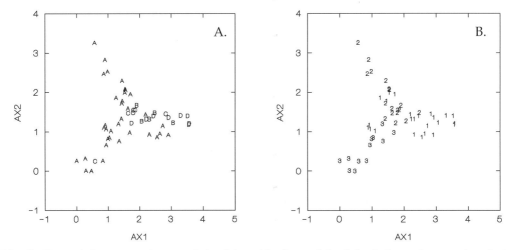

Fig. 3. Detrended correspondence analysis of the spider fauna of the *Spergulo-Corynephoretum* sites. Each symbol represents one trap. **(A)** Symbols are applied to the average cover of lichen and moss: A: average cover of both lichen and moss ≤ 0.5; B: average cover of moss > 0.5, of lichen < 0.5; C: average cover of lichen > 0.5, of moss < 0.5; D: average cover of both lichen and moss > 0.5. **(B)** Symbols are applied to the neighbouring habitats: 1: neighbouring heathland, 2: neighbouring grassland, 3: neighbouring *Pinus sylvestris*-forest.

humilis). Most of the species have a middle position in the ordination diagram (Fig. 4).

(ii) - The species of the second group correlate significantly and positively with the cover of moss and herbs. Most of these species are typical of heathland but were also found regularly in the sandy sites (e.g. *Trochosa terricola, Agroeca lusatica, Centromerita concinna*). Some species are stenotopic of the sandy habitat (e.g. *Zelotes petrensis, Alopecosa fabrilis*). All species of this group are positioned on the right side of the ordination diagram (Fig. 4).

(iii) - Four species of the third group correlate negatively with the vegetation cover: *Oedothorax apicatus, Arctosa perita, Archaeodictyna ammophila* and *Yllenus arenarius* are restricted to the initial stage of the *Spergulo-Corynephoretum* and live in the open sand. These species are positioned on the left side of the ordination diagram (Fig. 4).

Some species significantly prefer sandy sites which are surrounded by heathland (e.g. *Typhochrestus digitatus, Agroeca lusatica, Drassyllus pusillus*; data not shown). Most of them are typically living in heathland or prefer places with an advanced stage of succession within the inland dunes.

The size of the habitat was of no importance to most of the species in this study. The individual numbers of two species (*Xysticus kochi, Dictyna major*) correlated significantly and positively with the size of the habitat.

What role does the geographical location play? Some of the stenotopic spider species occur exclusively or predominantly at eastern or western localities, respectively (Table 2): 12 species were mainly found along the rivers Elbe and Oder in the east, 8 species were mainly found along the Ems in the west. In the west several species were found regularly in dune habitats which otherwise typically occur in heathland or forest or which are common pioneer species. In the east these species live in their typical habitats but do not find their way into the sandy sites.

DISCUSSION

The spider communities of inland dunes in the lowlands of Northern Germany can generally be distinguished from the spider communities of the neighbouring habitats. One cannot, however, speak of a uniform species community which is characteristic and unique to open

Table 1. Correlations between abundance of individuals and vegetation cover. Significant correlation coefficients are indicated. Limits of significance: P = 0.05: r = 0.273 (x); P = 0.01: r = 0.354 (xx); P = 0.001: r = 0.443 (xxx).

Species	Total veg. cover	Cover of lichen	Cover of grass	Cover of herbs	Cover of moss
Haplodrassus signifer	0.550 (xxx)	0.524 (xxx)	0.310 (x)	.	.
Araeoncus humilis	0.570 (xxx)	0.835 (xxx)	0.644 (xxx)	.	.
Alopecosa schmidti	0.590 (xxx)	0.879 (xxx)	0.413 (xx)	.	.
Drassyllus praeficus	0.567 (xxx)	0.801 (xxx)	0.555 (xxx)	.	.
Thanatus arenarius	0.496 (xxx)	0.682 (xxx)	0.728 (xxx)	.	.
Xysticus striatipes	0.588 (xxx)	0.836 (xxx)	0.642 (xxx)	.	.
Zelotes longipes	0.510 (xxx)	0.584 (xxx)	.	.	.
Trichopterna cito	0.495 (xxx)	0.543 (xxx)	0.388 (xx)	.	.
Zelotes electus	0.362 (xx)	0.610 (xxx)	.	.	.
Pardosa monticola	0.423 (xx)	0.322 (x)	0.506 (xxx)	.	.
Hypsosinga albovittata	0.284 (x)	0.504 (xxx)	.	.	.
Erigone atra	0.394 (xx)	0.644 (xxx)	.	.	.
Bathyphantes gracilis	.	0.446 (xxx)	.	.	.
Typhochrestus digitatus	0.694 (xxx)	0.513 (xxx)	.	.	0.442 (xx)
Aelurillus v-insignitus	0.684 (xxx)	0.738 (xxx)	.	.	0.302 (x)
Centromerita concinna	0.391 (xx)	.	.	0.617 (xxx)	0.422 (xx)
Centromerus sylvaticus	0.417 (xx)	.	.	0.464 (xxx)	0.304 (x)
Tapinocyba praecox	0.374 (xx)	.	.	0.340 (x)	0.323 (x)
Alopecosa barbipes	0.315 (x)	.	.	0.379 (xx)	.
Walckenaeria monoceros	0.341 (x)	0.439 (xx)	.	0.539 (xxx)	.
Trochosa terricola	0.339 (x)	.	.	0.458 (xxx)	0.355 (xx)
Drassyllus pusillus	.	.	.	0.425 (xx)	0.312 (x)
Hahnia nava	.	.	.	0.376 (xx)	0.429 (xx)
Agroeca lusatica	.	.	.	0.516 (xxx)	.
Zelotes petrensis	.	.	.	0.543 (xxx)	.
Micaria silesiaca	.	.	.	0.373 (xx)	.
Alopecosa fabrilis	.	.	.	.	0.546 (xxx)
Phrurolithus festivus	.	.	.	0.311 (x)	.
Agelena labyrinthica	.	.	.	0.275 (x)	.
Xysticus kochi	.	.	0.321 (x)	.	.
Oedothorax apicatus	.	.	.	.	-0.293 (-x)
Arctosa perita	-0.386 (-xx)	-0.43 (-xx)	.	.	-0.327 (-x)
Archaeodictyna ammophila	-0.306 (-x)	.	.	.	.
Yllenus arenarius	-0.294 (-x)	.	.	.	.

inland dunes. The concrete vegetation and environmental factors of the site and its surroundings determine the species composition of the spider community. Vegetation cover and kind of the neighbouring habitat have been documented as the most distinguishing features. The composition of the spider community seems to follow the stage of succession of the plant community: some of the species are restricted to the initial stage of open sand and sparse vegetation cover, some species prefer a dense cover of lichen, others are restricted to a dense cover of moss and to immigrated herbs which indicate an advanced stage of succession.

The atlantic climate in the west of the lowlands seems to accelerate the succession of the inland dunes and to reduce the extreme character of the habitat. Along the Ems several common species of other habitat types were found regularly in the dunes. In the east these species stay in their typical habitats, probably because the sandy sites in the east are drier and hotter than those in the west.

Concerning the spiders, the inland dunes of the lowlands of Northern Germany are intermediate between the coastal dunes along the North Sea and the Baltic Sea on the one hand, and the inland dunes in the east of Germany on

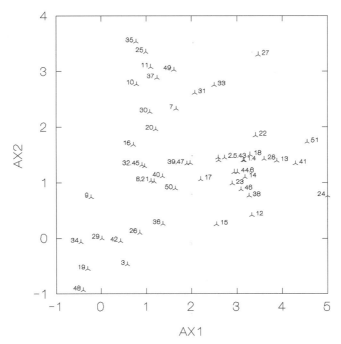

Fig. 4. Correspondence analysis of the spider fauna of the *Spergulo-Corynephoretum* sites. Ordination of species scores along the first and second axis of DCA. Numbering of the species:

1*Acartauchenius scurrilis*, 2*Aelurillus v-insignitus*, 3*Agelena labyrinthica*, 4*Agroeca lusatica*, 5*Alopecosa barbipes*, 6*Alopecosa fabrilis*, 7*Alopecosa schmidti*, 8*Araeoncus humilis*, 9 *Archaeodictyna ammophila*, 10*Arctosa perita*, 11*Bathyphantes gracilis*, 12*Berlandina cinerea*, 13*Centromerita concinna*, 14*Centromerus sylvaticus*, 15*Cheiracanthium virescens*, 16*Dictyna major*, 17*Drassyllus praeficus*, 18*Drassyllus pusillus*, 19*Enoplognatha serratosignata*, 20*Erigone atra*, 21*Erigone dentipalpis*, 22*Hahnia nava*, 23*Haplodrassus signifer*, 24*Haplodrassus silvestris*, 25*Hypsosinga albovittata*, 26*Lepthyphantes tenuis*, 27*Micaria fulgens*, 28*Micaria silesiaca*, 29*Oedothorax apicatus*, 30*Pachygnatha degeeri*, 31*Pardosa monticola*, 32*Pelecopsis parallela*, 33*Phrurolithus festivus*, 34*Porrhomma microphthalmum*, 35*Sitticus distinguendus*, 36*Sitticus saltator*, 37*Steatoda albomaculata*, 38*Tapinocyba praecox*, 39*Thanatus arenarius*, 40*Trichopterna cito*, 41*Trochosa terricola*, 42*Troxochrus scabriculus*, 43*Typhochrestus digitatus*, 44*Walckenaeria monoceros*, 45*Xysticus kochi*, 46*Xysticus sabulosus*, 47*Xysticus striatipes*, 48*Yllenus arenarius*, 49*Zelotes electus*, 50*Zelotes longipes*, 51*Zelotes petrensis*

the other hand. There are species typically found in coastal dunes, e.g. *Philodromus fallax, Sitticus distinguendus, Haplodrassus dalmatensis, Erigone arctica* (Hänggi et al. 1995; Schultz & Finch 1996). In Northern Germany most of these species are restricted to the inland dunes in the west. Some specimens are also found in the investigation area along the Oder, which is not far from the Baltic Sea. On the other hand there are species with a clear centre of distribution in the east of Europe, e.g. *Yllenus arenarius, Alopecosa schmidti, Archaeodictyna ammophila, Berlandina cinerea* (Buchar 1975; Weiss & Marcu 1979; Prószynski 1978, 1986; Grimm 1985; Thaler & Buchar 1994). In the lowlands of Northern Germany they are found along the

Oder and Elbe. Some of them reach their western distribution limits there.

ACKNOWLEDGEMENTS

Very many thanks to Dr. Peter Lühmann for reliably looking after the pitfall traps in the dune on the Oder and to Martina Hüls for kindly checking the English of the manuscript.

REFERENCES

Bauchhenß, E. 1995. Die epigäische Spinnenfauna auf Sandflächen Nordbayerns (Arachnida: Araneae). *Zoologische Beiträge* N.F. 36 (2), 221-250.

Braun, R. 1956. Zur Spinnenfauna von Mainz und Umgebung, mit besonderer Berück-

Table 2. Occurrence of several stenotopic and typical spider species in the inland dunes of five geographical regions. The regions are ordered from west to east. x: 1-3 individuals; xx: more than 3 individuals in the sandy sites.

Species	Ems	Weser	Lünebg.	Elbe	Oder
Species which are stenotopic or typical of sandy sites along the Elbe/Oder					
Alopecosa schmidti	.	.	.	xx	xx
Alopecosa trabalis	.	.	.	.	xx
Archaeodictyna ammophila	.	x	.	xx	xx
Argenna subnigra	.	.	.	xx	.
Berlandina cinerea	.	.	.	xx	.
Cheiracanthium gratum	.	.	.	.	xx
Drassyllus praeficus	.	.	.	xx	x
Enoplognatha serratosignata	.	.	.	xx	.
Hypsocephalus dahli	.	.	.	.	xx
Hypsosinga albovittata	.	.	x	xx	x
Micaria dives	x	.	.	.	xx
Ozyptila westringi	.	.	.	xx	.
Philodromus fallax	.	.	.	.	xx
Thanatus arenarius	.	.	.	xx	xx
Xysticus sabulosus	.	.	xx	xx	.
Xysticus striatipes	.	.	.	xx	.
Yllenus arenarius	.	xx	.	xx	.
Species which are stenotopic or typical of sandy sites along the Ems					
Ceratinopsis romana	xx	.	.	.	.
Erigone arctica	xx	.	.	.	.
Haplodrassus dalmatensis	xx	.	.	.	.
Micaria fulgens	xx	.	.	.	x
Micaria silesiaca	xx	.	.	.	.
Ostearius melanopygius	xx	.	.	.	.
Porrhomma montanum	xx	.	.	.	.
Sitticus distinguendus	xx	.	.	.	x
Zelotes petrensis	xx	.	.	.	.
Common species of heathland or forests, mainly in the sandy sites in the west of Northern Germany					
Drassodes cupreus	xx	.	x	.	x
Gnaphosa leporina	xx	.	x	.	.
Lepthyphantes pallidus	xx	.	.	x	.
Microneta viaria	xx	.	.	.	x
Pardosa lugubris	xx	x	.	x	x
Pardosa nigriceps	xx	.	.	.	.
Pirata hygrophilus	xx	.	.	.	.
Pisaura mirabilis	xx	.	.	.	.
Tapinocyba insecta	xx	.	.	.	.
Walckenaeria acuminata	xx	.	x	x	x
Walckenaeria cucullata	xx	.	.	x	.
Walckenaeria cuspidata	xx	.	.	.	x
Walckenaeria dysderoides	xx	.	x	x	.
Xerolycosa nemoralis	xx	.	.	x	.
Zora spinimana	xx	.	x	x	.

sichtigung des Gonsenheimer Waldes und Sandes. *Jahrbuch des Nassauer Vereins für Naturkunde* 92, 50-79.

Buchar, J. 1975. Arachnofauna Böhmens und ihr thermophiler Bestandteil. *Vestnik Ceskoslovenske spolecnosti zoologicke* 39, 241-250.

Deutscher Wetterdienst 1965. *Klima-Atlas von Niedersachsen*. Offenbach/M.

Finch, O.-D. 1997. Die Spinnen (Araneae) der Trockenrasen eines nordwestdeutschen Binnendünenkomplexes. *Drosera* '97 (1), 21-40.

Hänggi, A., Stöckli, E. & Nentwig, W. 1995.

Lebensräume mitteleuropäischer Spinnen. Centre Suisse de Cartographie de la Faune, Neuchâtel. [*Miscellanea Faunistica Helvetiae* 4, 1-459]

Jongman, R.H., ter Braak, C.J.F. & Tongeren, O. F.R. v. 1987. *Data analysis in community and landscape ecology.* Pudoc, Wageningen.

Lademann, J. 1995. Die Spinnenfauna (Araneae) unterschiedlicher Vegetationseinheiten eines norddeutschen Sandtrockenrasens. Diplomarbeit University of Bremen.

Leist, N. 1994. Zur Spinnenfauna zweier Binnendünen um Sandhausen bei Heidelberg (Arachnida: Araneae). *Beihefte Veröffentlichungen Naturschutz und Landschaftspflege Baden-Württemberg* 80, 283-324.

Merkens, S. 2000. Die Spinnenzönosen der Sandtrockenrasen im norddeutschen Tiefland im West-Ost-Transekt - Gemeinschaftsstruktur, Habitatbindung, Biogeographie. Dissertation Universität Osnabrück.

Prószynski, J. 1978. Distributional patterns of the palaearctic Salticidae (Araneae). *Symposia of the Zoological Society of London* 42, 335-343.

Prószynski, J. 1986. Remarques sur la composition de la faune européenne, sa répartition et son origine basées sur les études des Salticidae. *Mémoires de la Société Royale Belge d'Entomologie* 33, 165-170.

Sachs, L. 1978. *Angewandte Statistik. Statistische Methoden und ihre Anwendungen.* Springer, Berlin.

Schultz, W. & Finch, O.-D. 1996. *Biotoptypenbezogene Verteilung der Spinnenfauna der nordwestdeutschen Küstenregion. Charakterarten, typische Arten und Gefährdung.* Cuvillier, Göttingen.

Thaler, K. & Buchar, J. 1994. Die Wolfspinnen von Österreich 1: Gattungen *Acantholycosa, Alopecosa, Lycosa* (Arachnida, Araneida: Lycosidae) - Faunistisch-tiergeographische Übersicht. *Carinthia* II, 104(2), 357-375.

Weiss, I. & Marcu, A. 1979. [Soil surface spiders and harvestmen from the river dunes reserve from Hanu Conachi (district Galati)] (in Romanian). *Studii si comunicari Muzeul Brukenthal, Stiinte naturale (Sibiu)* 23, 251-254.

European Arachnology 2000 (S. Toft & N. Scharff eds.), pp. 223-228.
© Aarhus University Press, Aarhus, 2002. ISBN 87 7934 001 6
(Proceedings of the 19th European Colloquium of Arachnology, Århus 17-22 July 2000)

Distinctiveness of the epigeic spider communities from dune habitats on the Danish North Sea coast

PETER GAJDOŠ[1,2] & SØREN TOFT[2]

[1]*Institute of Landscape Ecology, Slovak Academy of Sciences, Bratislava, Branch Nitra, Akademicka 2, POB 23B, SK-94901 Nitra, Slovakia* (nrukgajd@savba.sk)
[2]*Department of Zoology, University of Aarhus, Building 135, DK-8000 Århus C, Denmark* (soeren.toft @biology.au.dk)

Abstract

Pitfall traps were operated through a full year in dune and heathland habitats of the Hanstholm Reserve close to the North Sea coast of NW Jutland, Denmark. Transect lines were laid out in the dune adjacent to the beach, and dunes 400-700 m from the shore. The following main habitat types could be distinguished: yellow dune, grey dune, *Empetrum/Calluna* heathland, low pine plantation, sandy areas with sparse vegetation and *Sphagnum* bog. The full-year catches of every trap were analysed with Canonical Correspondence Analysis (CCA), using topography and vegetation (species, coverage and height) as environmental variables.

The greatest faunistic differences were between the near-beach (yellow dune) communities and the rest, in spite of great differences in habitat conditions (e.g. humidity) and structure of the vegetation. Several (even some dominant) species were unique to the yellow dunes and were not found in the grey dunes only a few hundred meters inland.

Key words: Araneae, coastal dunes, habitat mosaic, habitat selection

INTRODUCTION

The spider fauna of coastal dune habitats in Northern Europe have been thoroughly studied in recent years (Finland: Perttula 1984; Sweden: Almquist 1973; England: Duffey 1968; Germany: Schultz 1995; Schultz & Finch 1996; Belgium: Bonte et al. 2000). Clausen (1987) reported on dune and dune heath spiders from the Danish island Læsø in the Kattegat Sea. The extensive dune system along the Danish North Sea coast has not been similarly investigated before from an ecological point of view. In 1997 we therefore initiated a study of the dune spiders in the Hanstholm Reserve, situated at the 'shoulder' of northwestern Jutland. Due to their location, these dunes are more exposed to the prevailing (north)westerly winds and occa-

sional storms than any other places along the Danish west coast. A pauperised fauna could therefore be expected compared to less exposed areas.

Dunes are dynamic landscapes regularly subjected to the modifying forces of wind (Ranwell 1972). Topographically they are hilly, often with steep slopes. This means that dune systems will always be mosaics of vegetation types varying in environmental conditions and being in different stages of successional development. Typical coastal dune profiles (cf. Duffey 1968; Ranwell 1972) are usually modified by local conditions. At Hanstholm no foredune is present because autumn and winter storms erode into the yellow dune, which is heavily influenced by shifting sand. Behind the yellow dune is a dune slack/dune heath ranging 400-600 m inland. At the site of investigation two high grey dunes raises here. Their slopes present a diversity of habitats: typical grey dune vegetation, windbreaks with bare sand, low pine plantation (planted to prevent windbreaks), dune heath, and at the lowest parts of the dune slack there are seasonally water-logged *Sphagnum* bogs. Our aim was to characterise the epigeic spider communities of this mosaic area and analyse how the changes in species composition relate to changes in habitat. In this preliminary report we present only some major patterns, while leaving a detailed analysis of the data to be published elsewhere.

METHODS

Sampling

A total of 88 pitfall traps were laid out in a series of 10 transects and were all operated through a whole year (11 May 1997 to 11 May 1998). The traps consisted of a plastic flower pot forming a stable hole in the ground, into which was fitted a removable plastic beaker (diam. 11 cm) with preservative. The latter contained a 2-3% formalin solution mixed ca. 4:1 with ethylene glycol, with a few drops of detergent added. The addition of ethylene glycol served to avoid winter freezing and complete desiccation during summer. The traps were covered by a small wooden roof (12 * 12 cm) in order to prevent rain water filling the traps. Emptying was done every two weeks during the warm season and every four weeks during winter, with fresh preservative supplied on each occasion.

Sites

The transects all started from the top of a dune and were placed in four directions perpendicular to each other. They were not exactly in the four compass directions, however. The deviation was due to the fact that the dunes were shaped by the prevailing WNW-winds. For simplicity we refer to the main compass directions, but in reality the transects are slightly

clockwise displaced (ca. 30°). The length of the transects and the number of traps in each transect varied depending on how the habitats changed in the different directions. We refer to three trap sites, corresponding to three dune tops. Trap site A was in the yellow dunes by the shore of the North Sea. The top of the dune (18 m a.s.l.) was only 5 m from a steep slope to the beach, eroded by the sea. The two trap transects started at this dune top. One ran north, parallel with the coastline at 5-10 m distance; the other ran east, inland from the coast. Sites B and C were further inland, 500 m and 650 m from the coast, respectively, and should be classified as grey dunes, except where winds had broken up the dune and created bare depressions. Both sites started on tops situated 28 m a.s.l. (known locally as Kobbelsbakke and Bøjebakke, respectively). Site B had a covering of low pines of varying density on parts of the N and W transects. Otherwise, most of the transects crossed a mosaic of bare sandy spots, grey dune vegetation, dwarfshrub heath, and mixtures of these vegetation types. The two south transects were extended somewhat further than the others into a low heathland bog.

On one occasion (20 July 1997) we recorded the habitat characteristics of the immediate surroundings (diam. 1 m) of each trap. The vegetation height and the coverage of the dominant plant species (>5%) or unvegetated area was noted. We also estimated the approximate altitude a.s.l. height a.s.l. of each trap. The direction of the slope of each trap was recorded (N, E, S, W) and the steepness of the slope (0 = plain, 1 = weak slope, 2 = steep slope).

For some presentations, the trap sites have been categorised into six main vegetation types, though the surroundings of some traps were intermediate between these:

-White dune: > 50% *Ammophila arenaria* was present, often as a monoculture. Most trap-sites were strongly influenced by shifting sand; white dunes occurred only at site A.

-Grey dune: mixture of *Ammophila arenaria*, *Carex arenaria* and other plants;

-Bare dune: > 50% bare sandy area or covered

with low (~1 cm) lichens or moss;

-Heath: *Empetrum nigrum* was the dominant dwarf shrub; only few places was *Calluna vulgaris* present to any extent;

-Bog: low-lying, more or less flooded during winter and spring. Vegetation consisted of >50% *Sphagnum* with *Molinia coerulea*, *Erica tetralix* or *Narthecium ossifragum*; deep (ca. 10 cm) turf layer;

-Pine plantation (*Pinus mugho*, ½-2 m high, ca. 50 years old) on N and W slopes of site B. The ground was covered with dead needles or a thick layer of moss.

Analysis

Spiders were identified to species level and a matrix with the total catch of each species in each of the 88 traps was produced. A similar matrix was made for the topographical and vegetation recordings (only dominant species included). The two files served as species- and environmental factor files for analysis in the CANOCO 4 program (ter Braak & Smilauer 1998). A canonical correspondence analysis (CCA) was used to compare the species distributions in relation to the traps and habitat factors.

RESULTS

Faunistics

A total of 22200 individuals were identified from 170 species. Several species were recorded for the first time from Denmark: *Evansia merens*, *Porrhomma egeria*, *Walckenaeria capito*, *Sitticus distinguendus*, *Cheiracanthium campestre* and *Micaria lenzi*. Other rare or previously rarely recorded species are: *Maro lepidus*, *Trichopterna thorelli*, *Hypsosinga sanguinea*, *Clubiona genevensis*, *Haplodrassus moderatus*, *Micaria aenea*, *M. dives* and *Synageles venator*.

Canonical correspondence analysis (CCA)

In the sites-plot (Fig. 1A) the species composition of the traps from the yellow dune transects (site A) is clearly distinguished from that of the traps of the two 'inland' sites (sites B and C). The classification according to main vegetation type reveals some separation of the communities of these habitats, but also some intergrading. However, there is a clear gradient along the first CCA-axis, with the more vegetated (and presumably more humid) habitats to the left and the bare, dry habitats to the right. The yellow dune traps and the bare inland dune traps have similar positions along the first axis, probably due to the influence of shifting sand at site A. They are clearly separated along the second axis, which may reflect a humidity gradient. Table 1 presents the overall capture statistics in relation to the habitat types. The yellow dune is the habitat with most exclusive species (i.e. species found only in this habitat). The distinctiveness of the yellow dune spider community is particularly revealed by the dominance of exclusive species: it amounts to nearly 10% against c. 0.8% at most in the other habitats.

Fig. 1B indicates which species and habitat factors are mainly responsible for the differences. *Hypomma bituberculatum*, *Pelecopsis nemoralis*, *Troxochrus scabriculus*, *Clubiona frisia*, *Sitticus distinguendus*, *Micaria dives*, *Tibellus maritimus* and *Marpissa nivoyi* are species that are completely or nearly completely restricted to the yellow dune habitat. Similarly, *Arctosa perita*, *Alopecosa fabrilis*, *A. accentuata*, *Sitticus saltator* and *Aelurillus v-insignitus* are more or less restricted to the bare 'inland' dune habitat. The heath habitat has hardly any species of its own; the dominant species here are mostly habitat generalists that extend their occurrence into the various adjoining habitats, both the drier and the more humid ones. The trap-plot (Fig. 1A) shows a surprising overlap between the traps in the pine plantation and the heath bog. Many of the exclusive species of the pine habitat are tree-dwelling (e.g. *Drapetisca socialis*, *Lepthyphantes obscurus*, *Zygiella atrica*), but only singletons were caught because these species are not recorded well with pitfall traps. The similarity with the bog and heath fauna is due to the dominance of euryhygric and euryoecious generalists like *Centromerita concinna*, *Lepthyphantes mengei*, *Pardosa nigriceps*, *Trochosa*

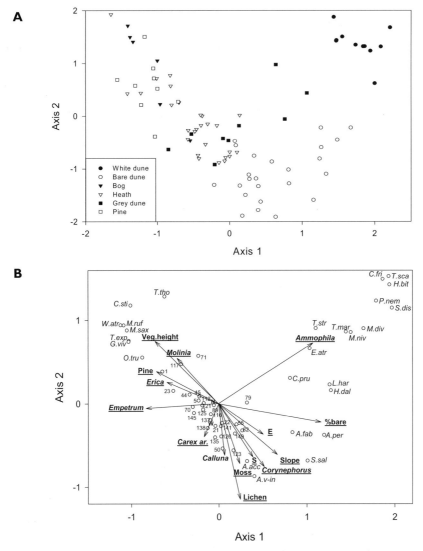

Fig. 1. Canonical Correspondence Analysis of dune spider communities from the Hanstholm Reserve, Denmark. **(A)** Trap plot, traps categorised according to habitat of trap surroundings. **(B)** Species and environmental factor plot. Species with >100 individuals (or >10 if of high indicator value). Abbreviated species names: A.acc *Alopecosa accentuata,* A.fab *Alopecosa fabrilis,* A.per *Arctosa perita,* A.v-in *Aelurillus v-insignitus,* C. fri *Clubiona friesia,* C.sti *Crustulina sticta,* E.atr *Erigone atra,* C.pru *Centromerus prudens,* G.viv *Gongylidiellum vivum,* H.bit *Hypomma bituberculatum,* H.dal *Haplodrassus dalmatensis:* L.har *Leptothrix hardyi,* M.div *Micaria dives,* M.niv *Marpissa nivoyi,* M.ruf *Macrargus rufus,* M.sax *Meioneta saxatilis,* O.tru *Ozyptila trux,* P.nem *Pelecopsis nemoralis,* S.dis *Sitticus distinguendus,* S.sal *Sitticus saltator,* T.exp *Tallusia experta,* T.mar *Tibellus maritimus,* T.sca *Troxochrus scabriculus,* T.str *Thanatus striatus,* T.tho *Trichopterna thorelli,* W.atro *Walckenaeria atrotibialis,* 1 *Ero furcata,* 5 *Euryopis flavomaculata,* 21 *Centromerita bicolor,* 22 *Centromerita concinna,* 23 *Centromerus dilutus,* 44 *Lepthyphantes ericaeus,* 45 *Lepthyphantes mengei,* 50 *Macrargus carpenteri,* 55 *Meioneta rurestris,* 70 *Peponocranium ludicrum,* 71 *Pocadicnemis pumila,* 79 *Stemonyphantes lineatus,* 82 *Tapinocyba praecox,* 89 *Walckenaeria antica,* 112 *Alopecosa pulverulenta,* 117 *Pardosa pullata,* 121 *Trochosa terricola,* 123 *Hahnia nava,* 125 *Agroeca proxima,* 126 *Scotina gracilipes,* 135 *Drassodes cupreus,* 137 *Drassyllus pusillus,* 138 *Gnaphosa leporina,* 141 *Haplodrassus signifer,* 145 *Micaria pulicaria,* 149 *Zelotes longipes.*

Table 1. Capture statistics for the spider sampling program in Hanstholm Reserve, NW Jutland , Denmark (May 1997 – May 1998).

	White dune	Bare	Grey dune	Heath	Bog	Pine	Total
No. pitfall traps	13	24	9	26	7	9	88
No. individuals per trap	157	223	250	334	341	177	
No. species per trap	36.8	35.5	40	42.3	44.7	40.3	
No. species per habitat	102	92	122	88	87	99	170
Exclusive species	11	4	2	8	9	8	
Exclusive species %	9.16	0.07	0.09	0.23	0.50	0.82	

terricola and others. The bog had some rarities but otherwise few specialists (e.g. *Trochosa spinipalpis*), probably due to its limited size. It was dominated by hygrophilic generalists (*Centromerus dilutus*, *Peponocranium ludicrum*, *Pocadicnemis pumila*, apart from those mentioned above).

Slope had some effect on the distribution of spiders (Fig. 1B), but this is probably due to its correlations with % bare ground, because bare habitats tended to occur mainly where the slope was steep and it was difficult to maintain a stable vegetation. The aspect of the slope had very little influence; in particular N- and W-facing slope vectors are situated close to the origin of the CCA-plot. Vegetation height is highly correlated with pines; also the bog vegetation was quite rich and on average higher than the heath and bare dune, especially where there were *Vaccinium* or *Erica*.

DISCUSSION

The recording of 170 spider species in a one-year sampling program from a very restricted area gives no indication of a pauperised fauna in this highly wind-exposed dune system. This number is c. one third of all spider species known in Denmark, in spite of the fact that a very special set of habitats were studied. On the contrary, the high number of new and rare species provides evidence of a high conservation value of the Hanstholm Reserve. Some of

these rarities were captured in the yellow and bare dunes, others in the bog or the heathland. Thus, the mosaic nature of the coastal dune landscape contributes significantly to the biodiversity. This underscores the importance of preserving extensive areas of relatively undisturbed habitat where the natural mosaic is still present, as in the Hanstholm Reserve. Fortunately the reserve is highly protected, e.g. from 1st April to 15th July public access is prohibited.

It remains to be established whether the dune fauna of the western coast of Jutland is generally so rich and diverse. Some of the peculiar habitat types, e.g. the bog, could easily disappear if and when areas are developed for tourist utilization. Because the results of this study particularly emphasise the nature value of the yellow dune, which is the habitat utilised by bathing guests all along the coast, there is a need for studies of the dunes that are more exposed to the wear of tourism. In particular, it is important to establish to what extent the exclusive species of the yellow dune compared to habitat generalist species tolerate such exploitation of the dunes.

ACKNOWLEDGEMENTS

We are grateful to Thy Skovdistrikt for allowing us access to part of the reserve throughout the year. Peter Gajdoš was supported by a grant from the Danish Research Academy.

REFERENCES

Almquist, S. 1973. Spider associations in coastal sand dunes. *Oikos* 24, 444-457.

Bonte, D., Maelfait, J.-P. & Hoffmann, M. 2000. Seasonal and diurnal migration patterns of the spider (Araneae) fauna of coastal grey dunes. *Ekológia (Bratislava)* 19 Suppl. 4, 5-16.

Clausen, I.H.S. 1987. Spiders (Araneae) from Nordmarken on the island of Læsø in Denmark. Faunistic notes, habitat description, and comparison of sampling methods. *Entomologiske Meddelelser* 55, 7-20.

Duffey, E. 1968. An ecological analysis of the spider fauna of sand dunes. *Journal of Animal Ecology* 37, 641-674.

Perttula, T. 1984. An ecological analysis of the spider fauna of the coastal sand dunes in the vicinity of Tvärminne Zoological Station, Finland. *Memoranda pro Societas Fauna Flora Fennica* 60, 11-22.

Ranwell, D.S. 1972. *Ecology of salt marshes and sand dunes.* Chapman and Hall, London.

Schultz, W. 1995. Verteilungsmuster der Spinnenfauna (Arthropoda, Arachnida, Araneida) am Beispiel der Insel Norderney und weitere friesischer Inseln. Thesis, Carl von Ossietzky Universität, Oldenburg.

Schultz, W. & Finch, O.-D. 1996. *Biotoptypenbezogene Verteilung der Spinnenfauna der nordwestdeutschen Küstenregion – Charakterarten, typische Arten and Gefährdung.* Cuvillier Verlag, Göttingen.

ter Braak, C.J.F. & Smilauer, P. 1998. *CANOCO 4.* Centre for Biometry, Wageningen.

European Arachnology 2000 (S. Toft & N. Scharff eds.), pp. 229-236.
© Aarhus University Press, Aarhus, 2002. ISBN 87 7934 001 6
(Proceedings of the 19th European Colloquium of Arachnology, Århus 17-22 July 2000)

The spider fauna of balks

MARIA WOLAK

Department of Zoology, University of Podlasie, ul. Prusa 12, 08-110 Siedlce, Poland
(wolak@ap.siedlce.pl)

Abstract

Results of studies of the spider fauna of three balks and a rye field adjacent to one of them in a mosaic of agrocoenoses in Eastern Poland are presented. Spiders were collected by pitfall trapping and sweep netting in the years 1998 and 1999. In total, 7589 specimens representing 94 species were collected. Spider diversity depended on width and age of the balk and on vegetation structure. More spider species occurred in the older balk, which had fewer plant species but a denser vegetation cover than in the younger one, which was covered with more diverse but less compact vegetation. Seasonal activity of the dominant spider species *Pachygnatha degeeri* was different in the balk and adjacent field. Balks, as important components of mosaic landscapes, play a significant role in enriching spider diversity in arable areas.

Key words: agricultural areas, Araneae, balks, biodiversity, refugial habitats

INTRODUCTION

Agricultural practices decrease the abundance of beneficial arthropods. The role of landscape diversification is to provide reservoirs that are strategically placed and which act as safe havens and sources of immigrants to the crop fields (Sunderland & Samu 2000).

In Eastern Poland the agricultural landscape has maintained its diverse character. It forms a mosaic of small fields (usually between 5 ha and 1 ha), separated by balks and small woods. Balks are narrow (less than 1 m) strips covered with herbs and grasses, slightly raised above the field level. A balk separates two fields and is unmanaged, although it may be affected by the agrotechnical treatments carried out in the adjacent fields. Because of their structural stability and vegetational composition balks might play a similar role for invertebrate animals as waste lands or meadows. Although balks are characteristic components of the Polish agricultural landscape, their importance as refugial habitats is not known.

Many investigations in other European countries have focused on the importance of unmanaged areas within arable lands. According to Barthel & Platcher (1995) a more diverse spider fauna in the agricultural landscape requires: larger field margins, fallow lands as sources of colonization, connectedness of uncultivated areas, and a stable vegetation structure of the margins. Bergthaler (1996) found that already during the first year of succession hedgerows functioned as an important refuge for the invertebrate fauna. Tóth & Kiss (1999) noticed that the presence of margins increased species richness of epigeic spiders in winter wheat fields. Kemp & Barrett (1989) found that predators were more abundant in uncultivated areas, especially successional corridors, than in soybean fields. The establishment of uncropped areas within the agricultural landscape might provide net ecological and economic benefits. Thomas et al. (1991) proposed the creation of 'island' habitats in the farmland. They noticed that during the first year of establishment the new habitats provided overwintering refuge sites for many species of Araneae, Carabidae and Staphylinidae. Experiments with the introduction of sown weed strips into crop fields were undertaken in Switzerland with positive results for the diversity of spiders and other arthropods (Lys & Nentwig 1994; Frank & Nentwig 1995).

Spiders and other invertebrates of balks have not previously been studied in Poland. The aim of this study was to determine the characteristics of balks which influence spider biodiversity. The role of balks as refugial habitats is considered.

MATERIAL AND METHODS
Study area
The spider fauna in three balks and the rye field adjoining one of them in the years 1998-99 was studied. The study area was situated in the village of Zbuczyn, Eastern Poland. Balk I separated a conventionally grown strawberry field and an organically grown wheat field. Balk II separated the strawberry field and organic potato fields. The width of the strawberry field between the two balks was about 6 m. Balks I and II were about 30 years old and about 40 cm wide. Both were covered with diverse vegetation composed of 33 plant species. Vegetation cover in balk I was not very dense. Apart from grass, some high and branched plants (e.g. *Artemisia vulgaris*) were present there. The vegetation in balk II was not as dense as in balk I and was composed mainly of low herbs. Balk III was situated at another farm, about 2 km from balks I and II, and separated two rye fields. Areas of cultivated fields were about 1 ha. The balk was about 80 years old and 70 cm wide. The flora of this habitat was composed of 13 plant species. Grass was a main component of the vegetation. Its long and bent blades formed a compact green mass.

Methods
Samples were taken from April till October in

Table 1. Number of species, Shannon-Weaver and Equitability index for all study sites. B1 balk I, B2 balk II, B3 balk III, C rye field.

	1998			1999	
	B1	B2	B3	B3	C
Species richness	43	39	52	58	45
Shannon index	2.69	2.70	2.25	1.95	1.71
Equitability	0.72	0.74	0.57	0.47	0.45

1998 (in balk II only from May), and from March till October in 1999. Epigeic spiders were collected by pitfall traps (plastic cups of 7 cm upper diameter, containing ethylene glycol and a few drops of detergent). Ten traps with a distance of 2 m between each one were installed at each site. The traps functioned for two weeks of every month. Spiders of the herbaceous layer were collected once a month by a sweep net of 40 cm diameter. One sampling consisted of 4 x 25 sweeps. The material was identified to species level or, in the case of young individuals, to genus level. For analysis Shannon-Weaver diversity index (H) and species similarity Sørensen index (So) were applied.

RESULTS
In total, spiders of 15 families and 94 species represented by 7589 specimens were recorded. Numbers of spider species caught by pitfall traps and sweep net in all study areas, as well as Shannon index and equitability, are given in Table 1. An analysis of epigeic spiders showed differences between balks. In balks I and II similar numbers of species were found in pitfall traps (34 and 30, respectively) but the total number of individuals was much higher in balk I (748) than in balk II (293). As regards the number of specimens, *Pachygnatha degeeri* and *Pardosa palustris* were most abundant in balk I, while *Oedothorax apicatus* was most numerous in balk II. The spider fauna of balk III was more diverse and a larger number of species (46) and individuals (1761) was recorded than in the two other balks. *P. degeeri, Centromerita*

Balk I 1998

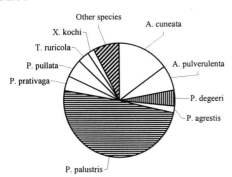

Balk II 1998

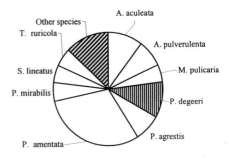

Balk III 1998

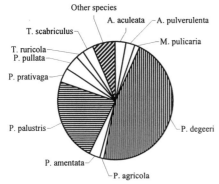

Balk III 1999

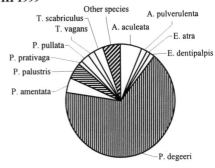

Rye field 1999

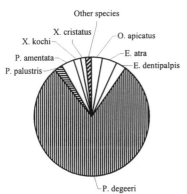

Fig. I. Dominance structure of the spider communities of balks and adjacent rye field as revealed by pitfall trapping in May.

bicolor, Centromerus sylvaticus, Pardosa palustris and *O. apicatus* were the most abundant species. In the herbaceous layer, *Mangora acalypha* was the most numerous in all studied balks. The number of individuals was twofold higher (50) in balk I than in balk II (25).

A comparison of balk III in 1998 and 1999 revealed only small differences in epigeic spider composition but the number of individuals was much higher (2469) in the second year of the study. Foliage-dwelling spider communities were different in the two years studied. In

1999 twice the number of species was recorded. In the rye field agrobiont species were most numerous (*P. degeeri, O. apicatus, Erigone dentipalpis* and *E. atra*). The field and the adjacent balk had 21 species in common.

The dominance structure of epigeic spiders recorded in May (Fig. 1) revealed that the spider fauna of balk III was quite different from that of balks I and II. Moreover, there were differences between balk I and II, in spite of their closeness. In balk III, *P. degeeri* dominated in both years (47% and 67% respectively). In 1998 the second dominating species was *P. palustris* (23%), whereas in the next year of study this spider accounted for only 5%. In the rye field the dominant spider species was also *P. degeeri* (79%).

The Sørensen similarity index between balk I and balk II was high (0.72). Lower values were recorded for balk I and III (0.65), and for balk II and III (0.66). Only small changes of spider communities in balk III in 1998 and 1999 were noted (So = 0.70). The value of this index for the rye field and balk III was low (0.48).

The number of individuals of the dominant spider species *P. degeeri* during the 1999 study season showed two peaks in the balk but only one peak in the adjacent rye field (Fig. 2); the species disappeared completely after harvest of the field.

DISCUSSION

This study revealed that the richness of the spider fauna of balks depends on the width of the balk and on its vegetation structure. While width is important for the general abundance of spiders, vegetation structure may influence spider composition. In balk III, which was almost twice as wide as the other two, a far larger number of specimens was found. The size of the area itself seems to be important: the wider the balk, the more space to inhabit. Barthel & Platcher (1995, 1996) stated that the size of uncultivated margins is an important factor for spider composition. Narrower margins might be more affected by mechanical treatments of the neighbouring cultivated areas

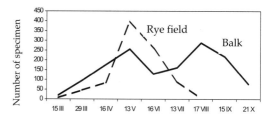

Fig. 2. Seasonal activity of *Pachygnatha degeeri* in balk III and adjacent rye field in 1999.

than wider margins, where an undisturbed central zone can be established. Stability of balks depends, among other things, on their age. More spider species were recorded in the 80-year-old-balk than in the 30-year-old ones. Similar relations were found by Frank & Nentwig (1995) in sown weed strips within cultivated fields and in field boundaries. Although strips were only one or two years old they were similar to balks in their width (1.5 m) and plant composition (25 weed species). The older strips contained significantly more spider species than the younger ones, and also the number of species in the ten-year-old boundary was significantly higher than in weed strips.

Vegetation structure is one of the essential factors for spiders. Dense and compact vegetation provides shade and humidity, appropriate conditions especially for small spiders of the families Linyphiidae and Theridiidae. These spiders, exposed to loss of water more than larger ones, find hiding places in numerous, tiny spaces of such habitats (Duffey 1975). Linyphiids were most abundant in balk III in which the vegetation provided good support for sheet webs. Apart from the Linyphiidae, spiders of five other families were registered here. According to Frank & Nentwig (1995) the dense grass cover reduces the mobility of spiders. In contrast, in balks I and II spiders representing 11 and 9 families respectively were found. It seems that less compact and more diverse vegetation provide better conditions for actively hunting and larger web-building spiders (more insolation and support for three-dimensional webs). Scheidler (1990) who studied the influence of habitat structure and vegetation architecture on spiders, found higher spider densities on broad plants with many branches than on plants with only few branches and a rather narrow architecture. He stated also that for some spiders the plant architecture may play the dominant role, while for other spider species special structures like leaves or buds are most important. Frank & Nentwig (1995) also recorded a more diverse spider fauna in areas covered with richly structured vegetation.

Although balk III was characterized by a lower Shannon diversity index than balks I and II, records of *Porrhomma errans*, *Allomengea vidua* and *Argiope bruennichi* in this site should be stressed. The two first species are known only from a few localities in Poland (Staręga 1983; Staręga & Stankiewicz 1996). *A. bruennichi*, although frequent in Western Europe (Nyffeler 1982; Barthel & Platcher 1995, 1996; Jmhasly & Nentwig 1995; Bergthaler 1996) and even in Western Poland (Radkiewicz & Jerzak 1991; Kuźniak 1998), is a relatively rare species in Eastern Poland and protected in the whole country.

Balks may serve as overwintering sites, especially for agrobiont spiders. Under a vegetation layer reduced temperature extremes can be recorded (Thomas et al. 1992) which might increase winter survival compared to open fields. This seems to be essential, especially for arthropods as spiders which are not able to dig well into the soil, and which showed the lowest overwintering abundance in cereal areas (Lys & Nentwig 1994). A pattern of seasonal activity of *Pachygnatha deggeeri* in balk III (Fig. 2) showed an increase in the number of individuals twice a year, in spring and in autumn. Similar data were obtained by Palmgren (1974). The second peak should not be related to the sexual activity of this species but rather to the migration to overwintering sites (Toft 1979). A number of individuals of this spider had only one peak in spring in the rye field and then it disappeared completely during harvest. The spiders might have migrated from the field to the balk, as this species is known as eurychro-

nous and adult specimens overwinter (Schaefer 1977). Frank & Nentwig (1995) considered *P. degeeri* a species with distinct preference for the field boundary. They listed as agrobiont species *Erigone atra*, *E. dentipalpis*, *Oedothorax apicatus*, *Pardosa agrstis* and *P. palustris* showing dispersal from weed strips into fields (i.e. in spring more individuals were in weed strips, but as the season progressed more individuals were in the fields). This means that these spider species overwintered in the strips. Also Łuczak (1979) pointed out that densely vegetated biotopes provided particularly attractive sites for hibernation.

Some spider species probably originating from the adjoining balk were found in the field, so one can speculate that spiders migrate from balks to the fields and vice versa. Because of the small width of the balks, the extent of this migration need not be large. Jmhasly & Nentwig (1995) observed that in many cases the spider density and web cover were larger close to the strips, so they supposed that weed strips might increase the spider population in the adjacent winter wheat.

The conclusion of this study is that balks, in spite of their small width, are important habitats for the spider fauna within Polish agricultural areas. Their presence might increase biodiversity in arable areas, because their spider fauna is much more diverse than in the cultivated fields. It is highly recommended to maintain or restore the Polish type of agricultural landscape with plenty of balks, woods, water bodies and small fields. Sustainable agriculture promotes diversified agroecosystems and extensive production in small-area farms so that it could be profitable. Gravesen & Toft (1987) mentioned the creation of grass strips along field margins and hedges as one way of 'manipulating' predator populations for productive and environmental benefits.

REFERENCES

Barthel, J. & Platcher, H. 1995. Distribution of foliage-dwelling spiders in uncultivated areas of agricultural landscapes (Southern

Bavaria, Germany) (Arachnida, Araneae). In: *Proceedings of the 15th European Colloquium of Arachnology* (V. Rüžička ed.), pp. 11-21. Institute of Entomology, České Budějovice.

Barthel, J. & Platcher, H. 1996. Significance of field margins for foliage-dwelling spiders (Arachnida, Araneae) in an agricultural landscape of Germany. *Revue Suisse de Zoologie* Hors série 2, 45-59.

Bergthaler, G. 1996. Preliminary results on the colonisation of a newly planted hedgerow by epigeic spiders (Araneae) under the influence of adjacent cereal fields. *Revue Suisse de Zoologie* Hors serie 2, 61-70.

Duffey, E. 1975. Habitat selection in man-made environments. *Proceedings of the 6th International Arachnological Congress*, pp. 53-67. Amsterdam.

Frank, T. & Nentwig, W. 1995. Ground dwelling spiders (Araneae) in sown weed strips and adjacent fields. *Acta Oecologica* 16, 179-193.

Gravesen, E. & Toft, S. 1987. Grass fields as reservoirs for polyphagous predators (Arthropoda) of aphids (Homopt., Aphididae). *Journal of Applied Entomology* 104, 461-473.

Jmhasly, P. & Nentwig, W. 1995. Habitat management in winter wheat and evaluation of subsequent spider predation on insect pests. *Acta Oecologica* 16, 389-403.

Kemp, J. & Barret, G. 1989. Spatial patterning: impact of uncultivated corridors on arthropod populations within soybean agroecosystems. *Ecology* 70, 114-128.

Kuźniak, S. 1998. Tygrzyk paskowany (*Argiope bruennichi*) w Przemęckim Parku Krajobrazowym. *Biuletyn Parków Krajobrazowych Wielkopolski* 3, 136-139.

Lys, J.-A. & Nentwig, W. 1994. Improvement of overwintering sites for Carabidae, Staphylinidae and Araneae by strip-management in a cereal field. *Pedobiologia* 38, 238-242.

Łuczak, J. 1979. Spiders in agrocoenoses. *Polish Ecological Studies* 5, 151-200.

Nyffeler, M. 1982. Field studies on ecological role of the spiders as insect predators in Agroecosystems (Abandoned grassland, meadows and cereal fields). Thesis, Swiss Federal Institute of Technology, Zürich.

Palmgren, P. 1974. Die spinnenfauna Finnlands und Ostfennoskandiens IV. Argiopidae, Tetragnathidae und Mimetidae. *Fauna Fennica* 24, 36-69.

Radkiewicz, J. & Jerzak, L. 1991. O stanowiskach pająka tygrzyka paskowanego na obszarze Polski. *Chrońmy Przyrodę Ojczystą* 47, 89-91.

Schaefer, M. 1977. Winter ecology of spiders (Araneida). *Zeitschrift für Angewandte Entomologie* 83, 113-134.

Scheidler, M. 1990. Influence of habitat structure and vegetation architecture on spiders. *Zoologischer Anzeiger* 225, 333-340.

Staręga, W. 1983. Wykaz krytyczny pająków (Aranei) Polski. *Fragmenta Faunistica* 27, 149-268.

Staręga, W. & Stankiewicz, A. 1996. Beiträge zur Spinnenfauna einiger Moore Nordostpolens. *Fragmenta Faunistica* 39, 345-361.

Sunderland, K.D. & Samu, F. 2000. Effects of agricultural diversification on the abundance, distribution, and pest control potential of spiders: a review. *Entomologia Experimentalis et Applicata* 95, 1-13.

Thomas, M., Wratten, S.D. & Sotherton, N. 1991. Creation of 'island' habitats in farmland to manipulate populations of beneficial arthropods: predator densities and emigration. *Journal of Applied Ecology* 28, 906-917.

Thomas, M., Mitchel, H. & Wratten, S.D. 1992. Abiotic and biotic factors influencing the winter distribution of predatory insects. *Oecologia* 89, 78-84.

Toft, S. 1979. Life histories of eight Danish wetland spiders. *Entomologiske Meddelelser* 47, 22-32.

Tóth, F. & Kiss, J. 1999. Comparative analyses of epigeic spider assemblages in northern Hungarian winter wheat fields and their adjacent margins. *Journal of Arachnology* 27, 241-249.

Appendix. List of spider species and numbers collected on study habitats using pitfall traps (PF) and sweeping net (SN).

Spider species	Balk 1 PF (1998)	SN	Balk 2 PF	SN	Balk 3 PF	SN	Balk 3 PF (1999)	SN	Rye field PF	SN
Mimetidae										
Ero furcata (Villers)			1							
Theridiidae										
Achaearanea lunata (Clerck)	1							1		
Enoplognatha ovata (Clerck)		2		2				1		
Neottiura bimaculata (Linnaeus)							2	7		
Robertus arundineti (O.P. Cambridge)	3				4		6	2		
Robertus lividus (Blackwall)		1								
Theridion impressum (L.Koch)		4						1		6
Theridion mystaceum (L. Koch)				3						
Theridion sp. (juv.)		5				4		7		13
Linyphiidae										
Allomengea vidua (L. Koch)	1			1						
Araeoncus humilis (Blackwall)				1	11		8			
Bathyphantes gracilis (Blackwall)	2				14		27		8	
Bathyphantes parvulus (Westring)							5		2	
Centromerita bicolor (Blackwall)	53		35		165		118		10	
Centromerus aequalis (Westring)									1	
Centromerus sylvaticus (Blackwall)	72		5		157		15			
Dicymbium brevisetosum (Locket)							8			
Diplocephalus latifrons (O.P. C.)							2	1		
Diplostyla concolor (Wider)							4			
Erigone atra (Blackwall)	5				17		69		154	
Erigone dentipalpis (Wider)	9		1	1	57		97	1	161	
Erigone longipalpis (Sundevall)					1					
Gnathonarium dentatum (Wider)						1	1			
Lepthyphantes angulipalpis (Westring)						1	1			
Lepthyphantes pallidus (O.P. C.)						1	1			
Linyphia triangularis (Clerck)						1	1			
Macrargus rufus (Wider)								2		
Meioneta rurestris (C.L.Koch)				5	4		6		16	
Meioneta tenera (Menge)					1		2			
Micrargus herbigradus (Blackwall)										1
Microlinyphia pusilla (Sundevall)				1				2		
Microlinyphia sp. (juv.)		2								
Microneta viaria (Blackwall)									1	
Neriene clathrata (Sundevall)					1		1	1		
Oedothorax apicatus (Blackwall)	44		66		137		169		634	
Oedothorax fuscus (Blackwall)					4		3		2	
Oedothorax gibbosus (Blackwall)							1			
Oedothorax retusus (Westring)					9		7		2	
Pocadicnemis pumila (Blackwall)							1			
Porrhomma errans (Blackwall)							1			
Porrhomma pallidum (Jackson)							1			
Porrhomma pygmaeum (Blackwall)									1	
Silometopus reussi (Thorell)										
Stemonyphantes lineatus (Linnaeus)	24		40		3		15		7	
Tapinocyba insecta (L. Koch)					3		1			
Tiso vagans (Blackwall)			2		5		35		1	
Troxochrus scabriculus (Westring)	12		1		53		43		2	
Walckenaeria obtusa (Blackwall)							1			
Tetragnathidae										
Pachygnatha clercki (Sundevall)	3				19		12		5	
Pachygnatha degeeri (Sundevall)	177		37		747		1420	1	888	
Tetragnatha extensa (Linnaeus)				1		1		4		
Tetragnatha pinicola (L. Koch)		3				1		14		3
Araneidae										
Aculepeira ceropegia (Walckenaer)		2		4		3	1	1	1	1
Araneus diadematus (Clerck)				1			1			
Araneus marmoreus (Clerck)						1			1	
Araneus quadratus (Clerck)		4						2		
Araneus sp. (juv.)								1		
Araniella cucrbitina (Clerck)		5			4					1
Araniella opisthographa (Kulczyński)					2					1

Appendix, continued.

Spider species	1998 Balk 1 PF	SN	Balk 2 PF	SN	Balk 3 PF	SN	1999 Balk 3 PF	SN	Rye field PF	SN
Araniella sp. (juv.)						3				
Argiope bruennichi (Scopoli)										
Cyclosa oculata (Walckenaer)	1									
Hypsosinga pygmea (Sundevall)										
Mangora acalypha (O. P. Cambridge)	16		5		6		26		10	
Lycosidae										
Alopecosa aculeata (Clerck)	1		4		13		27		2	
Alopecosa cuneata (Clerck)	22		1		4		2			
Alopecosa pulverulenta (Clerck)	12		5		31		26			
Alopecosa sp. (juv.) Simon	31		12		6		4			
Pardosa agrestis (Westring)	16		3		4		29			
Pardosa agricola (Thorell)	1				3				6	
Pardosa amentata (Clerck)	40		20		32		34		31	
Pardosa lugubris (Walckenaer)					1				2	
Pardosa palustris (Linnaeus)	115		11		140		161		79	
Pardosa prativaga (L. Koch)	15		3		14		31		3	
Pardosa pullata (Clerck)	7				19		24		2	
Pardosa sp. (juv.)	8	1			13		6		3	
Pirata latitans (Blackwall)							2			
Pirata piraticus (Clerck)					1					
Pirata tenuitarsis (Simon)	1								1	
Trochosa ruricola (De Geer)	33		7		15		12		17	
Trochosa terricola (Thorell)	2				8					
Trochosa sp. (juv.)	6		3	3	8		3			
Xerolycosa nemoralis (Westring)					3					
Pisauridae										
Pisaura mirabilis (Clerck)	1	3	2		6		2		1	
Hahnidae										
Hahnia pusilla (C. L. Koch)	1									
Dictynidae										
Argenna subnigra (O.P. Cambridge)									1	
Cicurina cicur (Fabricius)	1									

Spider species	1998 Balk 1 PF	SN	Balk 2 PF	SN	Balk 3 PF	SN	1999 Balk 3 PF	SN	Rye field PF	SN
Liocranidae										
Agroeca brunnea (Blackwall)				3						
Agroeca proxima (O.P. Cambridge)	1			1						
Phrurolithus festivus (C. L. Koch)			1		1					
Gnaphosidae										
Drassyllus lutetianus (L. Koch)	1			1						
Drassyllus pusillus (C. L. Koch)	1				1		1			
Haplodrassus sylvestris (Blackwall)			1		1					
Micaria pulicaria (Sundevall)	2		3		3		3			
Micaria sp. (juv.)			2		2					
Zelotes sp. (juv.)	2				2					
Zoridae										
Zora spinimana (Sundevall)							2			
Philodromidae										
Philodromus aureolus (Clerck)										1
Philodromus sp. (juv.)		1								
Thomisidae										
Ozyptila scabricula (Westring)	1									
Ozyptila trux (Blackwall)					2					
Ozyptila sp. (juv.)		1			1					
Xysticus cristatus (Clerck)	16		1		10		3			
Xysticus kochii (Thorell)	2		1		6		8		10	
Xysticus sp. (juv.)	3	4	7	4	8	3	13		12	14
Salticidae										
Euophrys frontalis (Walckenaer)							1			
Total of individuals	798		318		1790		2543		2140	
Total of species	43		39		52		58		45	

European Arachnology 2000 (S. Toft & N. Scharff eds.), pp. 237-242.
© Aarhus University Press, Aarhus, 2002. ISBN 87 7934 001 6
(Proceedings of the 19th European Colloquium of Arachnology, Århus 17-22 July 2000)

The epigeic spider fauna of single-row hedges in a Danish agricultural landscape

SØREN TOFT[1] & GABOR L. LÖVEI[2]

[1]*Department of Zoology, University of Aarhus, Building 135, DK-8000 Århus C, Denmark*
(soeren.toft@biology.au.dk)
[2]*Danish Institute of Agricultural Science, Research Centre Flakkebjerg, Department of Crop Protection,*
DK-4200 Slagelse, Denmark (gabor.lovei@agrsci.dk)

Abstract

To characterise arthropod biodiversity supported by one type of non-cultivated habitat patches in a cultivated Danish landscape, ground-active arthropods were collected by pitfall traps in three single-row hedgerow types near Bjerringbro, central Jutland, Denmark. Three each of hawthorn (*Crategus monogyna*), rowan (*Sorbus intermedia*), or white spruce (*Picea glauca*) hedgerows were sampled twice yearly, in early (June) and late summer (late August) using 20 pitfall traps per habitat patch (10 in centre, 10 at edge). A total of 71 spider species were identified among 1422 individuals: 33 species (515 individuals) were found in hawthorn hedges, 52 species (653 individuals) in rowan, and 48 species (254 individuals) in spruce. Principal Component Analysis clearly separated the spider assemblages by tree species of the hedge. There was no difference between edge and central traps neither at the assemblage nor at the species level. Most species captured were characteristic of non-cultivated land (*Diplostyla concolor*, *Diplocephalus latifrons*, *Oxyptila praticola*, *Zelotes pusillus*), or associated with more permanent grassland rather than cultivated crops (*Pardosa prativaga*, *Pachygnatha degeeri*). Species typical of cultivated agricultural fields were infrequent (*Erigone atra*, *Bathyphantes gracilis*, *Meioneta rurestris*, *Oedothorax apicatus*) or missing altogether (eg. *Araeoncus humilis*). Thus, the narrow single-row hedges were faunistically very little influenced by the cultivated matrix habitat enclosing them.

Key words: biodiversity, hedgerows, spider assemblages, habitat affinity, rowan, hawthorn, spruce

INTRODUCTION

Biodiversity in the agricultural landscape has traditionally been associated with non-cultivated areas, even though a considerable number of species can be found on cultivated land (Meszaros 1984; regarding spiders: Toft 1989; Vangsgaard 1996). With the increasing human pressure on such habitats, the significance of biodiversity that can be supported by agricultural land is bound to increase. Little is known, however, of the level of biodiversity that can be supported, and how uncultivated areas interact with the cultivated land. An improved understanding of these is necessary for a more efficient management of biodiversity as well as of the ecosystem services they provide (Daly 1999). The study of predatory arthropods such as spiders is obviously relevant in this respect.

Because of methodological differences, studies of hedgerow spiders can be grouped into those considering the fauna of higher vegetation (see Nährig 1991; Ysnel & Canard 2000), and those considering the ground-active fauna (Blick 1989; Henatsch & Blick 1993; Møller-Nielsen 1990; Reinke & Irmler 1994; Bergthaler 1996). Hedgerows obviously add to the biodiversity of the arable landscape through the shrub and tree living species that would otherwise be absent. Hedges also add, however, to the fauna of ground-active spiders, especially woodland species. Earlier studies on the epigeic fauna indicated that: A) hedgerows are inhabited by an assemblage distinct from those of fields as well as woodlands (Blick 1989; Møller-Nielsen 1990; Reinke & Irmler 1994); B) they are often dominated by wood-land or forest-edge species rather than by field species (Blick 1989); and C) the spider assem-blages show high species richness because woodland and open-land species mix into a composite assemblage. Reinke & Irmler (1994) found only one type of hedge with dominance of the field species *Erigone atra* (Bl.): low hedges on sandy soil, poor in ground vegetation. Though some of the hedges studied previously were quite young, they were composed of at least three rows of trees or shrubs and thus of a considerable width (Blick 1989: 6-8 m; Møller-Nielsen 1990: 4.5 - 5 m; Bergthaler 1996: 3.5 m). As soon as the canopy closes, a 'forest floor' habitat is created in the centre of the hedge, the ground being covered by leaf litter rather than grasses or herb vegetation. The spider fauna of old single-row hedges in which the woodland character of the habitat is not obvious, was ex-pected to be more influenced by influx from surrounding fields, but has to our knowledge not previously been analysed.

We studied the assemblages of ground-active spiders in three types of single-row hedges in central Jutland, Denmark, and from the analysis of one year's results, we describe a fairly high degree of species richness. More-over, the fauna was not highly influenced by that of the fields. Apart from an unpublished thesis (Møller-Nielsen 1990) on three 13 year old 3-rowed hedges, this is the first account of hedgerow spiders in Denmark.

STUDY AREA AND METHODS

Our study took place in the area of Bjerringbro, central Jutland, Denmark. Nine hedgerows were selected for study; three of hawthorn (*Crategus monogyna*) (localities: Lådnehøje,

Aidt1, Aidt2), three of rowan (*Sorbus interme-dia*) (Sahl, Gerning, Aldrup), and three of spruce (two of white spruce *Picea glauca*, one of sitka spruce *Picea sitchensis*) (Sahl1, Sahl2, Lådnehøje). They were old, well established hedges of the single-row type. The total width was variable as some hedges had a wide grass covered base (total hedge width, range of means at the three locations: hawthorn, 2.4 – 4.0 m; rowan, 2.6 – 5.0 m; spruce, 3.0 – 3.3 m). One hedge differed in some of these respects: the sitka hedge was planted alternately in two rows, it was rather dense with branches to the ground, thus creating a vegetation-free centre. Other peculiarities of single hedges will be mentioned along with the results. The nine sample locations were at a distance of 200 m – 10 km from each other, within a 4 km x 10 km area. Individual hedgerows were sampled twice yearly, in early (June 1999) and late sum-mer (early September 1999), using 20 pitfalls per habitat patch. Ten of the traps were set at the edge, and ten in the centre of the hedgerow, at a distance of 10 m between individual traps. Neighbouring traps alternated with respect to position. Edge traps were situated only 10-20 cm from the adjacent cultivated field.

Individual pitfall traps were plastic cups of 10 cm diameter, filled with about 200 ml of 70% ethylene glycol solution and a drop of deter-gent. Traps were sunk into the ground so that their rim was level with the soil surface. Every trap was covered with a galvanised metal square cover to protect the trap contents from rain and disturbance by frogs, birds or small mammals. Traps were set for one week at a time. Trap catches were sieved in the field and transferred into glass vials containing 70% ethyl alcohol. Trap catches with small mam-mals or frogs as well as displaced or raised traps (33 of a total of 360) were not included in the evaluation. In the laboratory, the samples were sorted under a microscope, and ground beetles and spiders were put into separate vials and stored in 70% ethyl alcohol until identifica-tion.

Identification was made by ST. Spider tax-

Fig. 1. The results of Principal Component Analysis on untransformed spider capture data in nine hedgerows near Bjerringbro, Jutland, Denmark. E: edge traps, C: centre traps. Aidt1, Aidt2, Ald, Ger, Låd, Sahl1, Sahl2 are abbreviations for sites (see text).

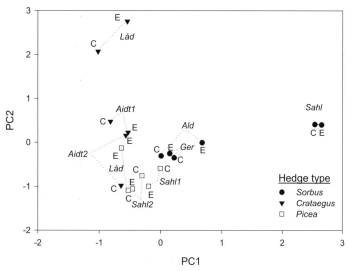

Fig. 2. Habitat affinity relations of the 19 most common spider species according to PCA (Cf. Fig. 1.).
For abbreviations of species names: see appendix 1.

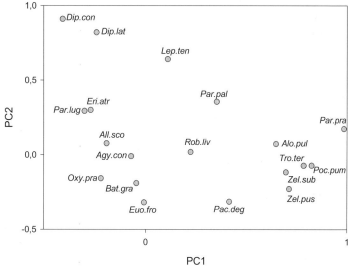

onomy follows Platnick (1993). The total (early + late summer) catches of 10 traps (centre or edge) of each hedgerow were summed, giving 18 sites. The species/sites matrix was analysed with a Principal Component Analysis (PCA) on untransformed numbers using the CANOCO 4 program (ter Braak & Smilauer 1998; cf. Jongman et al. 1987). All species were included in the analysis, but only the more abundant ones (≥ 10 individuals) are shown in the species plot.

RESULTS

In the captures from the nine hedgerows, a total of 72 spider species were identified among 1563 individuals (Appendix 1). In *Sorbus* hedgerows, 722 individuals of 52 species were captured, followed by *Crataegus* (548 individuals of 33 species). *Picea* hedgerows had the smallest number of individuals (293 spiders) but not the lowest species richness (49 species). The Principal Component Analysis gave a clear separation of hedges according to tree species (Fig. 1): *Sorbus* hedges were distinguished from the other two types along the first axis; *Crataegus* and *Picea* hedges separated along the second axis.

Fig. 2 shows which species were responsible for the separations of hedge types: *Pardosa prativaga*, *Trochosa terrestris*, and *Pocadicnemis pumila* were especially abundant in the *Sorbus* hedges. *Diplostyla concolor*, *Diplocephalus latifrons* were particularly abundant in the *Crataegus* hedges. The dominant species of the *Picea* hedges were more abundant elsewhere (notice that the origin of the graph is at the lower left); *Euophrys frontalis* were mostly found here.

Centre and edge positions of the same hedge had a high similarity in their spider faunas (Fig. 1). For the two most abundant species, *Pardosa prativaga* and *Diplostyla concolor*, centre and edge catches were tested against each other (Kruskal-Wallis ANOVA) for single hedges where they occurred in high numbers (spring period only). No significant differences were found (*P. prativaga* in *Sorbus* hedges: Aldrup, H = 1.93, P = 0.17; Sahl, H = 0.13, P = 0.72; Gerning, H = 0.03, P = 0.87. D. *concolor* in *Crataegus* hedge, Lådnehøje, H = 0.44, P = 0.51).

One each of the *Sorbus* and *Crataegus* hedges deviated in their spider fauna from the other two of their kind (Fig. 1). In both cases the hedgerow trees had low hanging branches that shaded the central part of the hedge, which was more or less devoid of ground vegetation. On top of that the *Sorbus* hedge at Sahl was bordered to one side by a permanent meadow. The two-rowed sitka hedge at Lådnehøje did not differ from the other spruce hedges.

DISCUSSION
Species richness
Møller-Nielsen (1990) recorded 105 spider species (among c. 9000 individuals) in 70 pitfall traps catching continuously for four months (June - September) in three young hedges. This is equivalent to the 100 species found by Blick (1989) in Germany in an only slightly larger sample. We found 72 species among only 1423 individuals collected within two weeks. All together these results indicate a relatively high species richness of the hedgerow spider fauna.

A study of newly planted hedgerows in Austria (Bergthaler 1996), even though longer in duration and collecting more spiders, found fewer species (44). Several of the dominant species in Bergthaler's (1996) study were associated with agricultural fields, but species characteristic of forests also started to appear.

Habitat relations
The habitat relations of the species have been evaluated from the data compiled by Hänggi et al. (1995) and other sources. The eight most abundant species on the dominance list can be characterized as follows: *Pardosa prativaga* is widespread in many types of open habitats, particularly in marshes and meadows; occurrence in cereal fields depending on adjacent permanent grassland. As a matter of fact, the high dominance of *P. prativaga* was mainly due to an extreme abundance of this species at the *Sorbus* hedge at Sahl; the adjacent permanent meadow was obviously the source habitat of the species. *Diplostyla concolor* occurs in both forested and open habitats, reaching highest abundance in forests, forest edges, and hedgerows. *Oxyptila praticola* is a species of forest edges, hedgerows and open shrubland. *Diplocephalus latifrons* is primarily a forest and forest-edge species, with no association to agricultural land. *Zelotes pusillus* is species of dry grassland and heaths. *Erigone atra* is a habitat generalist though most abundant in open habitats. It is often the most abundant species in Northern European agricultural fields (Sunderland 1987; Blick et al. 2000). *Pachygnatha degeeri* is widespread in meadows, grass fields and sometimes in agricultural fields. *Bathyphantes gracilis* is also a habitat generalist abundant in meadows, marshes and agricultural fields. Only *E. atra*, *B. gracilis* and sometimes *P. degeeri* may be among the dominant agricultural species (Sunderland 1987; Blick et al. 2000), which may owe their occurrence in the hedges to the adjacent agricultural habitats. Thus, the hedge fauna was dominated partly by species of permanent open habitats, partly by species characteristic of forest edges and

even by true forest species. The most abundant agrobionts were rather low in dominance. Several species that are typical of Danish and European cereal fields (cf. Toft 1989; Vangsgaard 1996; Blick et al. 2000) were infrequent (e.g. *Meioneta rurestris*, *Oedothorax apicatus*, *Lepthyphantes tenuis*) or missing altogether (e.g. *Araeoncus humilis*).

In conclusion, we found that the spider fauna of single-row hedgerows depended on the tree species of the hedge, with *Sorbus* and *Picea* hedges (both with dense grassy herb layer) richest in species. Habitat structure is plausibly an important determinant of species richness, but we do not have data on that. Spruce hedgerows were the poorest habitat for ground beetles (Lövei, unpublished), but spruce hedgerows had a thick, grassy ground vegetation layer, and this could be the reason for a higher spider species richness in spruce vs. rowan hedges. Thus, Asteraki et al. (1992) found that herbicide-removal of hedgerow vegetation affected linyphiid spiders. The most unexpected finding was that in spite of their narrowness, the spider fauna of these hedgerows was dominated by species originating from permanent open-land habitats, forest or forest-edge species, and species characteristic of agricultural fields had low dominance, were rare or absent, even at the edge of the hedgerows.

ACKNOWLEDGEMENTS

This work was done within the framework of the 'Boundaries in the landscape' project, financed by the Danish Research Agency. We thank Jette Lilholt, Ulla Sandberg, and Lena Pedersen for technical assistance, and Henny Rasmussen for help with the illustrations.

REFERENCES

Asteraki, E.J., Hanks, C.B. & Clements, R.O. 1992. The impact of the chemical removal of the hedge-base flora on the community structure of carabid beetles (Col., Carabidae) and spiders (Araneae) of the field and hedge bottom. *Journal of Applied Entomology* 113, 398-406.

Bergthaler, G.J. 1996. Preliminary results on the colonization of a newly planted hedgerow by epigeic spiders (Araneae) under the influence of adjacent cereal fields. *Revue Suisse de Zoologie* vol. hors serie, 61-70.

Blick, T. 1989. Die Beziehungen der epigäischen Spinnenfauna von Hecken zum Umland. *Mitteilungen der Deutschen Gesellschaft für Allgemeine und Angewandte Entomologie 7*, 84-89.

Blick, T., Pfiffner, L. & Luka, H. 2000. Epigäische Spinnen auf Ackern der Nordwest-Schweiz im mitteleuropäischen Vergleich (Arachnida: Araneae). *Mitteilungen der Deutschen Gesellschaft für Allgemeine und Angewandte Entomologie 12*, 267-276.

Daily, G. C. 1999. Developing a scientific basis for managing Earth's life support systems. *Conservation Ecology* 3(2): 14. [online] URL: http://www.consecol.org/vol3/iss2/art14

Hänggi, A., Stöckli, E. & Nentwig, W. 1995. *Lebensräume Mitteleuropäischer Spinnen.* CSCF/SZKF, Neuchâtel, Switzerland.

Henatsch, B. & Blick, T. 1993. Zur Tageszeitlischen Laufaktivität der Laufkäfer, Kurzpflügelkäfer und Spinnen in einer Hecke und einer angrenzenden Brachfläche (Carabidae, Staphylinidae, Araneae). *Mitteilungen der Deutschen Gesellschaft für Allgemeine und Angewandte Entomologie.* 8, 529-536.

Jongman, R.H.G., ter Braak, C.J.F. & van Tongeren, O.F.R. 1987. *Data analysis in community and landscape ecology.* Pudoc, Wageningen.

Meszaros, Z. (ed.) 1984. Results of faunistical studies in Hungarian maize stands. *Acta Phytopathologica Academiae Scientiarum Hungariae* 19, 65-90.

Møller-Nielsen, L. 1990. En undersøgelse af løbebille- (Carabidae) og eddekoppefaunaen (Araneae) i trerækkede løvtræshegn. Unpublished MSc Thesis, University of Aarhus.

Nährig, D. 1991. Charakterisierung und Bewertung von Hecken mit Hilfe der Spinnenfauna. *Zoologische Beiträge* N.F. 33, 253-263.

Platnick, N.I. 1993. *Advances in spider taxonomy 1988-1991.* New York Entomological Society.

Reinke, H.-D. & Irmler, U. 1994. Die Spinnenfauna (Araneae) Schlegswig-Holsteins am Boden und in der bodennahen Vegetation. *Faunistisch-Ökologische Mitteilungen* Suppl. 17, 1-148.

Sunderland, K.D. 1987. Spiders and cereal aphids in Europe. *Bulletin SROP/WPRS* 10 (1), 82-102.

ter Braak, C.J.F. & Smilauer, P. 1998. *CANOCO 4*. Centre for Biometry, Wageningen.

Toft, S. 1989. Aspects of the ground-living spider fauna of two barley fields in Denmark: species richness and phenological synchronization. *Entomologiske Meddelelser* 57, 157-168.

Vangsgaard, C. 1996. Spatial distribution and dispersal of spiders in a Danish barley field. *Revue Suisse de Zoologie* hors serie 2, 671-682.

Appendix 1. List of spiders captured in three different hedgerow types near Bjerringbro, central Jutland, Denmark. The captures from three locations per hedgerow type were combined. Numbers in parentheses indicate dominance rank of species in hedgerow type (only species with ≥ 10 individuals). Underlining marks the abbreviations used on Fig. 2.

Species	Hedgerow type Sorbus	Crataegus	Picea	Total
Gnaphosidae				
Haplodrassus signifer (C.L.K.)	1	0	2	3
Zelotes latreillei (Simon)	2	1	1	4
Zelotes longipes (L.K.)	0	1	0	1
Zelotes pusillus (C.L.K.)	(3) 28	3	(3) 21	52
Zelotes subterraneus (C.L.K.)	(9.5) 10	1	5	16
Zelotes sp. juv.	4	0	6	10
Micariidae				
Micaria pulicaria (Sund.)	1	1	1	3
Clubionidae				
Clubiona terrestris Westr.	0	1	0	1
Clubiona sp. juv.	2	0	1	3
Liocranidae				
Agroeca proxima (O.P.-C.)	0	0	3	3
Zoridae				
Zora spinimana (Sund.)	0	0	3	3
Thomisidae				
Oxyptila praticola (C.L.K.)	(4) 25	(2) 74	4	103
Thomisidae sp. juv.	1	1	2	4
Salticidae				
Euophrys frontalis (Walck.)	2	0	8	10
Lycosidae				
Pardosa agrestis (Westr.)	0	0	2	2
Pardosa palustris (L.)	7	4	1	12
Pardosa pullata (Cl.)	2	2	1	5
Pardosa prativaga (L.K.)	(1) 353	(4) 41	(1) 44	438
Pardosa amentata (Cl.)	0	1	1	2
Pardosa lugubris (Walck.)	3	6	2	11
Pardosa nigriceps (Thor.)	1	0	0	1
Alopecosa pulverulenta (Cl.)	8	4	6	18
Trochrosa ruricola (Deg.)	1	0	0	1
Trochosa terricola Thor.	8	1	5	14
Lycosidae sp. juv.	18	8	8	34
Agelenidae				
Tegenaria atrica C.L.K.	0	1	0	1
Agelenidae sp. juv.	0	1	0	1
Hahniidae				
Hahnia nava (Bl.)	0	0	4	4
Theridiidae				
Crustulina guttata (Wider)	1	0	0	1
Robertus lividus (Bl.)	7	3	9	19
Robertus neglectus (O.P.-C.)	1	0	0	1
Tetragnathidae				
Pachygnatha degeeri Sund.	(8) 11	3	(4) 17	31
Pachygnatha clercki Sund.	0	1	0	1

Species	Hedgerow type Sorbus	Crataegus	Picea	Total
Linyphiidae				
Walckenaeria acuminata Bl.	4	0	4	8
Walckenaeria antica (Wider)	0	0	1	1
Walckenaeria dysderoides (Wider)	2	0	6	8
Dicymbium brevisetosum Locket	2	1	0	3
Dismodicus bifrons (Bl.)	3	0	0	3
Oedothorax apicatus (Bl.)	3	1	1	5
Gonatium rubens (Bl.)	2	0	1	3
Pocadicnemis pumila (Bl.)	(5) 17	0	(5) 12	29
Minyriolus pusillus (Wider)	0	0	1	1
Troxochrus scabriculus (Westr.)	2	0	0	2
Tiso vagans (Bl.)	0	1	1	2
Micrargus herbigradus (Bl.)	2	0	0	2
Gongylidiellum vivum (O.P.-C.)	1	0	0	1
Tapinocyba praecox (O.P.-C.)	0	0	1	1
Tapinocyba insecta (L.K.)	1	0	1	2
Tapinocyba pallens (O.P.-C.)	4	2	1	7
Savignia frontata (Bl.)	2	0	0	2
Diplocephalus cristatus (Bl.)	0	0	1	1
Diplocephalus latifrons (O.P.-C.)	2	(3) 47	4	53
Diplocephalus picinus (Bl.)	1	0	0	1
Erigone atra (Bl.)	(6) 14	(5) 18	4	36
Erigone dentipalpis (Wider)	1	0	0	1
Agyneta conigera (O.P.-C.)	7	2	3	12
Agyneta subtilis (O.P.-C.)	3	0	1	4
Meioneta rurestris (C.L.K.)	0	0	1	1
Microneta viaria (Bl.)	4	0	0	4
Ostearius melanopygius (O.P.-C.)	0	0	1	1
Porrhomma microphthalmum (O.P.-C.)	1	0	0	1
Centromerus sylvaticus (Bl.)	3	0	4	7
Centromerus dilutus (O.P.-C.)	0	0	2	2
Centromerita bicolor (Bl.)	1	2	0	3
Poeciloneta globosa (Wider)	1	0	0	1
Stemonyphantes lineatus (L.)	0	1	0	1
Bathyphantes gracilis (Bl.)	8	(8) 10	(6) 12	30
Bathyphantes parvulus (Westr.)	3	0	1	4
Diplostyla concolor (Wider)	(2) 57	(1) 252	(2) 36	345
Bolyphantes alticeps (Sund.)	0	1	0	1
Lepthyphantes mengei Kulcz.	5	0	2	7
Lepthyphantes tenuis (Bl.)	(9.5) 10	(7) 13	2	25
Lepthyphantes insignis O.P.-C.	1	0	1	2
Lepthyphantes ericaeus (Bl.)	1	0	1	2
Lepthyphantes angulipalpis(Westr.)	1	0	0	1
Neriene clathrata (Sund.)	0	1	0	1
Allomengea scopigera (Grube)	(7) 12	(6) 14	2	28
Linyphiidae sp. juv.	44	23	21	88
Total	722	548	293	1563

European Arachnology 2000 (S. Toft & N. Scharff eds.), pp. 243-252.
© Aarhus University Press, Aarhus, 2002. ISBN 87 7934 001 6
(Proceedings of the 19th European Colloquium of Arachnology, Århus 17-22 July 2000)

The riparian spider fauna (Araneae) of the river Gaula, Central Norway: implications for conservation efforts

KJETIL AAKRA

Museum of Natural History and Archaeology, Department of Natural History, NTNU, N-7491 Trondheim, Norway Present address: *Regionsmuseet for Indre Midt-Troms, Boks 82, N-9059 Storsteinnes, Norway* (kjetil.aakra@consultant.online.no)

Abstract

The riparian spider fauna of the river Gaula near Trondheim, Central Norway, has been investigated in spring/summer 1994 by means of pitfall traps, limited sieving and hand picking. Sites varied from sand and gravel banks completely devoid of vegetation, to sand/silt deposits with or without vegetation and *Salix triandra/Alnus incana* forests on sandy soil. These special habitats turned out to harbour a very special and remarkable spider fauna. Two species were recorded for the first time in Fennoscandia: *Arctosa stigmosa* and *Caviphantes saxetorum*, both being the northernmost records in Europe. Two species were new to Norway: *Singa nitidula* and *Myrmarachne formicaria*, and 10 species were new regional records.

The spider fauna of the river banks consists of the following ecological groups based on their known ecology in Norway: (1) riparian species, both psammophilous and lithophilous, (2) hygrophilous species, (3) pioneer species, (4) ubiquitous species, and (5) accidental guests from nearby habitats. Group (1) represents about 29% of the total number of species. The isolated occurrence of rare riparian species is paralleled by riparian beetles (Coleoptera). Central Norway harbours one of the richest and most diverse riparian faunas of Northern Europe.

The river Gaula is protected by law against hydroelectric exploitation but about 65 km of the river banks have been destroyed by the construction of flood preventing walls and both riparian beetles and spiders have disappeared from many sites. Most of the riparian species are vulnerable to human disturbance and changes in the flooding pattern of the river and four are included in the Norwegian Red List with the status 'Declining, care demanding'. Protection measures include reduction of silt/sand removal, adaptation of flood preventing walls to suit the riparian fauna, reduction of leisure activities on particularly vulnerable localities and the designation of certain particularly valuable sites as Nature Reserves or Protected Sites. Clear-cutting and management of vegetation in order to maintain the open areas should also be considered.

Key words: Riparian spiders, central Norway, biological conservation

INTRODUCTION

Undisturbed river bank ecosystems are rare and threatened in Europe today (e.g. Plachter 1986), including Norway (DN 1999a), and it has become increasingly urgent to document the fauna and flora restricted to and dependent upon this habitat type. The flora is in general well-known, while the study of the riparian invertebrate fauna is still in its infancy (Tischler 1993). With regard to spiders, some of the first ecological studies on the riparian spider fauna were the works by Schenkel (1932), Knülle (1953) and Casemir (1962). Most of the recent research has been carried out in central parts of continental Europe (e.g. Beyer 1995; Beyer & Grube 1997; Framenau 1995; Hugenschütt 1996;

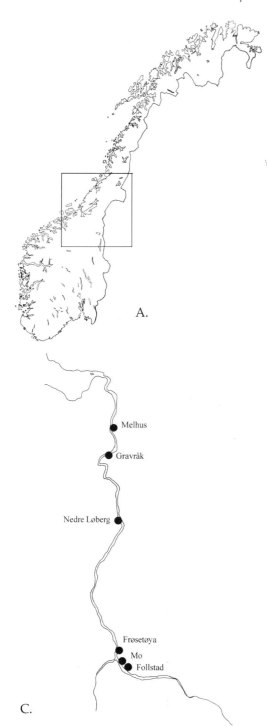

A.

B.

C.

Fig. 1. (A) The study area in Central Norway. **(B)** Major rivers in the Trøndelag counties. **(C)** Study sites along the river Gaula.

Ruzicka & Hajer 1996; Steinberger 1996), while in northern Europe riparian invertebrates have been virtually neglected. Only beetles have received any attention in Fennoscandia (e.g. Andersen 1970, 1983, 1984, 1997; Andersen & Hanssen 1993). These investigations revealed that Central and North Norwegian river systems harbour a rich and extremely varied beetle fauna consisting of many rare species. Several of these are highly sensitive to human influence (Andersen & Hanssen 1994) and are now red-listed (DN 1999b). In fact, Norway has one of the richest riparian beetle faunas of northern Europe (Andersen 1983) and it was suspected that other riparian invertebrates would also exhibit a large diversity (Andersen & Hanssen 1994).

Andersen & Hanssen (1994) used, amongst other methods, pitfall traps in their study and an interesting by-product was a comparatively large number of spiders. These were determined by the present author and the results were found to have important implications for future management and for the conservation status of the river system of Gaula.

MATERIALS AND METHODS

Most of the material presented here originated from pitfall traps consisting of plastic cups, 66 mm wide and 95 mm deep, dug into the substrate. The catching period started on 13 May 1994 and lasted to 21 August, although the traps were not operated for exactly the same period on all localities. No attempt will therefore be made to compare the catches from the different sites. The traps were emptied on 24 May, 20 June and 24 July.

Limited hand collecting and sieving were also carried out on a few other localities. These methods were not standardised and information on sampling intensity is not available. The resulting material is presented in Table 2.

STUDY AREA

The study sites were situated along the lower part of the river Gaula in Central Norway (Fig. 1). This is the least disturbed major river in the region. Total length of the riverbanks is 114.5 km, catchment area is 3651 km², and about 70% of the total area lies between 300-900 m. above sea level (Stølen 1992). Below the Gaulfossen waterfall the river is heavily influenced by agriculture and construction activities. Some 65 km of river bank has been destroyed by the construction of flood-preventing walls, including as much as 50% below Gaulfossen (Stølen 1992). A highway and railroad track follow the river for most of its length, exerting a considerable influence on the river's topography. The river is protected against hydroelectric development (Stølen 1992).

The lower portion of the river was chosen for study because it contains the highest number of substrate types and habitats. The following localities along the river were studied (Fig. 1C), the first four using pitfall traps and the other three using manual collection methods (habitat descriptions and quality assessment with regard to beetles from Andersen & Hanssen 1994):

Mo (south)
Heterogeneous river banks consisting of gravel and stone deposits near the river and sand/silt deposits close to a former tributary. The heterogeneous state of the deposits is caused by previous disturbance through digging activities. There are numerous temporary ponds and open sand banks separated by small stands of *Alnus* spp., *Salix triandra* and *Myricaria germanica*. Ten pitfall traps were used.

Mo (north)
Open stone and gravel deposits. Higher elevations covered partly with *Rhacomitrium* spp. and *Astragalus* spp. tufts on sand and gravel. Sand/silt deposits mostly overgrown by *Salix triandra*. Seven pitfall traps were used.

Frøsetøya
Large deposit containing open stone and gravel in the lower part of the 'island', partly grown with *Myricaria germanica*. Smaller occurrences of sand/silt deposits with a little vegetation. Vegetation-free sand banks are currently small and possibly not large enough to sustain many riparian beetle species (Andersen & Hanssen 1994). Eight pitfall traps were used.

Follstad
Large, open and sparsely grown sand deposit situated next to a forested deposit at higher elevation. Several grain sizes of sand present and patches with some silt. Frequently flooded. Number of pitfall traps was not recorded.

Gravråk
Old deposit with sand banks at higher elevations surrounded by deciduous forests (*Alnus* and *Salix triandra*). The upper part contains a small pond created by previous extraction of material. The open sand banks are old, at least 40 – 50 years.

Melhus
Relatively large sand/silt deposits situated between the river and a tributary. A well-grown *Salix* forest to one side. One of the best investigated sites along the river with regard to beetles and one of the most important localities for riparian species. Extraction of material in the 1970s and –80s did not destroy the locality, but the reduced elevation of the river itself has led to the sand banks becoming overgrown. It is therefore likely that many riparian species are close to extinction here.

Table I. Spiders collected by pitfall traps at river Gaula, Norway. <u>Underlined</u>: riparian species.

Species	♂/♀	New to	Locality
Araneidae			
<u>Singa nitidula L. Koch, 1844</u>	1/0	Norway	Mo (south)
Clubionidae			
Clubiona lutescens Westring, 1851	1/1	Central Norway	Frøsetøya, Melhus, Mo (south)
Gnaphosidae			
<u>Micaria nivosa L. Koch, 1866</u>	5/8		Follstad, Frøsetøya, Mo
Hahniidae			
Cryphoeca silvicola (C. L. Koch, 1834)	1/0		Mo (north)
Linyphiidae			
Allomengea scopigera (Grube, 1859)	3/1		Mo
Bathyphantes gracilis Blackwall, 1841	1/0		Mo (north)
B. nigrinus (Westring, 1851)	1/2		Mo (north)
<u>Caviphantes saxetorum (Hull, 1916)</u>	0/1	Fennoscandia	Mo (south)
Dicymbium nigrum (Blackwall, 1834)	0/1		Mo (north)
D. tibiale (Blackwall, 1836)	6/0		Mo
Diplocephalus cristatus Blackwall, 1833	4/0		Mo
Diplostyla concolor (Wider, 1834)	0/1		Mo (north)
Dismodicus bifrons (Blackwall, 1841)	2/1		Follstad, Mo (south)
Erigone atra Blackwall, 1833	11/2		Follstad, Mo
E. dentipalpis (Wider, 1834)	3/0		Follstad, Frøsetøya, Mo (south)
Erigonella hiemalis (Blackwall, 1841)	1/0	Central Norway	Follstad
Gongylidium rufipes (Linnaeus, 1758)	2/1		Follstad, Mo
Hypomma bituberculatum (Wider, 1834)	4/1		Follstad
Kaestneria pullata (O.P.-Cambridge, 1863)	2/2	Central Norway	Follstad, Mo (south)
Leptohoptrum robustum (Westring, 1851)	2/2		Mo (nord), Mo (south)
Neriene clathrata (Sundevall, 1830)	0/1		Frøsetøya
<u>Oedothorax agrestis (Blackwall, 1853)</u>	1/0	Central Norway	Mo (nord)
O. apicatus (Blackwall, 1850)	3/0	Central Norway	Mo
<u>O. retusus (Westring, 1851)</u>	192/158		Follstad, Frøsetøya, Mo
Pocadicnemis pumila (Blackwall, 1841)	2/0		Follstad, Mo (north)
<u>Porrhomma pygmaeum (Blackwall, 1834)</u>	1/4	Central Norway	Mo
Savignia frontata Blackwall, 1833	4/7		Mo
<u>Silometopus reussi (Thorell, 1871)</u>	3/0	Central Norway	Mo (north)
<u>Tapinocyba insecta (L.Koch, 1869)</u>	0/1	Central Norway	Mo (south)
Tenuiphantes alacris (Blackwall, 1853)	1/0		Follstad
<u>Troxochrus scabriculus (Westring, 1851)</u>	8/13	Central Norway	Frøsetøya, Mo
Walckenaeria cuspidata (Blackwall, 1833)	2/0		Mo (north)
W. nudipalpis (Westring, 1851)	0/1		Mo (north)
<u>W. vigilax (Blackwall, 1853)</u>	2/0	Central Norway	Follstad
Lycosidae			
<u>Arctosa cinerea (Fabricius, 1777)</u>	4/0		Mo
<u>A. stigmosa (Thorell, 1875) X</u>	31/12	Fennoscandia	Follstad, Frøsetøya, Mo
<u>Pardosa agricola (Thorell, 1856)</u>	73/24		Follstad, Frøsetøya, Mo
P. amentata (Clerck, 1757)	75/35		Follstad, Frøsetøya, Mo
P. lugubris (Walckenaer, 1802)	1/0		Mo
P. riparia (C. L. Koch, 1833)	1/1		Frøsetøya
Trochosa terricola Thorell, 1856	4/4		Follstad, Mo (syd)
Salticidae			
Myrmarachne formicaria (De Geer, 1778)	1/0	Norway	Mo (north)
Thomisidae			
Xysticus audax (Schrank, 1803)	1/0		Mo (south)
X. cristatus (Clerck, 1757)	1/0		Follstad
X. ulmi (Hahn, 1832)	1/0		Mo (south)
TOTAL	460/285		

Table 2. Spider collected by sieving and hand picking at river Gaula, Norway. <u>Underlined</u>: riparian species.

Species	♂/♀	Locality
Clubionidae		
Clubiona lutescens Westring, 1851	1/1	Melhu
Linyphiidae		
Bathyphantes nigrinus (Westring, 1851)	1/0	Gravråk
Erigone atra Blackwall, 1833	0/1	Nedre Løberg
E. dentipalpis (Wider, 1834)	6/3	Gravråk, Melhus, Nedre Løberg
Hypomma bituberculatum (Wider, 1834)	3/2	Melhus, Nedre Løberg
Leptorhoptrum robustum (Westring, 1851)	0/1	Melhus
Neriene peltata (Wider, 1834)	1/0	Gravråk
<u>Oedothorax agrestis (Blackwall, 1853)</u>		Gravråk, Melhus
<u>O. retusus (Westring, 1851)</u>	14/39	Gravråk, Melhus, Nedre Løberg
<u>Porrhomma pygmaeum (Blackwall, 1834)</u>	1/0	Nedre Løberg
<u>Troxochrus scabriculus (Westring, 1851)</u>	0/6	Gravråk, Nedre Løberg
Lycosidae		
<u>Arctosa cinerea (Fabricius, 1777)</u>	1/1	Gravråk
<u>A. stigmosa (Thorell, 1875)</u>	1/1	Melhus
<u>Pardosa agricola (Thorell, 1856)</u>	3/2	Gravråk, Melhus
P. amentata (Clerck, 1757)	4/2	Gravråk
Tetragnathidae		
Pachygnatha clercki Sundevall, 1823	0/1	Gravråk
TOTAL	34/61	

Nedre Løberg

Large open sand deposits surrounded by *Myricaria germanica*. Mostly coarse sand but with some finer deposits along the vegetation belt. Beyond the *Salix* stands there are stone and gravel deposits. A few flats of silt close to an artificial gravel formation. Previous extraction of material is evident but a few riparian beetles have managed to re-establish populations here.

RESULTS

Altogether 745 specimens belonging to 45 species from 8 families were collected by pitfall traps (Table 1), and 98 specimens belonging to 16 species from 4 families (Table 2) were taken by manual methods. The latter method yielded only two species not taken by the pitfall traps. Linyphiids dominated the material, being represented by 30 species, followed by lycosids with 7 species and thomisids with 3 species. All other families were represented by a single species. In terms of individuals the linyphiids made up some 62% of the pitfall material and the lycosids some 36%. The dominance of the linyphiids was largely due to the preponderance of one species, *Oedothorax retusus*, which made up 47% of the total pitfall material. The species was also the most numerous in the manually collected mate-

rial. Apart from *Pardosa amentata*, *P. agricola* and *Arctosa stigmosa*, the other species were taken in relatively small numbers.

A great number of faunistic and ecologically interesting species were discovered, including two new to Fennoscandia, two new to Norway and 10 not previously recorded from the central parts of Norway. All but one of the species new to Norway/Fennoscandia (*Myrmarachne formicaria*) are riparian. As many as 13 species may be classified as riparian species (following the definition of Andersen 1983). The most important new records are presented in Aakra (2000).

The following ecological groups of species may be recognised (based on published and unpublished Norwegian material), although it should be noted that the habitat preferences of some species may differ from that recorded in other regions of Europe:

- Riparian species, both psammophilous (sand preferents) and lithophilous (gravel preferents). They include *Arctosa cinerea*, *A. stigmosa*, *Pardosa agricola*, *Singa nitidula*, *Micaria nivosa*, *Caviphantes saxetorum*, *Oedothorax agrestis*, *O. restusus*, *Silometopus reussi*, *Tapinocyba insecta*, *Troxochrus scabriculus* and *Walckenaeria vigilax..* With the exception of *A. stigmosa*, *C. saxetorum* and *S. nitidula* these species have also been found in other habitat types (mainly sandy beaches) in southern parts of Norway, but in Central Norway comparable habitats are scarce or absent and the species are therefore considered riparian within the region.

- Hygrophilous species, which may be found in a variety of moist habitats, including *Clubiona lutescens, Allomengea scopigera, Bathyphantes gracilis, B. nigrinus, Gongylidium rufipes, Hypomma bituberculatum, Savignia frontata, Pardosa amentata* and *Trochosa terricola*.
- Pioneer species, *Erigone atra* and *E. dentipalpis*.
- Ubiquitous species, which are found in a wide variety of habitats in Norway, such as *Xysticus cristatus, Walckenaeria cuspidata* and *W. nudipalpis*.
- Accidental guests from nearby habitats, mainly forest species like *Pardosa lugubris s. str., Cryphoeca silvicola, Dicymbium tibiale* and *Tenuiphantes alacris*.

Group (1) represents about 29% of the total number of species. In other words, the riparian fauna of Gaula consists of a comparatively large number of species which are unlikely to be found in other habitats in the region. It is also obvious that the river banks, including the *Alnus* and *Salix triandra* forests, are important for a wide range of hygrophilous species.

DISCUSSION

Studies suggest that riparian spiders have behavioural adaptations to cope with frequent inundations in what is a very variable and harsh environment (Cooke & Merrett 1967; Siepe 1985). The continental studies also show that river banks harbour a very rich and diverse spider fauna with a large number of extremely rare species, many of which are not or only occasionally encountered in other habitats (Steinberger 1996).

The results presented here fully corroborate this notion. From a comparatively modest spider material, 29% were classified as riparian and at least five of them are rare or uncommon in Norway with no, or only very few, previous records. The vulnerability of these species is best highlighted by examining the species' habitat requirements, distribution and general

conservation status in some detail. The noteworthy species may be divided into two groups: species rare and uncommon throughout their known range; and more frequently recorded species which are, however, rarely found in central Norway where comparatively few appropriate habitats are available.

Arctosa cinerea

The known distribution of this species in Norway has been described by the author (Aakra 2000). Distribution in central Norway, including older records, is rather extensive. There is also one older, unverified record from coastal sand dunes in south-eastern Norway (Strand 1898). Unfortunately, the species has disappeared from at least one known locality along Gaula, a serious indication of the species' vulnerability to habitat changes (see Framenau 1995). In Sweden, the species occurs on sandy lake and sea shores north of Norrbotten (Holm 1947). It is only known from coastal parts of southern and central Finland and sandy inland habitats of northern Finland (Krogerus 1932; Palmgren 1939). The species is widely distributed in Europe at least as far south as Italy, but it has clearly been declining in numbers in at least Switzerland and Germany (see Framenau 1995).

A. cinerea is strongly psammophilous, preferring sites with a mixture of sand and shingle (Framenau 1995) in both coastal sand dunes as well as along rivers and by estuaries. It was given the Red List category 'Declining, care demanding' by Aakra & Hauge (2000).

Arctosa stigmosa

This is perhaps the most surprising arachnological discovery in recent times in Norway, being the first record from Fennoscandia and the northernmost in Europe. A single specimen has also been found by the river Orkla to the west of Gaula. It is likely that *A. stigmosa* can be found in the other major river systems in the Trøndelag counties, e.g. where *A. cinerea* has been found outside saline habitats.

The species was previously known from scattered localities in central parts of continental Europe, the closest known occurrences are from Baltikum (Relys 1994; Mikhailov 1997) and Poland (J. Kupryjanowicz pers. comm.). Known European records are from France, Romania, Bulgaria, Hungary, Slovakia, southern Germany, Switzerland (Denis 1937; Buchar 1968; Maurer & Hänggi 1990; Blick & Scheidler 1991; Blick et al. 2000; Framenau 1995; Steinberger 1996), and the species ranges eastwards through southern Siberia (Mikhailov 1997), China and Korea to Japan (Paik 1994 sub *A. subamylacea* (Bösenberg & Strand, 1906). In continental parts of Europe it is very rare and local (Buchar & Thaler 1995; Steinberger 1996).

Little is known of the ecology of this species which is apparently confined to sand and gravel-covered river banks. It is almost certainly not halotolerant like *A. cinerea*. It does not appear to construct a burrow in the sand, but probably spends the day in litter close to the river banks. Preliminary studies suggests that *A. stigmosa* is nocturnal (Aakra 2000). The species is probably just as vulnerable to habitat deterioration as *A. cinerea* and was given the same Red List category by Aakra & Hauge (2000).

Caviphantes saxetorum

Another first record for Fennoscandia. Based on available literature *C. saxetorum* is rare throughout its range. The closest known occurrence is Scotland (Cooke & Merrett 1967; Locket et al. 1974) and the species is otherwise known from central parts of continental Europe (Georgescu 1973; Wunderlich 1975, 1979; Maurer & Hänggi 1989; Steinberger 1996; Thaler 1993), Poland (Starega 1972), Russia (Mikhailov 1997) and North America (Crawford 1990).

The principal habitat appears to be sand banks along rivers (Locket et al. 1974; Steinberger 1996) but there are also records from dry ruderal sites (Wunderlich 1975; Maurer & Hänggi 1989). Apparently the species occupies the spaces between rocks and sand which would classify it as microcavernicolous (Cooke & Merrett 1967; Wunderlich 1979). Presence of sand seems to be a requirement for this species, probably with

high moisture levels. The species was given the status 'Declining, care demanding' in the Norwegian Red List (Aakra & Hauge 2000).

Myrmarachne formicaria

New to Norway and the northernmost record in Europe. This species is probably not a riparian element. In Sweden it is known to occur as far north as Värmland (L.J. Jonsson pers. comm.), whereas in Finland it is only known from the southern coastal region (Palmgren 1943).

Heimer & Nentwig (1991) report the species to be found in a variety of habitats: The current record is in accordance with Tullgren (1944) who indicated a preference for spaces beneath flat rocks on warm beaches.

Singa nitidula

This is the first and hitherto only known record from Norway. The species is apparently rare in both Sweden (L.J. Jonsson and T. Kronestedt pers. comm.) and Finland (Palmgren 1974) where it occurs from Uppland and Dalarne and north to the end of the Botnian Bay, respectively. It was recently placed on the Swedish Red List under 'Data Deficient' (Gärdenfors 2000) and the Norwegian Red List (Aakra & Hauge 2000) as 'Declining, care demanding'. *S. nitidula* is usually found in low vegetation and litter along streaming water (Palmgren 1974) and can therefore be classified as riparian.

Species commonly found along the river

A relatively large number of species are common along the rivers of central Norway (this paper, unpublished data), but for several of them (e.g. *Silometopus reussi,Tapinocyba insecta, Troxochrus scabriculus, Walckenaeria vigilax*) there are no other records in this area and they may be restricted here to river banks. Other species are almost certainly found in other habitats as well, but preliminary data suggests that sand and gravel banks along rivers constitute their main habitat in central Norway (e.g. *Micaria nivosa, Oedothorax agrestis, O. retusus, Pardosa agricola*). It is therefore obvious that the river banks are important not only for rare and stenotopic species, but also for more wide-spread and eurytopic species. Focus should therefore not be restricted to the rare and faunistically interesting species, but extended to the whole river bank community and its role in the ecosystem.

Conservation of the riparian invertebrate fauna

Although the river Gaula is protected by law against hydroelectric developments, other human activities pose a severe threat to the riparian invertebrate fauna. Specifically, the river banks are not protected against construction of flood preventing walls, leisure activities or other common human disturbance factors. New roads are planned which will be situated close to the river. Furthermore, the nearby flood-influenced forests and meadows are often chopped down or converted to agricultural land, a practice which deprives the riparian species of their winter quarters (Andersen & Hanssen 1994). In other words, the legislated protection of the river systems does not, in practice, fulfil its purpose for the riparian invertebrates. These animals are dependent upon the river's regular flooding, which is in contrast to the human interest in taming the river by the construction of flood preventing walls. The continued survival of the riparian species depends on future management decisions. Several solutions to this problem were proposed by Andersen & Hanssen (1994) , including:
- cessation of sand and silt extraction,
- placement of flood reducing walls only when the effect on deposition of sand and silt will be minimal,
- creation of suitable habitats for the species by removing parts of the forests and alluvial vegetation.

In light of the rare and remarkable species recently discovered along the river, I propose a few additional conservation measures:
- designation of the largest sand bank areas as Nature Reserves,
- prohibition of extensive leisure and construction activities in these sites,
- monitoring procedures and further research

along Gaula and other major river systems in order to assess the status and range of the rare species.

Whether the rare riparian invertebrate species along Gaula can be preserved for future generations remains to be seen. Given its uniqueness in a north European context, the number of rare and vulnerable species (both beetles and spiders) found there, and the apparent decline of some species in recent years, it is obvious that such precautions cannot be implemented too soon. If no actions are taken by the local and/or national authorities, the future of species like *Arctosa cinerea, A. stigmosa* and *Caviphantes saxetorum* may seem bleak, indeed.

ACKNOWLEDGEMENTS.

Many thanks to Johan Andersen for fruitful and inspiring discussions regarding riparian invertebrates and their conservation; to Dr. Torbjørn Kronestedt for confirming the identity of *Arctosa stigmosa,* and to Oddvar Hanssen and Frode Ødegaard for providing the material and information on the localities along Gaula. I also thank Søren Toft, Domir DeBakker and Ambros Hänggi for their helpful suggestions which improved the manuscript considerably.

REFERENCES

Andersen, J. 1970. Habitat choice and life history of Bembidiini (Col., Carabidae) on river banks in Central and Northern Norway. *Norsk entomologisk Tidskrift* 17, 17-65.

Andersen, J. 1983. Towards an ecological explanation of the geographical distribution of riparian beetles in western Europe. *Journal of Biogeography* 10, 421-435.

Andersen, J. 1984. Gaula, et vassdrag med en unik elvebreddfauna (Gaula, a river system with a unique riparian fauna). *Insekt-Nytt* 9, 21-27.

Andersen, J. 1997. Habitat distribution of riparian species of Bembidiini (Col., Carabidae) in South and Central Norway. *Fauna Norvegica* Serie B 44, 11-25.

Andersen, J. & Hanssen, O. 1993. Geographical distribution of the riparian species of the tribe Bembidiini (Col., Carabidae) in South and Central Norway. *Fauna Norvegica* Serie B 40, 59-69.

Andersen, J. & Hanssen, O. 1994. Invertebratfaunaen på elvebredder - ett oversett element. 1. Biller (Coleptera) ved Gaula i Sør-Trøndelag. *NINA Oppdragsmelding* 326, 1-23.

Beyer, W. 1995. Untersuchungen zur Spinnenfauna (Araneida) im überflutungsbeeinflussten Deichvorland des Unteren Odertales. Diplomarbeit, FB Biol., FU Berlin.

Beyer, W. & Grube, R. 1997. Einfluss des Überflutungsregimes auf die epigäische Spinnen- und Laufkäferfauna an Uferabschnitten im Nationalpark 'Unteres Odertal' (Arach.: Araneida, Col.: Carabidae). *Verhandlungen der Gesellschaft für Ökologie* 27, 349-356.

Blick, T. & Scheidler, M. 1991. Kommentierte Artenliste der Spinnen Bayerens (Araneae). *Arachnologische Mitteilungen* 1, 27-80.

Blick, T. & A. Hänggi, unter Mitarbeit von K. Thaler (2000). Checkliste der Spinnentiere Deutschlands, der Schweiz und Österreichs (Arachnida: Araneae, Opiliones, Pseudoscorpiones, Scorpiones, Palpigradi). Vorläufige Version vom 7. Juli 2000. http://www.arages.de/checklisten.html.

Buchar, J. 1968. Zur Lycosidenfauna Bulgariens (Arachn., Araneae). *Věstník československé Společnosti zoologické* 32, 116-130.

Buchar, J. & Thaler, K. 1995. Die Wolfspinnen von Österreich 2: Gattungen *Arctosa, Tricca, Trochosa* (Arachnida, Lycosidae). *Faunistisch-Tiergeographische Übersicht.* Carinthia II 185, 481-498.

Casemir, H. 1962. Spinnen vom Ufer des Altrheins bei Xanten/Niederrhein. Gewässer Abwässer 30/31, 7-35; Düsseldorf.

Cooke, J. A. L. & Merrett, P. 1967. The rediscovery of *Lessertiella saxetorum* in Britain (Araneae, Linyphiidae). *Journal of Zoology* 151, 323-328.

Crawford, R.L. 1990. Discovery of *Caviphantes saxetorum* in North America; Status of *Scironis tarsalis* (Araneida, Linyphiidae). *Journal of Arachnology* 18, 235-236.

Denis, J. 1937. Une station nouvelle de *Dolomedes plantarius* et remarques sur *Arctosa stigmosa*. *Bulletin de la Societe D'Historie naturelle de Toulouse* 71, 451-456.

DN (Direktoratet for Naturforvaltning) 1999a. *Kartlegging av naturtyper - verdisetting av biologisk mangfold. March 1999.* DN-håndbok. Direktoratet for Naturforvaltning.

DN (Direktoratet for Naturforvaltning) 1999b. *Nasjonal rødliste for truete arter i Norge 1998. Norwegian Red List 1998.* DN-rapport 1999-3, 1-161. Direktoratet for Naturforvaltning.

Framenau, V. 1995. Populationsökologie und Ausbreitungsdynamik von *Arctosa cinerea* (Araneae, Lycosidae) in einer alpinen Wildflublandschaft. Diplomarbeit am Fachbereich Biologie der Phillips-Universität Marburg. Fachgebiet Naturschutz.

Georgescu, M. 1973. Le developpement postembryonnaire de l'espece cavernicola *Lessertiella dobrogica* Dumitrescu & Miller (Araneida, Micryphantidae). *Travaux Institutului de Speologie 'Emil Racovita'* 12, 63-73.

Gärdenfors, U. 2000. *Rödlistade arter i Sverige - The 2000 Swedish Red List of Swedish Species.* ArtDatabanken, SLU, Uppsala.

Heimer, S. & Nentwig, W. 1991. *Spinnen Mitteleuropas: Ein Bestimmungsbuch.* Verlag Paul Parey, Berlin.

Holm, Å. 1947. *Svensk Spindelfauna. 3. Egentliga spindlar. Araneae. Fam. 8 - 10, Oxyopidae, Lycosidae aoch Pisauridae.* Entomologiske Föreningen i Stockholm.

Hugenschütt, V. 1997. Bioindikationsanalyse von Uferzonationskomplexen der Spinnen- und Laufkäfergemeinschaften (Arach.: Araneida, Col.: Carabidae) an Fliegewässern des Drachefelser Ländchens. Arch. Zool. Publ. 2: 1-350; Wiehl (Martina Galunder-Verl.). [= Diss. Univ. Bonn] s.a. Autorreferat in *Arachnologische Mitteilungen* 13, 56-58.

Knülle, W. 1953. Zur Ökologie der Spinnen an Ufern und Küsten. *Zeitschrift für Morphologie und Ökologie der Tiere* 42, 117-158.

Krogerus, R. 1932. Über die Ökologie und Verbreitung der Arthropoden der Triebsandgebiete an den Küsten Finnlands. *Acta Zoologica Fennica* 12, 1-308.

Locket, G.H., Millidge, A.F. & Merrett, P. 1974. *British Spiders* Vol. III. Ray Society, London.

Maurer, R. & Hänggi, A. 1989. Für die Schweiz neue und bemerkenswerter Spinnen (Araneae) III. *Mitteilungen schweizerische entomologische Gesellschaft* 62, 175-182.

Maurer, R. & Hänggi, A. 1990. *Katalog der Schweizerischen Spinnen*. Documenta Faunistica Helvetica 12. Centre Suisse de cartographie de la Faune, Neuchâtel.

Mikhailov, K.G. 1997. *Catalogue of the spiders of the territories of the former Soviet Union (Arachnida, Aranei)*. Zoological Museum of the Moscow State University, Moscow.

Paik, K.Y. 1994. Korean spiders of the Genus *Arctosa* C.L. Koch, 1848 (Araneae: Lycosidae). *Korean Arachnology* 10, 36-65

Palmgren, P. 1939. Die Spinnenfauna Finnlands. I. Lycosidae. *Acta Zoologica Fennica.* 25, 1-86.

Palmgren, P. 1943. Die Spinnenfauna Finnlands. II. Pisauridae, Oxyopidae, Salticidae, Clubionidae, Anyphaenidae, Sparassidae, Ctenidae, Drassidae. *Acta Zoologica Fennica* 36, 1-112.

Palmgren, P. 1974. Die Spinnenfauna Finnlands und OstFennoskandiens. IV. Argiopidae, Tetragnathidae und Mimetidae. *Fauna Fennica* 24, 1-70.

Plachter, H. 1986. Die Fauna der Kies- und Schotterbänke dealpiner Flüsse und Empfehlungen für ihren Schutz. *Ber. Akad. Naturschutz Landschaftspflege* 10, 119-147.

Relys, V. 1994. Unkommentierte Liste der Spinnen Litauens (Araneae). *Arachnologische Mitteilungen* 7, 1-19.

Ruzicka, V. & Hajer, J. 1996. [Arachnofauna on the Labe river banks from Usti nad Lebem to Male Brezno]. *Fauna Bohemiae Septentrionalis* 21, 85-92.

Schenkel, E. 1932. Spinnen am Ufer der Untertrave. In: *Das linke Untertraveufer*, pp. 410-421.

Siepe, A. 1985. Einflub häufiger Überflutungen auf die Spinnen-Besiedlung am Oberrhein-Ufer. *Mitteilungen Deutsche Gesellschaft für Allgemeine und Angewandte Entomologie* 4, 281-284.

Starega, W. 1972. (Für die Fauna Polens neue und seltene Spinnearten (Aranei), nebst Beschreibung von *Lepthyphantes milleri* n. sp.). *Fragmenta Faunistica* 18, 55-98.

Steinberger, K.-H. 1996. Die Spinnenfauna der Uferlebensräume des Lech (Nordtirol, Österreich) (Arachnida: Araneae). – *Berichte Naturwissenschaftlicher-Medizinischer Verein in Innsbruck* 83, 187-210.

Strand, E. 1898. Einige fundorte für Araneiden im südlischen Norwegen. *Verhandlungen der Zoologische-Botanischen Gesellschaft in Wien* 48, 401-404.

Stølen, A. 1992. Miljøindikator: Endring av biodiversitet i elvekantvegetasjon langs større vassdrag. Numsedalslågen, Drammenselva, Gudbrandslågen, Gaula. UNIT-SMU meddelelse nr. 4/92: 1-30.

Thaler, K. 1993. Über wenig bekannte Zwergspinnen aus den Alpen – IX (Arachnida: Aranei, Linyphiidae, Erigoninae). *Revue Suisse de Zoologie* 100, 641-654.

Tischler, W. 1993. *Einführung in die Ökologie.* 4. Aufl. Fischer Stuttgart.

Tullgren, A. 1944. *Svensk Spindelfauna. 3. Aranèae (Salticidae, Thomisidae, Philodromidae och Eusparassidae).* Stockholm.

Wunderlich, J. 1975. Spinnen von Kaisersthul (Arachnida, Araneae). *Entomologica Germanica* 1, 381-386.

Wunderlich, J. 1979. Die Gattungen *Caviphantes* Oi 1960 und *Lessertiella* Dumitrescu & Miller 1962. *Senckenbergiana Biologica* 60, 85-89.

Aakra, K. 2000. Noteworthy records of spiders (Araneae) from central regions of Norway. *Norwegian Journal of Entomology* 47, 153-162.

Aakra, K. & Hauge, E. 2000. Provisional list of rare and potentially threatened spiders (Arachnida: Araneae) in Norway including their Proposed Red List status. *NINA Scientific Report* 42, 1-38.

REGIONAL FAUNISTICS

European Arachnology 2000 (S. Toft & N. Scharff eds.), pp. 255-259.
© Aarhus University Press, Aarhus, 2002. ISBN 87 7934 001 6
(Proceedings of the 19th European Colloquium of Arachnology, Århus 17-22 July 2000)

The spider fauna of Russia and other post-Soviet republics: a 2000 update

KIRILL G. MIKHAILOV

Zoological Museum MGU, Bolshaya Nikitskaya Str.6, Moscow 103009 Russia (kmk2000@online.ru)

Abstract

A brief review of the recent version (July 2000) of the *Catalogue of the spiders of the territories of the former Soviet Union* is provided. Calculations of species numbers are given for post-Soviet republics and physiographical areas, mostly in comparison with 1996 and earlier data. Totally, 2824 spider species belonging to 512 genera and 49 families are known from the former Soviet Union.

Key words: spiders, catalogue, faunistics, Russia, former Soviet Union

INTRODUCTION

The basic catalogue of spiders of the former Soviet Union (FSU) (Mikhailov 1997) - followed by three annual additions (Mikhailov 1998, 1999, 2000) - covers all literature data since the 18th century till July 2000. The last calculation of species numbers in the post-Soviet republics and main physiographical areas was made in 1996 (Mikhailov 1997).

The aim of this paper is to provide recent calculations (July 2000) of the numbers of spider taxa and to reflect changes in the knowledge of the FSU spiders, based on the aforementioned catalogue.

RESULTS

Until now, 2824 spider species belonging to 512 genera and 49 families have been reported from the FSU territories (Table 1). Since the 1996 evaluation, the main increase in species composition is recorded in Salticidae (+41 species), then successively in Linyphiidae (+23), Agelenidae (+9), Theridiidae (+7), Gnaphosidae and Dictynidae (+6 each). During 1989-1996, the increase of species number was 510, or ap-

proximately 73 species annually. In 1996-2000, the respective indices are 130 and 33. Nevertheless, my estimation of the richness of the total FSU spider fauna remains as 3400-3500 species. The reason of a certain delay in the FSU spider study is connected with a decline of activity of Russian and especially other post-Soviet arachnologists, mainly in Linyphiidae (+28 species annually in 1989-1996 vs. +ca.6 in 1996-2000).

Main spider families are treated here (Tables 2 & 3) as being represented by 1.5% or more of the whole FSU species number. As earlier, Linyphiidae shows the highest diversity in species and genera in the FSU (Table 2). An enormous increase of generic diversity in Linyphiidae in 1996-2000 (+24 genera vs. +23 species) is a consequence of splitting in this family, especially in the former genus *Lepthyphantes*. The second place of Salticidae (instead of Gnaphosidae in 1996) was already predicted (Mikhailov 1997).

Analysis of the spider fauna of the post-Soviet republics (Table 3) reveals almost the same relations as in 1996: Russia supports the highest diversity, followed by Ukraine, Ka-

Table 1. Genus/species composition of spider families known from the FSU territories in July 2000 (comparative data of 1996–1989 are given in brackets).

Family	No. of genera			No. of species		
	2000	1996	1989	2000	1996	1989
Atypidae	1	1	1	5	4	2
Ctenizidae	2	2	2	3	3	3
Dipluridae	1	1	1	1	1	1
Nemesiidae	2	2	3	11	11	10
Filistatidae	4	4	1	7	7	6
Sicariidae	1	1	1	1	1	1
Scytodidae	2	2	1	5	5	2
Leptonetidae	1	1	0	1	1	0
Pholcidae	6	5	5	18	16	15
Segestriidae	1	1	1	4	4	4
Dysderidae	6	6	4	91	90	51
Oonopidae	2	2	2	3	3	2
Palpimanidae	1	1	1	2	3	3
Mimetidae	2	2	2	6	6	6
Eresidae	2	2	2	6	6	6
Oecobiidae	4	4	5	7	7	8
Hersiliidae	1	1	1	3	3	3
Uloboridae	4	4	4	6	6	6
Nesticidae	4	3	3	13	12	10
Theridiidae	22	18	17	132	125	116
Theridiosomatidae	1	1	1	2	2	2
Linyphiidae	237	213	153	873	850	654
Tetragnathidae	7	7	8	43	43	30
Araneidae	18	18	12	113	108	114
Lycosidae	17	17	14	263	247	210
Pisauridae	3	3	3	12	12	12
Agelenidae	12	7	4	54	45	44
Cybaeidae	1	1	1	6	6	3
Argyronetidae	1	1	1	1	1	1
Desidae	1	1	1	6	6	3
Hahniidae	5	5	4	17	17	10
Dictynidae	14	14	8	59	53	49
Amaurobiidae	3	3	2	8	8	9
Titanoecidae	2	2	2	19	21	13
Zoropsidae	1	1	1	1	1	1
Oxyopidae	1	1	1	7	7	6
Anyphaenidae	2	3	2	4	4	5
Liocranidae	7	7	4	29	24	19
Clubionidae	2	2	2	99	98	83
Corinnidae	3	3	4	5	4	5
Zodariidae	3	3	3	23	23	23
Cithaeronidae	1	1	0	1	1	0
Prodidomidae	2	2	1	2	2	1
Gnaphosidae	28	28	23	294	286	206
Zoridae	1	1	1	7	7	6
Heteropodidae	3	3	3	5	5	5
Philodromidae	6	6	4	74	73	61
Thomisidae	16	16	17	168	164	146
Salticidae	45	41	38	307	266	211
TOTAL 49 families	512	474	375	2827	2698	2187

zakhstan, and Azerbaijan. Spider faunas of Lithuania, Moldavia and Armenia are not sufficiently studied. The main increase in species richness is recorded in Georgia (+130 species), Kirghizia (+106) and Russia (+100). The last index can be explained by the immense territory under study, whereas the first two derive from the revision of Georgian spiders (Mkheidze 1997) and a check-list made in Kirghizia (Zonshtein et al. 1996). In addition, the spider study in Russia declines only minimally compared to the other post-Soviet republics.

Among physiographical areas (Fig. 1, Table 4), the most diverse in spiders are the Russian Plain, the mountains of South Siberia, the Caucasus, and the mountains of Middle Asia. The main increases in species number during the last five years are reported from West Siberia, mountains of South Siberia, Fennoscandia, the Caucasus, the Urals, and the mountains of Middle Asia (+114, +99, +87, +82, +67, and +60 species, respectively). The first index owes several contributions to the faunas of Tuva, the Altais, and Transbaikalia. Numerous doubtful records (not entered) are peculiar to the Carpathians. Adding them, the whole species diversity will be 540, not 428 species as indicated in Table 4!

Table 2. Genus/species composition of main spider families in the FSU territories in July 2000 (in %) (comparative data of 1996 are given in brackets).

Family	% of genera	% of species
Linyphiidae	46.3 (45.1)	30.9 (31.6)
Salticidae	8.8 (8.7)	10.9 (9.9)
Gnaphosidae	5.5 (5.9)	10.4 (10.6)
Lycosidae	3.3 (3.6)	9.3 (9.2)
Thomisidae	3.1 (3.4)	6.0 (6.1)
Theridiidae	4.3 (3.8)	4.7 (4.7)
Araneidae	3.5 (3.8)	4.0 (4.0)
Clubionidae	0.4 (0.4)	3.5 (3.6)
Dysderidae	1.2 (1.3)	3.2 (3.3)
Philodromidae	1.2 (1.3)	2.6 (2.6)
Dictynidae	2.7 (3.0)	2.1 (2.0)
Agelenidae	2.3 (1.5)	1.9 (1.7)
Tetragnathidae	1.4 (1.5)	1.5 (1.6)

Table 3. Species composition of main spider families in the post-Soviet republics in July 2000 (in %) (comparative data of 1996 are given in brackets). Abbreviations: Li - Linyphiidae, Sa - Salticidae, Gn - Gnaphosidae, Ly - Lycosidae, To - Thomisidae, Te - Theridiidae, Ar - Araneidae, Cl - Clubionidae, Dy - Dysderidae, Ph - Philodromidae.

Republic	Li	Sa	Gn	Ly	To	Te	Ar	Cl	Dy	Ph	Total species		Total families	
Russia	37.4	9.0	9.1	9.7	5.6	5.2	7.1	4.4	0.8	3.1	1974	(1874)	37	(37)
Estonia	39.9	6.9	6.9	9.2	4.5	7.5	5.7	3.9	0	2.6	509	(506)	25	(25)
Latvia	35.6	8.0	8.5	10.7	4.7	8.0	7.5	4.2	0	3.0	402	(401)	22	(21)
Lithuania	26.2	5.5	7.8	14.0	6.3	6.3	9.6	6.6	0.4	3.7	271	(241)	22	(21)
Byelorussia	35.4	7.0	6.1	9.5	6.3	8.0	7.5	4.6	0	2.9	412	(383)	26	(26)
Ukraine	27.2	8.6	9.8	9.2	5.7	7.8	5.8	3.6	1.9	3.0	830	(808)	37	(37)
Moldavia	19.2	7.2	6.9	11.0	10.6	7.9	7.9	5.8	2.4	2.7	292	(291)	29	(29)
Georgia	18.2	8.8	5.0	9.7	9.7	7.2	8.8	3.7	7.3	3.1	456	(326)	37	(34)
Azerbaijan	15.9	15.4	9.8	8.6	7.5	8.4	5.9	3.9	5.0	2.9	559	(500)	37	(36)
Armenia	29.1	12.6	7.9	3.2	8.7	2.4	7.1	7.1	7.9	4.7	127	(118)	19	(19)
Kazakhstan	17.0	19.2	14.1	9.7	8.2	5.0	5.8	3.2	0.3	4.2	719	(679)	34	(34)
Uzbekistan	8.7	17.5	10.7	9.1	11.3	5.8	8.1	3.6	1.0	5.2	309	(290)	33	(33)
Turkmenia	9.0	23.6	14.3	6.4	8.2	4.0	6.1	1.6	2.9	2.7	377	(353)	38	(39)
Kirghizia	21.8	18.1	11.2	5.0	8.2	6.5	7.5	1.7	1.1	2.8	464	(358)	31	(29)
Tajikistan	8.4	19.7	11.6	10.0	10.3	6.1	5.5	1.9	2.3	4.5	310	(293)	34	(34)

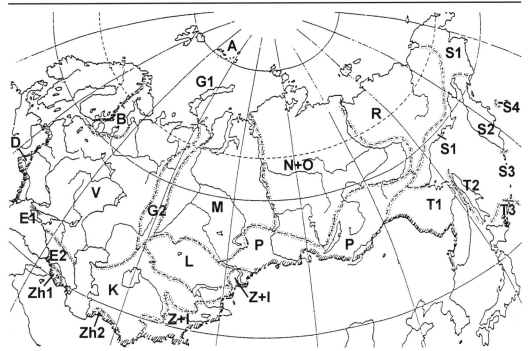

Fig. 1. Physiographical areas of the FSU. Abbreviations see Table 4.

Table 4. Spider families/species composition in the FSU physiographical areas in July 2000 (in %) (comparative data of 1996-1989 are given in brackets). Abbreviations: Li - Linyphiidae, Sa - Salticidae, Gn - Gnaphosidae, Ly - Lycosidae, To - Thomisidae, Te - Theridiidae, Ar - Araneidae, Cl - Clubionidae, Dy - Dysderidae, Ph - Philodromidae.

Code	Area	Li	Sa	Gn	Ly	To	Te	Ar	Cl	Dy	Ph	Total species	Total families
A	Atlantic-Arctic area	100	-	-	-	-	-	-	-	-	-	2 (1-1)	1 (1-1)
B	Fennoscandia	40.5	7.0	7.6	9.5	4.1	7.8	6.4	3.2	0	2.5	516 (429-385)	26 (25-23)
V	Russian Plain	31.2	9.0	11.5	8.6	5.5	7.1	5.1	3.3	1.1	3.2	1026 (1001-936)	36 (35-36)
G1	Novaya Zemlya	85.7	-	4.8	-	-	4.8	-	-	-	-	21 (21)	4 (4)
G2	Urals	36.5	7.9	8.7	9.2	6.1	7.2	4.9	3.2	-	3.2	750 (683)	26 (24)
D	Carpathians	21.0	9.1	5.1	11.7	6.8	8.6	9.1	5.1	0.9	2.8	428 (421-435)	34 (34-35)
E1	Crimea	10.8	12.9	10.2	10.2	10.2	8.5	9.1	4.1	3.8	2.9	342 (311-308)	31 (31-31)
E2	Caucasus	19.7	12.2	10.1	9.2	7.3	6.6	5.5	3.5	6.7	2.8	834 (752)	40 (40)
Zh1	Armenian Upland	28.2	14.1	9.6	3.0	8.2	2.2	5.9	7.4	8.2	4.4	135 (127)	19 (19)
Zh2	Kopetdagh Mts.	8.3	19.6	13.8	4.6	10.4	5.4	5.8	2.5	2.1	2.9	240 (221)	36 (37)
Z+I	Mountains of Middle Asia	19.5	17.2	13.1	7.9	7.9	5.5	6.2	2.5	1.0	3.2	833 (773)	38 (40)
K	Deserts of Middle Asia	6.5	26.3	16.0	6.5	8.3	4.7	5.0	2.4	2.4	4.4	338 (318-291)	34 (35-37)
L	Kazakhstan hills	2.8	20.3	18.2	9.1	11.9	4.2	11.2	3.5	-	9.8	143 (129-103)	18 (18-15)
M	West Siberia	37.6	10.3	6.5	12.1	6.5	5.4	7.0	2.9	-	4.0	554 (440-243)	22 (21-21)
N+O	Middle Siberia	49.2	6.8	6.3	9.5	5.8	5.0	4.3	2.8	-	3.6	634 (624-532)	22 (22-24)
P	Mountains of South Siberia	31.9	10.6	11.8	12.4	6.5	5.4	4.8	3.5	-	4.8	912 (813-436)	24 (24-23)
R	Northeastern Siberia	57.2	6.3	6.8	7.1	3.8	4.5	3.5	1.8	-	3.3	397 (395-277)	17 (17-16)
S1	Continental Far North-East	59.5	4.8	5.3	7.0	2.9	3.9	3.9	2.4	-	2.9	415 (411)	18 (17)
S2	Kamchatka	51.7	2.2	3.9	11.0	3.9	5.0	5.5	3.9	-	3.9	182 (184)	15 (15)
S3	N-Kuriles	70.0	5.0	1.7	8.3	-	5.0	1.7	5.0	-	3.3	60 (54)	8 (7)
S4	Commander Islands	75.0	-	-	10.0	-	10.0	-	-	-	-	20 (19)	4 (3)
T1	Continental southern Far East	31.1	11.1	5.0	8.1	6.7	5.7	8.7	7.8	-	3.7	566 (507)	28 (25)
T2	Sakhalin	48.8	6.5	3.6	4.1	6.2	4.7	6.5	6.8	-	3.3	338 (343)	20 (20)
T3	S-Kuriles	32.2	12.8	3.4	6.0	8.7	7.4	6.0	7.4	-	4.7	149 (144)	18 (18)

A small decrease in the number of spider families in some Middle Asian areas and republics derives from the entering of only identified and described species in 2000, whereas in 1996 several indications of undescribed species were counted as well (see footnotes in Mikhailov 1997).

A greater part of linyphiids is peculiar to boreal, arctic, and northern island zones (A, G1, S4, S3, S1, R, S2, N+O, T2, B). Salticids dominate only in southern, mainly arid areas (K, L, Zh2, E1).

The above data make it obvious that a faunistic study of spiders of Russia and the FSU is not complete.

REFERENCES

Mikhailov, K.G. 1997. *Catalogue of the spiders of the territories of the former Soviet Union (Arachnida, Aranei).* Zoological Museum of the Moscow State University (Archives 37), Moscow.

Mikhailov, K.G. 1998. *Catalogue of the spiders (Arachnida, Aranei) of the territories of the former Soviet Union. Addendum 1.* KMK Scientific Press, Moscow.

Mikhailov, K.G. 1999. *Catalogue of the spiders (Arachnida, Aranei) of the territories of the former Soviet Union. Addendum 2.* Zoological Museum MGU, Moscow.

Mikhailov, K.G. 2000. *Catalogue of the spiders (Arachnida, Aranei) of the territories of the former Soviet Union. Addendum 3.* Zoological Museum MGU, Moscow.

Mkheidze, T.S. 1997. *Spiders of Georgia (taxonomy, ecology, zoogeographical review).* Tbilisi Univ. Publ. House, Tbilisi [in Georgian, with Russian summary].

Zonshtein, S.L., Gromov, A.V., Zyuzin, A.A. & Ovchinnikov, S.V. 1996. Class Arachnida - terrestrial arachnids. Order Araneae - spiders. In: *Kadastr geneticheskogo fonda Kyrgyzstana. Vol.2. Vira, Bacteria, Animalia (Protozoa, Porifera, Coelenterata, Plathelminthes, Nemathelminthes, Acanthocephales, Annelida, Bryozoa, Mollusca, Tardigrada, Arthropoda,* pp.132-153. Biologo-pochvenniy Institut, Bishkek [in Russian].

European Arachnology 2000 (S. Toft & N. Scharff eds.), pp. 261-265.
© Aarhus University Press, Aarhus, 2002. ISBN 87 7934 001 6
(Proceedings of the 19th European Colloquium of Arachnology, Århus 17-22 July 2000)

Altitudinal and biotopic distribution of the spider family Gnaphosidae in North Ossetia (Caucasus Major)

KIRILL G. MIKHAILOV & ELENA A. MIKHAILOVA
Division of Invertebrates, Zoological Museum MGU, Bolshaya Nikitskaya Str. 6 Moscow 103009 Russia
(kmk2000@online.ru)

Abstract

An analysis of the spider fauna, its distribution and zoogeography of the family Gnaphosidae of North Ossetia is provided. More than 2100 specimens were collected in 1985 by pitfall trapping. A total of 40 species is reported from the area studied, which includes 29 biotopes in 4 mountain ranges. A biotopical arrangement of the species found is given and dominant species are indicated. Most of the species have Euro-Siberian and Euro-Kazakhstanian ranges. Several new species (as yet undescribed) were found.

Key words: Araneae, Gnaphosidae, Caucasus Major, altitudinal distribution, biotopic distribution

INTRODUCTION

Up-to-date detailed quantitative studies of Caucasian spiders have not yet been conducted properly. The aim of this project is to study herpetobiont spiders of model plots on the northern macroslope of the Caucasus Major including several parallel ridges with decreasing altitude and increasing xerophytization (Fig. 1). All these plots are situated in the North Ossetian State Reserve and its surroundings.

Three main stages of the project are planned:
(1) a study of the spider fauna at the family level (already made),
(2) analysis of the fauna, distribution, and zoogeography of separate spider families,
(3) a definitive analysis of spider species distribution.

METHODS

All the material was collected by pitfall traps during April–November 1985 in several parallel ridges of the Caucasus Major: Bokovoy, Tsei, Skalistiy, Pastbishchniy, and Kabardino-Sunzhenskiy Mt. Ridges (Fig. 1). Traps were placed in lines of 10 jars with formaldehyde in the following biotopes: 2 in Bokovoy Mt. Ridge (V series), 9 (8 for gnaphosids) in Tsei Mt. Ridge (Ts series), 6 in Skalistiy and Pastbishchniy mt. ridges (G series), 6 in Unal Kettle (nr. Skalistiy Mt. Ridge, K series), and 6 in Kabardino-Sunzhenskiy Mt. Ridge (S series). As a result, 29 biotopes including 5 steppe, 11 forest, 11 meadow, and 2 bushy ones were examined (Table 1). All biotopes are situated in low, middle, and high montane areas.

RESULTS

A total of ca. 18000 spider specimens of 26 families was collected. More than 2100 specimens of Gnaphosidae were captured (ca. 12.1% of the total) making it the second most abundant family after the Lycosidae.

Gnaphosids are most abundant in mountain steppes and in middle & high montane xerophytous communities (up to 38%), as well

Fig. 1. (A-C) Map of collecting sites in North Ossetia, Caucasus Major, Russia, 1985. **(B)** Rectangle in A enlarged. **(C)** rectangle in B enlarged. Sample series are indicated. Abbreviations: BK — Bokovoi Mt. Ridge, CM — Caucasus Major, LS — Lesistiy ("Woody") Mt. Ridge, PB — Pastbishchniy ("Pasturable") Mt. Ridge, SK — Skalistiy ("Rocky") Mt. Ridge, SN — Kabardino-Sunzhenskiy Mt. Ridge.

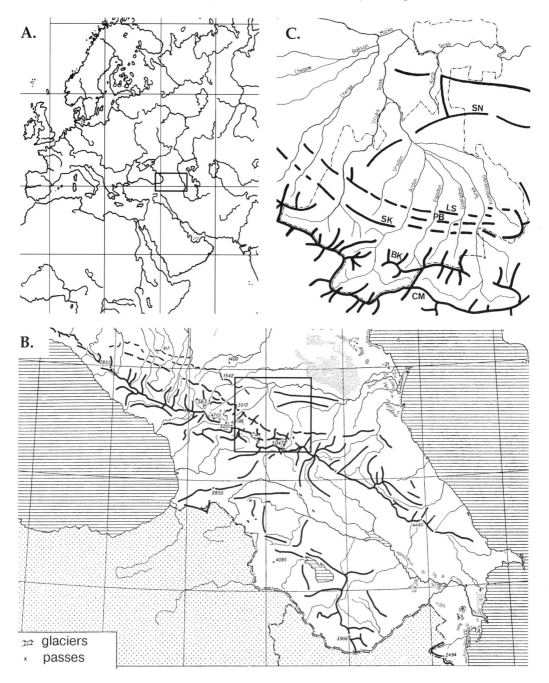

Table I. List of biotopes studied (North Ossetia, Caucasus, Russia, 1985)

Basic regions Biotope Subregions	Pitfall trap series	Altitude m a.s.l.
LOW MONTANE (Kabardino-Sunzhenskiy Mt. Ridge)		
Steppe:	S6	450
	S5	500
	S3	880
Forest:		
Quercus (young)	S4	500
Quercus	S1	600
Fagus	S2	570
MIDDLE MONTANE (Unal Kettle, Pastbishchniy, Tsei, Bokovoy Mt. Ridges)		
Xerophytous open communities:	K3	1170
Mountain steppes	K2	1200
Other xerophytous open communities	K6	1000
	K4	1100
	K1	1200
Forest:		
Quercus	G1	900
Quercus (young)	K5	1050
Quercus (sparse)	Ts9	1400
Broadleaved	V1	1500
Mesophytous meadow	Ts10	1350
	V2	1500
HIGH MONTANE (Skalistiy, Tsei Mt. Ridges)		
Forest belt		
Forest:		
Pinus	Ts8	2000
Pinus (young)	Ts7	2300
Pinus	Ts6	2300
Betula	G6	2000
Mesophytous meadow	Ts5	2550
Subalpine/Alpine belt		
Mesophytous subalpine meadow	G5	2300
	G2	2550
	Ts4	2750
Small bushes	G4	2500
	Ts2	3000
Xerophytous community	G3	2500
Alpine meadow	Ts3	3000

in high montane *Betula* forest (where Agelenidae is dominant). Generally, the Gnaphosidae and Lycosidae prefer open communities.

A total of 40 gnaphosid species belonging to 12 genera were found (Table 2). Of them, 13 species are connected with the low montane belt, 31 with middle-height mountains, and 20 with high mountains. Seven species were found in all belts, and 11 species in two belts.

Twenty-seven (67.5%) species were found in steppic and xerophytous biotopes, and 15 (37.5%) species were found only in them. In the alpine/subalpine zone, a mixture of species with wide ranges and endemics is found.

The altitudinal and biotopical distribution of Gnaphosidae is poorly studied in the Caucasus Major northern macroslope. Only Ovtsharenko (1979) provides data on 13 species. But he studied the wider area (from North Ossetia to Black Sea coast) and did not collect in the steppes or in *Quercus* and *Pinus* forests. Of his list, only *Zelotes hermani* (Chyzer, 1897) and *Drassyllus vinealis* (Kulczynski, 1897) were not found by us. Both species were reported from *Fagus* forests poorly represented in North Ossetia.

Distributional data were compiled from different sources united in a handwritten card catalogue partly published by Mikhailov (1997). From the viewpoint of zoogeography (Table 3), most of the gnaphosids (35%) are represented by Euro-Siberian (in a wide sense) and Euro-Kazakhstanian species. Twenty-five percent of species belong to Holarctic and trans-Palaearctic patterns. There was also a large proportion of European species (12.5%). The exact percentage of endemics is not clear, but it could be as high as 12.5%.

REMARKS ON ECOLOGY

As usual, males predominate in most of the samples. Only in low montane habitats (both in forests and in steppe) is the prevalence of females in summer time (June–August) recorded. The activity peaks of males are mainly in spring and autumn. Such a phenomenon can

as on alpine meadow. To a lesser extent they are also represented in low montane steppes, on meadows of the forest belt, and in subalpics. Gnaphosids are not abundant in forests (0–9 %, or 14–25 % in young forests). They are not found in low montane *Fagus* forest (where Linyphiidae and Dysderidae are dominant), or

Table 2. List of Gnaphosidae of North Ossetia (Caucasus Major) caught in pitfall traps. Abbreviations: Main belts: L low montane, M middle montane, H high montane;
Biotopes: (forests) Br broadleaved forest, P *Pinus* forest, P_j young *Pinus* forest, Q *Quercus* forest, Q_j young *Quercus* forest, Q_{sp} sparse *Quercus* forest; (open communities) alp alpine/subalpine belt, md meadows in forest belt, st steppes and xerophytous communities;
Range: (geographically) Alt Altaian (Altai Mts.), Baik Baikalian (Baikal Lake), Cauc Caucasia, EEu East European, Eu European, Hol Holarctic, Kaz Kazakhstanian, Kopetd Kopetdaghian (Kopetdagh Mts.), MAs Middle Asian, Med Mediterranean, Mong Mongolian, NCauc North Caucasian, Sib Siberian, trPal trans-Palearctic; (zonally) bor boreal, des deserticolous, nem nemoral, polyz polyzonal, st steppic; endem endemic.

	Main belt	Altitude m a.s.l.	Biotope Forest	Open	Range
1. *Berlandina cinerea* (Menge, 1868)	M	1170-1350		st	Eu-Kaz nem
2. *Callilepis nocturna* (Linnaeus, 1758)	H	2300	P_j		trPal polyz
3. *Drassodes lapidosus* (Walckenaer, 1802)	L, M, H	500-1200	P	st,md	trPal polyz
4. *Drassodes pubescens* (Thorell, 1856)	L, M, H	500-3000	Br, P	st, md, alp	Eu-Sib bor-nem
5. *Haplodrassus kulczynskii* Lohmander, 1942	L, M	450-1400	Q	st, md	Eu nem-st
6. *Haplodrassus signifer* (C.L. Koch, 1839)	M, H	1100-3000	P	st, md, alp	Hol polyz
7. *Haplodrassus cf. silvestris* (Blackwall, 1833)	L, M, H	500-2750	Q, Br	st, alp	endem?
8. *Haplodrassus umbratilis* (L.Koch, 1866)	M, H	1050-2750	Q_j	md, alp	Eu-Kaz nem-st
9. *Nomisia aussereri* (L. Koch, 1872)	M	1000-1200		st	Med-Mas st-des
10. *Poecilochroa conspicua* (L. Koch, 1866)	L, M	880-1050	Q_j	st	trPal nem
11. *Poecilochroa variana* (C.L. Koch, 1839)	M	1050-1350	Q_j	md	Eu-Mong nem
12. *Scotophaeus* sp. 1	M	1500		md	endem?
13. *Drasyllus praeficus* (L. Koch, 1866)	M	1170-1500		st, md	Eu-Kaz nem
14. *Drasyllus pumilus* (C.L. Koch, 1839)	L, M	450-1200		st	Eu nem
15. *Drasyllus pusillus* (C.L. Koch, 1833)	L, M, H	880-3000	Br, P	st, md, alp	trPal nem
16. *Zelotes aeneus* (Simon, 1878)	M	1000		st	Eu nem
17. *Zelotes atrocaeruleus* (Simon, 1878)	L	450-880		st	Eu-Kaz st
18. *Zelotes declinans* (Kulczynski, 1897)	M	1000		st	Eu-Kaz st
19. *Zelotes electus* (C.L. Koch, 1839)	H	2500		alp	Eu-Kaz nem
20. *Zelotes cf. erebeus* (Thorell, 1871)	L, M, H	600-2300	Q, P	st	endem?
21. *Zelotes gracilis* (Canestrini, 1868)	L	450		st	Eu st
22. *Zelotes longipes* (L. Koch, 1866)	L, M	450-1500	Br	st	Eu-Alt nem-st
23. *Zelotes petrensis* (C.L. Koch, 1839)	L, M, H	500-2750	Q_j, P	st, md, alp	Eu-Kaz nem
24. *Zelotes subterraneus* (C.L. Koch, 1833)	M, H	1050-2300	Q_j, Q_{cn}, Br, P	st	trPal polyz
25. *Zelotes* sp. 1	L, H	450-2500		st, alp	endem?
26. *Parasyrisca alexeevi* Ovtsharenko et al., 1995	M	ca. 1000-1200		st?	endem
27. *Gnaphosa caucasica* Ovtsharenko et al., 1992	H	2300-3000	P	alp	NCauc[1]
28. *Gnaphosa leporina* (L. Koch, 1866)	M, H	1500-2750		md, alp	Eu-Baik nem
29. *Gnaphosa lucifuga* (Walckenaer, 1802)	M	1000-1170		st	Eu-Kaz-MAs nem-st
30. *Gnaphosa lugubris* (C.L. Koch, 1839)	M, H	1200-2750		st, alp	Eu nem-st
31. *Gnaphosa mongolica* Simon, 1895	M	1000-1170		st	trPal st
32. *Gnaphosa montana* (L. Koch, 1866)	H	2300	P		Eu-Baik bor-nem
33. *Gnaphosa steppica* Ovtsharenko et al., 1992	M	1000-1200		st	EEu-Kaz st
34. *Gnaphosa taurica* Thorell, 1875	M	1350		md	EEu-MAs st
35. *Micaria dives* (Lucas, 1846)	M	1200		st	trPal nem-st
36. *Micaria formicaria* (Sundevall, 1831)	M, H	1100-2300	P_j	at, md	trPal nem-st
37. *Micaria fulgens* (Walckenaer, 1802)	L, M, H	500-2000	Q_j, Q_{cn}, P	st, md	Eu-Baik nem
38. *Micaria kopetdaghensis* Michailov. 1986	H	2000-2750	P	md, alp	Cauc-Kopetd
39. *Micaria pulicaria* (Sundevall, 1831)	H	2300		alp	Hol bor-nem
40. *Micaria silesiaca* L. Koch, 1875	M	1350		md	Eu-Baik bor

Table 3. Zoogeography of Gnaphosidae of North Ossetia (Caucasus Major).

	Range pattern	No. of species	%	Σ%
Widely distributed	Holarctic	2	5	25
	Trans-Palearctic	8	20	
Moderately-widely distributed	Euro-Siberian[1]	7	17.5	37.5
	Euro-Kazakhstanian	7	17.5	
	Euro-(Mediterranean)-Middle Asian	1	2.5	
Moderately distributed	European species	5	12.5	17.5
	East European[2]	2	5	
Locallydistributed	Caucaso-Kopetdaghian	1	2.5	17.5
	Caucasian	1	2.5	
	Endemics	1+4?	12.5	
TOTAL		40	100	100

[1]In a wide sense, including Euro-Baikalian, Euro-Mongolian and Euro-Altaian ranges.
[2]East European-Kazakhstanian etc.

be explained by the fact that a combination of dry and hot climatic conditions in summer is not very favourable for active males.

REMARKS ON TAXONOMY

1. *Zelotes aeneus* (Simon, 1878) may constitute a separate subspecies in the Caucasus differing by small details of embolus structure. Females are not distinguishable. A West European/Caucasian disjunction can be proposed for this species (the closest records are in Byelorussia; a record in Crimea is rather doubtful).

2. *Zelotes* sp.1: males are closer to *Z. atro-caeruleus* (Simon, 1878), whereas females — to *Z. apricorum* (L. Koch, 1876). This species found both in low and high montane habitats can be widely distributed in the Caucasus Major.

3. *Scotophaeus* sp. A single male found is close to *S. quadripunctatus* (Linneaeus, 1758) and *S. scutulatus* (L. Koch, 1866) differing by the details of male palp structure.

REFERENCES

Mikhailov, K.G. 1997. *Catalogue of the spiders of the territories of the former Soviet Union (Arachnida, Aranei).* Sbornik trudov Zoologicheskogo muzeya MGU 37, Moscow.

Ovtsharenko, V.I. 1979. Spiders of the families Gnaphosidae, Thomisidae, Lycosidae (Aranei) of the Caucasus Major. In: *Fauna i ekologiya paukoobraznykh.* Trudy Zoologicheskogo Instituta AN SSSR 85, 39-53. Leningrad [in Russian, with English summary].

European Arachnology 2000 (S. Toft & N. Scharff eds.), pp. 267-271.
© Aarhus University Press, Aarhus, 2002. ISBN 87 7934 001 6
(Proceedings of the 19th European Colloquium of Arachnology, Århus 17-22 July 2000)

Spider fauna of peat bogs in southwestern Finland

SEPPO KOPONEN

Zoological Museum, Centre for Biodiversity, University of Turku, FIN-20014 Turku, Finland
(sepkopo@utu.fi)

Abstract

The most common and abundant spider species on open peat bogs in mainland of SW Finland were *Arctosa alpigena, Pardosa hyperborea, P. sphagnicola, Alopecosa pulverulenta, Trochosa spinipalpis, Antistea elegans, Agyneta cauta* and *Pirata uliginosus*. Northern bog spiders can be divided into three groups: 1) species found only on the inland bogs (e.g. *Pardosa atrata* and *Mecynargus sphagnicola*), 2) also on the coastal bogs (e.g. *Pardosa hyperborea* and *Gnaphosa lapponum*) and 3) even on the archipelago bogs (*Arctosa alpigena* and *Gnaphosa microps*). Southern species on bogs include spiders 1) occurring only in the archipelago and coastal area (e.g. *Ozyptila gertschi* and *Satilatlas britteni*) and 2) species found also on mainland peat bogs in SW Finland (e.g. *Glyphesis cottonae* and *Zora parallela*).

Key words: spiders, peat bogs, abundance, distribution patterns, Araneae, Finland

INTRODUCTION

The proportion of bogs or peatlands is high in Finland. About a third of the land area of Finland was originally covered by peatlands, of which about half has been drained for forestry, farming and peat harvesting (Wahlström et al. 1996). Therefore peatland habitats (mires, bogs, fens), especially in southern Finland, can be listed as endangered habitats. Although the situation is not so serious as in Central Europe, many spider species living on bogs are nowadays considered threatened (e.g. Koponen 1985). In the recently updated Finnish Red Data Book (Rassi et al. 2001) six of the total 34 listed spider species are mire-dwellers (Koponen et al. 2001).

The spiders living on bogs have not been studied thoroughly in Northern Europe, Fennoscandia and the Baltic states. Some older publications can, however, be mentioned, e.g. Krogerus (1960), Koponen (1968, 1978, 1979), Palmgren (1977) and Vilbaste (e.g. 1980-81). Data on bog spiders can be found also in many faunistic works, e.g. Palmgren (1972), Lehtinen et al. (1979) and in several faunistic reports by Lohmander (e.g. 1956). In recent years, research on spiders of North European bogs has increased, at least in Norway, Lithuania and Finland. In the present paper, ground-living spider fauna on open peat bogs in southwestern Finland will be presented. Special attention is paid to the most abundant and common species as well as to other typical bog spiders. Also distributional patterns of some species are discussed.

STUDY AREA, MATERIAL AND METHODS

The nine study sites are situated in southwestern Finland (Fig. 1). The bogs studied are open peat bog sites (without trees) where *Sphagnum,*

Table I. The study sites in SW Finland (see Fig. I); Bog%: proportion of peatland in an area of 150 km^2 around the study site. Dist.km: distance from the Baltic Sea coast (km).

Site no.	Site name (Abbreviation)	Bog%	Dist. km
1	Masku, Karevansuo (MA)	5	10
2	Kustavi, Lautreski (KU)	2	0
3	Halikko, Sammalsuo (HA)	5	10
4	Vahto, Rehtsuo (VA)	20	20
5	Pöytyä , Kontolanrahka (PÖ)	15	50
6	Tammela , Torronsuo (TA)	25	75
7	Renko, Seitsemänlamminsuo (RE)	20	90
8	Huittinen, Isosuo (HU)	20	75
9	Ruovesi, Siikaneva (RU)	15	140

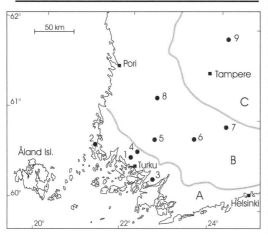

Fig. I. The study sites in SW Finland, cf. Table I. Key to bog zones: A: plateau bogs; B: concentric kermi bogs; C: eccentric kermi bogs (see Eurola et al. 1984).

Eriophorum and *Carex* species dominate; in addition, *Andromeda, Calluna, Ledum, Betula nana, Vaccinium oxycoccos* and *Rubus chamaemorus* can be found. Especially *Calluna, Ledum* and *Betula nana* grow on the hummocks. The distance of the bogs from the Baltic Sea coast is shown in Table 1. The isolation of the bogs is presented as 'bog%' which is the proportion of peatland in an area of 150 km^2 around each study site, based on information on 1:250000 maps (Table 1).

Material has been collected by pitfall traps.

Trapping periods and years as well as number of traps varied between sites, thus statistical analyses have not been done and the comparisons are based on percentages of abundant species on each bog studied. Nomenclature is after Platnick (1997) and the material is deposited in the Zoological Museum, University of Turku.

Besides the above study sites, some data on the occurrence of bog spiders, especially from the archipelago area, have been taken from earlier publications (Lehtinen et al. 1979; Koponen 1985).

RESULTS AND DISCUSSION
Fauna on peat bogs
The most abundant species of each study site are presented in Table 2. The percentages have been given only for the 'top-scorers' of each site (10-15 most abundant species; minimum 1-2.5% of the total number of specimens). The Karevansuo bog in Masku was studied most thoroughly; about 100 species were found during a two-year investigation. Its fauna represents well the SW-Finnish bog fauna: only five of the 45 listed 'top-scorers' of Table 2 were absent in the Karevansuo material, i.e. *Walckenaeria atrotibialis, Glyphesis cottonae, Pardosa atrata, Mecynargus sphagnicola* and *Agyneta decora*. Many of the abundant species (in Table 2) occur in Finland only on bogs; however, there are also species typical of moist habitats in general (e.g. *Antistea elegans, Bathyphantes gracilis* and *Dolomedes fimbriatus*) and some more eurytopic spiders (e.g. *Alopecosa pulverulenta, Zora spinimana, Agroeca proxima, Pardosa palustris* and *Cnephalocotes obscurus*).

The most dominant species on the nine bogs were *Pardosa hyperborea* (top dominant on three bogs), *P. sphagnicola, Arctosa alpigena* and *Pirata uliginosus* (each top dominant on two bogs).

The most common and abundant species on the studied peat bogs are shown in Table 3. Due to the collecting method (pitfall traps) lycosids dominated, the most frequent 'top-scorer' species being *Arctosa alpigena, Pardosa hyperborea, P. sphagnicola, Alopecosa pulverulenta*

Table 2. The most abundant ('top-scorer') species on peat bogs in SW Finland (% of the total number of specimens at each study site). +: found in small numbers; -: not found. For distribution and abbreviation of site names, see Fig. I and Table I.

| | 1 | 2 | 3 | 4 | 5 | 6 | 7 | 8 | 9 |
	MA	KU	HA	VA	PÖ	TA	RE	HU	RU
Pirata uliginosus (Thorell, 1856)	24.1	-	10.0	+	+	-	1.8	+	18.5
Pardosa hyperborea (Thorell, 1872)	21.8	-	+	51.0	21.2	44.6	3.7	38.3	9.1
Arctosa alpigena (Doleschall, 1852)	4.3	27.4	+	7.9	30.6	13.3	+	14.5	16.0
Trochosa spinipalpis (F.O.P.-Cambridge, 1895)	3.2	10.6	7.5	+	+	1.8	1.8	-	4.0
Agyneta cauta (O.P.-Cambridge, 1902)	3.0	-	1.4	-	1.4	+	2.2	-	-
Walckenaeria antica (Wider, 1834)	3.0	-	2.3	+	+	+	1.0	+	+
Pardosa sphagnicola (Dahl, 1908)	2.7	17.3	28.3	5.4	1.4	+	62.4	-	14.2
Alopecosa pulverulenta (Clerck, 1757)	2.5	2.9	+	1.7	8.7	+	3.8	1.0	+
Macrargus carpenteri (O.P.-Cambridge, 1894)	2.3	-	-	-	-	-	-	-	-
Lepthyphantes angulatus (O.P.-Cambridge, 1881)	2.2	-	-	+	-	+	2.9	-	+
Antistea elegans (Blackwall, 1841)	1.5	-	7.7	+	5.6	+	1.8	7.4	-
Maro lepidus Casemir, 1961	1.5	+	-	-	-	-	-	-	-
Drepanotylus uncatus (O.P.-Cambridge, 1873)	1.3	+	-	-	-	-	-	1.9	-
Pirata piscatorius (Clerck, 1757)	1.3	13.5	+	-	-	-	+	+	-
Centromerita concinna (Thorell, 1875)	1.3	-	-	+	-	+	-	1.6	4.0
Bathyphantes gracilis (Blackwall, 1841)	+	4.8	+	-	-	+	-	+	-
Pardosa pullata (Clerck, 1757)	+	3.7	2.4	3.1	+	-	-	+	3.3
Centromerus arcanus (O.P.-Cambridge, 1873)	+	1.9	+	+	+	-	2.9	-	-
Dolomedes fimbriatus (Clerck, 1757)	+	1.4	1.4	-	-	-	+	-	-
Gnaphosa nigerrima L. Koch, 1877	+	1.4	2.6	-	+	-	+	-	-
Xysticus cristatus (Clerck, 1757)	+	1.4	-	+	-	-	-	+	-
Euryopis flavomaculata (C.L. Koch, 1836)	+	-	3.5	+	2.4	11.5	+	-	-
Zora spinimana (Sundevall, 1833)	+	-	2.3	-	-	-	+	-	-
Scotina palliardi (L. Koch, 1881)	+	-	2.1	+	+	-	+	1.3	2.9
Pirata insularis Emerton, 1885	+	-	2.1	-	-	-	+	-	+
Meioneta affinis (Kulczynski,1898)	+	-	1.9	+	-	+	+	-	-
Walckenaeria atrotibialis O.P.-Cambridge, 1878	-	-	1.6	+	-	-	+	-	-
Pardosa palustris (Linnaeus, 1758)	+	-	-	2.8	-	+	-	1.0	-
Erigone atra Blackwall, 1833	+	-	-	2.5	-	-	-	-	-
Glyphesis cottonae (La Touche, 1945)	-	+	-	2.0	-	-	-	-	-
Gnaphosa lapponum (L. Koch, 1866)	+	-	-	2.0	+	6.5	-	2.6	-
Pachygnatha degeeri Sundevall, 1830	+	-	-	1.7	3.5	-	-	+	-
Porrhomma microphthalmum (O.P.-Cambridge, 1871)	+	-	-	1.7	-	+	-	-	-
Pardosa atrata (Thorell, 1873)	-	-	-	+	13.2	3.6	-	21.9	-
Robertus arundineti (O.P.-Cambridge, 1871)	+	-	+	+	1.0	-	-	-	+
Walckenaeria nudipalpis (Westring, 1851)	+	+	-	+	1.0	+	+	+	2.5
Drassodes pubescens (Thorell, 1856)	+	+	-	+	+	2.2	+	+	+
Haplodrassus signifer (C.L. Koch, 1839)	+	-	+	+	+	1.4	-	-	-
Agroeca proxima (O.P.-Cambridge, 1871)	+	-	+	+	+	1.4	+	-	-
Maro minutus O.P.-Cambridge, 1906	+	-	-	-	-	1.4	-	+	-
Gnaphosa microps Holm, 1939	+	-	-	-	-	-	4.6	+	-
Mecynargus sphagnicola (Holm, 1939)	-	-	-	-	-	-	-	2.3	+
Centromerus levitarsis (Simon, 1884)	+	-	-	+	+	-	-	-	3.6
Cnephalocotes obscurus (Blackwall, 1834)	+	+	+	+	+	-	+	-	3.6
Agyneta decora (O.P.-Cambridge, 1871)	-	-	+	-	+	-	-	-	3.3
Number of specimens	3670	216	428	355	288	278	628	311	275
Number of species	98	33	36	53	34	28	40	28	34

Table 3. The most common and abundant spider species on peat bogs in SW Finland (frequency as dominants in nine 'top-scorer' lists (Table 2)).

Arctosa alpigena	7/9
Pardosa hyperborea	7/9
P. sphagnicola	7/9
Alopecosa pulverulenta	6/9
Trochosa spinipalpis	6/9
Antistea elegans	5/9
Agyneta cauta	4/9
Pardosa pullata	4/9
Pirata uliginosus	4/9
Centromerita concinna	3/9
Euryopis flavomaculata	3/9
Gnaphosa lapponum	3/9
Pardosa atrata	3/9
Scotina palliardi	3/9
Walckenaeria antica	3/9

and *Trochosa spinipalpis*. Also *Pardosa pullata, Antistea elegans, Agyneta cauta* and *Pirata uliginosus* were among dominants at least on four of the nine studied bogs (Table 3).

Many interesting bog spiders (some of them rare in Finland) not included in Table 2 were also found on some of the studied open peat bogs. These include: *Haplodrassus moderatus* (Kulczynski, 1897), *Agyneta suecica* Holm, 1950, *Aphileta misera* (O.P.-Cambridge, 1882), *Carorita limnaea* (Crosby & Bishop, 1927), *Mecynargus foveatus* (Dahl, 1912), *Meioneta mossica* Schikora, 1993, *Minicia marginella* (Wider, 1834), *Sintula corniger* (Blackwall, 1856), *Tallusia experta* (O.P.-Cambridge, 1871), *Taranucnus setosus* (O.P.-Cambridge, 1863), *Agroeca dentigera* Kulczynski, 1913, *Hygrolycosa rubrofasciata* (Ohlert, 1865), *Neon valentulus* Falconer, 1912, *Dipoena prona* (Menge, 1868), *Theonoe minutissima* (O.P.-Cambridge, 1879) and *Zora parallela* Simon, 1878. Most of them prefer bogs, or are restricted to bogs in Finland, although some are more eurytopic in Central Europe (cf. Hänggi et al. 1995).

Distributional notes

Peat bogs are habitats with a special microclimate, due not only to the moisture in the peat and *Sphagnum* layers but also to the open, sun-exposed surface (cf. Nørgaard 1951; Krogerus 1960). Bogs or mires are known as localities of northern species, often these species are eurytopic in the North and found in the southern part of their range only on bogs; Petersen (1954) called these 'mire species'. In the present material, *Pardosa hyperborea* and *Gnaphosa microps* are good examples of 'mire species'. *P. hyperborea* is common on peat bogs (except some coastal ones), whereas *G. microps* is rarely caught.

Depending on the southern distributional limit, the northern species found in the present study area can be divided into three groups: 1) species found only on inland bogs (zones B-C in Fig. 1), e.g. *Pardosa atrata, Mecynargus sphagnicola* and *Diplocentria rectangulata* (Emerton, 1915); 2) species found also on the coastal mainland bogs, at least on some of them, e.g. *Pardosa hyperborea, Gnaphosa lapponum* and *Walckenaeria capito* (Westring, 1861); 3) species found even on bogs in the archipelago, at least on larger bogs in the Åland Islands (cf. Lehtinen et al. 1979), e.g. *Arctosa alpigena, Gnaphosa microps, Robertus lyrifer* Holm, 1939 and *Sisicus apertus* (Holm, 1939).

Southern species are also known to live on peat bogs. These can be divided into two groups: 1) species found only in the archipelago and coastal bogs (cf. also Koponen 1985), e.g. *Ozyptila gertschi* Kurata, 1944, *Lepthyphantes ericaeus* (Blackwall, 1853), *Satilatlas britteni* (Jackson, 1912) and *Centromerus semiater* (L. Koch, 1879) (the last-mentioned found also in the easternmost Finland, but not in the SW-Finnish mainland); 2) species found also on inland peat bogs of the study area. e.g. *Zora parallela, Agroeca dentigera, Drassyllus pusillus* (C.L. Koch, 1833*), D. lutetianus* (L. Koch, 1866), *Neon valentulus, Glyphesis cottonae, Maro minutus, Sintula corniger* and *Taranucnus setosus*.

(Micro)climatic factors, connected with the continentality/oceanity of the site (Dist.km in Table 1), seem to be the main factors explaining distribution patterns of the bog spiders. This fits also with the bog zonation of SW Finland (cf. Eurola et al. 1984), especially the border between zones A and B (Fig. 1) seems to be a limit for some northern and southern species. The isolation of bog (bog% in Table 1) as well as the size of investigated open bog habitat seems to have less effect on the distribution of bog spiders in southwestern Finland.

ACKNOWLEDGEMENTS
I wish to thank Peter van Helsdingen and Jean-Pierre Maelfait for comments on the manuscript and Veikko Rinne for help in compiling the map.

REFERENCES

Eurola, S., Hicks, S. & Kaakinen, E. 1984. Key to Finnish mire types. In: *European mires* (P.D. Moore ed.), pp. 11-117. Academic Press, London.

Hänggi, A., Stöckli, E. & Nentwig, W. 1995. *Lebensräume mitteleuropäischer Spinnen.* Centre Suisse de Cartographie de la Faune, Neuchâtel. [*Miscellanea Faunistica Helvetiae* 4, 1-460]

Koponen, S. 1968. Über die Evertebrata-Fauna (Mollusca, Chilopoda, Phalangida, Araneae und Coleoptera) von Hochmooren in Südwest-Häme. *Lounais-Hämeen Luonto* 29, 12-22.

Koponen, S. 1978. On the spider fauna of mires in Kuusamo. *Acta Universitatis Ouluensis A 68 Biologia* 4, 209-214. [In Finnish with an English summary]

Koponen, S. 1979. Differences of spider fauna in natural and man-made habitats in a raised bog. *The National Swedish Environment Protection Board, Report PM* 1151, 104-108.

Koponen, S. 1985. On changes in the spider fauna of bogs. *Memoranda Societatis pro Fauna et Flora Fennica* 61, 19-22. [In Finnish with an English summary]

Koponen, S., Relys, V. & Dalius, D. 2001. Changes in structure of ground-living spider (Araneae) communities on peatbogs along transect from Lithuania to Lapland. *Norwegian Journal of Entomology* 48, 167-174.

Krogerus, R. 1960. Ökologische Studien über nordische Moorarthropoden. *Societas Scientiarum Fennica, Commentationes Biologicae* 21 (3), 1-238.

Lehtinen, P.T., Koponen, S. & Saaristo, M. 1979. Studies on the spider fauna of the southwestern archipelago of Finland II. The Aland mainland and the island of Eckerö. *Memoranda Societatis pro Fauna et Flora Fennica* 55, 33-52.

Lohmander, H. 1956. Faunistikt fältarbete 1955 (huvudsakligen Södra Värmland). *Göteborgs Musei Årstryck* 1956, 32-94.

Nørgaard, E. 1951. On the ecology of two lycosid spiders (*Pirata piraticus* and *Pardosa pullata*) from a Danish Sphagnum bog. *Oikos* 3 (1), 1-21.

Palmgren, P. 1972. Studies on the spider populations of the surroundings of the Tvärminne Zoological Station, Finland. *Societas Scientiarum Fennica, Commentationes Biologicae* 52, 1-133.

Palmgren, P. 1977. Notes on the spiders of some vanishing habitats in the surroundings of Helsingfors, Finland. *Memoranda Societatis pro Fauna et Flora Fennica* 53, 39-42.

Petersen, B. 1954. Some trends of speciation in the cold-adapted holarctic fauna. *Zoologiska Bidrag från Uppsala* 30, 233-314.

Platnick. N.I. 1997. *Advances in spider taxonomy 1992-1995.* New York Entomological Society, New York.

Rassi, P., Alanen, A., Kanerva, T. & Mannerkoski, I. 2001. *Red data book of Finland.* Ministry of Environment, Helsinki. [In Finnish with an English summary]

Vilbaste, A. 1980-81. The spider fauna of Estonian mires. *Eesti NSV Teaduste Akadeemia Toimetised, Biloogia* 29, 313-327 and 30, 7-17.

Wahlström, E., Hallanaro, E-L. & Manninen, S. 1996. *The future of the Finnish Environment.* Edita, Helsinki.

European Arachnology 2000 (S. Toft & N. Scharff eds.), pp. 273-278.
© Aarhus University Press, Aarhus, 2002. ISBN 87 7934 006 1
(Proceedings of the 19th European Colloquium of Arachnology, Århus 17-22 July 2000)

Comments to the checklist of Gnaphosidae and Liocranidae (Arachnida, Araneae) of the Baltic States, with remarks on species new to Lithuania

VYGANDAS RĖLYS[1] & DALIUS DAPKUS[2]

[1]*Department of Zoology, Vilnius University, Čiurlionio 21/27, LT - 2009, Vilnius, Lithuania*
(vygandas.relys@gf.vu.lt)

[2]*Department of Zoology, Vilnius Pedagogical University, Studentų 39, LT - 2034, Vilnius, Lithuania*

Abstract

Until the present time, a total of 48 Gnaphosidae and 11 Liocranidae species were registered in the Baltic States. The records of *Gnaphosa sticta* and *Zelotes apricorum* are doubtful. Old records of *Micaria formicaria* and *Scotophaeus scutulatus* cannot be verified by recent data. Despite the given distribution in some sources, no published data on localities of *Drassodes hypocrita* and *Drassodes villosus* are available. It is suggested that these six species should not be included in the regional species checklist. Following this, the recent checklist of the Baltic States comprises 42 species of Gnaphosidae and 11 of Liocranidae: 29 species of Gnaphosidae and 9 of Liocranidae known in Lithuania, 31/10 in Estonia and 32/3 in Latvia. Only a few new species known in Central Europe can be expected to occur in this region (*Poecilochroa conspicua, Apostenus fuscus, Liocranum rupicola*). Some species are found only in one of these countries. *Gnaphosa microps, Haplodrassus moderatus, Zelotes aeneus, Agroeca dentigera, Agroeca lusatica, Phrurolithus minimus* and *Scotina palliardi* are new to the Lithuanian spider fauna.

Key words: spiders, Gnaphosidae, Liocranidae, Estonia, Latvia, Lithuania, zoogeography

INTRODUCTION

Grimm (1985, 1986) published one of the most comprehensive studies on European Gnaphosidae and Liocranidae. Only sparse data on spiders found in Lithuania, Latvia and Estonia (former USSR republics) were available at the time of the publication of these works. Some information dealing with spiders in Latvia and Estonia was published in local scientific journals and was not accessible for the mentioned revisions. For more references on these omitted data see Vilbaste (1987) and Spungis & Relys (1997). The data from this region used in the revision of Liocranidae and Gnaphosidae by Grimm were mainly based on Vilbaste (1980).

Some information was taken from a synopsis by Tyshchenko (1971). New publications containing data about these families from the Baltic countries were published after 1986 (Vilbaste, 1987; Šternbergs, 1990). Recent studies of the Lithuanian spider fauna have revealed new data on these families. The main aim of this article is to present additional information on the distribution of Liocranidae and Gnaphosidae in the Baltic States.

MATERIAL

The data in the literature was analysed. Liocranidae and Gnaphosidae were well presented in the synopsis of the Estonian spiders

(Vilbaste, 1987). Publications dealing with the occurrence of these families in Latvia were also considered (Šternbergs, 1981, 1990). In addition, an article dealing with Latvian spiders published after 1990 (Šternbergs, 1995) was checked for new records of Gnaphosidae and Liocranidae species in this country. Unpublished diploma theses were checked at the Department of Zoology and Animal Ecology of the University of Latvia in Riga. The material of Gnaphosidae from the collection of M. Šternbergs seems to be lost, and all our efforts to find it were unsuccessful. All published data on these families in Lithuania were considered in this paper. Unpublished material collected in Lithuania during 1993-1999 was used as well. The nomenclature of spiders follows Platnick (1993).

RESULTS

The compiled list includes 48 species of Gnaphosidae and 11 species of Liocranidae registered in the Baltic States (33/10 in Lithuania, 33/3 in Latvia and 35/10 in Estonia). A wide number of species were registered only as single individuals, mainly females. Some of the registered species are known only from old records. Grimm (1985) mapped the two species *Drassodes hypocrita* and *Drassodes villosus* over the entire considered region. All published material on these families in the Baltic States was checked, but there was no original data on the localities of these two species in the region. Probably the distribution data of these spiders was taken from Tyshchenko (1971) who gave only approximate distribution data for these species. The list of Gnaphosidae and Liocranidae species registered in the Baltic States is given in the Table 1. Some of the doubtful records and species worthy of revision are separated in the table and discussed below.

Estonia

Vilbaste (1987) presented 35 Gnaphosidae and 10 Liocranidae species found in Estonia. *Zelotes longipes, Phaeocedus braccatus, Gnaphosa sticta, Gnaphosa leporina, Micaria silesiaca* and *Scotina gracilipes* were reported as female singletons. The other 8 Gnaphosidae and 5 Liocranidae species are known only as females. As to the three *Scotina* species, both sexes of only *Scotina palliardi* are registered. 1 female of *Scotina gracilipes* and 2 females of *Scotina celans* are known from Estonia. Only old records of *Trachyzelotes pedestris* and *Micaria formicaria* are available (Grube, 1859). According to the distribution pattern of *Zelotes apricorum* and *Zelotes subterraneus* in Europe (Grimm, 1985), the occurrence of four females of *Zelotes apricorum* in Estonia is doubtful. These specimens could belong to *Zelotes subterraneus* which is well known in the whole region. The identity of one female of *Gnaphosa sticta* inhabiting the northern regions (Koponen, pers. comm.) and not occurring in the southern part of Finland should be revised. This could have been *Gnaphosa microps*, well known from wetland habitats in Estonia and Lithuania. *Gnaphosa nigerrima* is included in the cheklist of Estonia as a synonym of *Gnaphosa lugubris*. They are currently regarded as separate species and one of these species might be overlooked. *Gnaphosa nigerrima* is well known in Lithuania and Finland and is suspected of occurring also in Estonia. Its inclusion in the list must be postponed, however, until new or revised records from Estonia are obtained. In any case, revision of the available material is required to solve these questions. We therefore suggest *Gnaphosa sticta, Zelotes apricorum, Trachyzelotes pedestris* and *Micaria formicaria* be omitted from the checklist. The precise data on *Gnaphosa nigerrima* and *Gnaphosa lugubris* as well as on two *Scotina* species are also doubtful.

Latvia

The analysis of data shows 33 species of Gnaphosidae and 3 species of Liocranidae registered in Latvia. *Gnaphosa bicolor, Haplodrassus soerenseni, Haplodrassus moderatus, Haplodrassus umbratilis, Zelotes petrensis*, and *Micaria romana* are reported as female singletons. Most of these species (except *M. romana*) are known in neighbouring Baltic countries and from both

Table 1. Gnaphosidae and Liocranidae of the Baltic States. ++: species registered in both sexes; +: registered in one sex; x: species not recorded during the last 40 years; ???: occurrence indicated by Grimm (1985) but no exact records available.

	Species	Estonia	Latvia	Lithuania
	Gnaphosidae			
1	*Berlandina cinerea* (Menge, 1872)			++
2	*Callilepis nocturna* (Linnaeus, 1758)	++	+	
3	*Drassodes lapidosus* (Walckenaer, 1802)	++	++	
4	*Drassodes pubescens* (Thorell, 1856)	++	++	++
5	*Drassyllus lutetianus* (L. Koch, 1866)	++	++	++
6	*Drassyllus praeficus* (L. Koch, 1866)	++	++	++
7	*Drassyllus pusillus* (C.L. Koch, 1839)	++	+	++
8	*Gnaphosa bicolor* (Hahn, 1833)	++	+	++
9	*Gnaphosa leporina* (L. Koch, 1866)	+	+	
10	*Gnaphosa lucifuga* (Walckenaer, 1802)		++	
11	*Gnaphosa lugubris* (C.L. Koch, 1839)	++	++	
12	*Gnaphosa microps* Holm, 1939	+		++
13	*Gnaphosa montana* (L. Koch, 1866)	+	++	x
14	*Gnaphosa muscorum* (L. Koch, 1866)		+	++
15	*Gnaphosa nigerrima* L. Koch, 1877			++
16	*Haplodrassus cognatus* (Westring, 1861)	+	++	++
17	*Haplodrassus dalmatensis* (L. Koch, 1866)			++
18	*Haplodrassus moderatus* (Kulczyński, 1897)	++	+	++
19	*Haplodrassus signifer* (C.L. Koch, 1839)	++	++	++
20	*Haplodrassus silvestris* (Blackwall, 1833)		++	++
21	*Haplodrassus soerenseni* (Strand, 1900)	++	+	++
22	*Haplodrassus umbratilis* (L. Koch, 1879)	+	+	++
23	*Micaria nivosa* L. Koch, 1866	++		
24	*Micaria fulgens* (Walckenaer, 1802)	+	++	++
25	*Micaria lenzi* Bösenberg, 1899			+
26	*Micaria pulicaria* (Sundevall, 1832)	++	++	++
27	*Micaria romana* L. Koch, 1866		+	
28	*Micaria silesiaca* L. Koch, 1875	+		+
29	*Micaria subopaca* Westring, 1861	+	++	x
30	*Phaeocedus braccatus* (L. Koch, 1866)	+	+	
31	*Poecilochroa variana* (C.L. Koch, 1839)	+	+	
32	*Scotophaeus quadripunctatus* (Linnaeus, 1758)	++	++	++
33	*Sosticus loricatus* (L. Koch, 1866)	+	++	x
34	*Trachyzelotes pedestris* (C.L. Koch, 1837)	x	+	
35	*Zelotes aeneus* (Simon, 1878)			++
36	*Zelotes clivicola* (L. Koch, 1870)	+	++	++
37	*Zelotes electus* (C.L. Koch, 1839)	++		++
38	*Zelotes exiguus* (Müller & Schenkel, 1895)			+
39	*Zelotes latreillei* (Simon, 1878)	+	++	++
40	*Zelotes longipes* (L. Koch, 1866)	+	+	++
41	*Zelotes petrensis* (C.L. Koch, 1839)	++	+	++
42	*Zelotes subterraneus* (C.L. Koch, 1833)	++	++	++
	Liocranidae			
1	*Agroeca brunnea* (Blackwall, 1833)	++	++	++
2	*Agroeca cuprea* Menge, 1873	+		++
3	*Agroeca dentigera* Kulczyński, 1913			++
4	*Agroeca lusatica* (L. Koch, 1875)	+		++
5	*Agroeca proxima* (O.P.-Cambridge, 1871)	+		++
6	*Agraecina striata* (Kulczyński, 1882)	+		++
7	*Phrurolithus festivus* (C.L. Koch, 1835)	++	++	++
8	*Phrurolithus minimus* C.L. Koch, 1839	+		++
9	*Scotina celans* (Blackwall, 1841)	+		
10	*Scotina gracilipes* (Blackwall, 1859)	+		
11	*Scotina palliardi* (L. Koch, 1881)	++	++	++
	Doubtful or old records			
	Gnaphosa sticta Kulczyński, 1908	+		
	Micaria formicaria (Sundevall, 1832)	x		
	Scotophaeus scutulatus (L. Koch, 1866)			x
	Zelotes apricorum (L. Koch, 1876)	+	++	
	Drassodes hypocrita (Simon, 1878)	???	???	???
	Drassodes villosus (Thorell, 1856)	???	???	???

sexes. Another three species, *Callilepis nocturna*, *Gnaphosa leporina* and *Gnaphosa muscorum*, are known in Latvia only from females. The occurrence of one male and one female of *Zelotes apricorum* is doubtful and revision of the material is necessary.

Lithuania

In comparison with Latvia and Estonia, some older data on Gnaphosidae from the Vilnius region are available (Pupiska, 1939). The main material of spiders was collected during the last 8 years (Relys, 1996, 2000). A total of 33 Gnaphosidae and 9 Liocranidae species were registered in Lithuania. *Gnaphosa montana*, *Micaria subopaca*, *Scotopaeus scutulatus*, and *Sosticus loricatus* are known only from old records. We suggest not including these species in the checklist before recent data are obtained. The absence of some species (*Callilepis nocturna*, *Phaeocedus braccatus* and *Poecilochroa variana*) which are common in the neighbouring countries could be explained by insufficient studies in specific habitats.

Gnaphosidae and Liocranidae species new to Lithuania

Seven species mentioned in this paper viz. *Gnaphosa microps*, *Haplodrassus moderatus*, *Zelotes aeneus*, *Agroeca dentigera*, *Agroeca lusatica*, *Phrurolithus minimus*, and *Scotina palliardi* have not previously been reported for Lithuania. Most of them were found during the investigations of Lithuanian peatbogs and peatlands using pitfall traps in 1999. A part of the data is given in papers by Koponen et al. (2001), Relys & Dapkus (2001). The data on *Agroeca lusatica* refer to the material collected in 1993 and 1998. In the present paper we will briefly summarise data about these new species in Lithuania.

Gnaphosa microps Holm, 1939

Species found in all investigated peatbogs: Čepkeliai (54°00'N, 24°30'E), Laukėnai (54° 00'N, 24°30'E), Baloša (54°54' N, 25°48' E) and Kertušas (53°56'N, 24°34'E). Main activity in July – August. Abundance in different peat-bogs varied markedly. Highest abundances in communities in peatbogs overgrown with pines.

Material (total catches): Čepkeliai, 15.04.-12.11.1999, 81♂ 6♀; Laukėnai, 18.04.-14.11.1999, 3♂ 1♀; Baloša, 15.04.-03.11.1999, 4♂; Kertušas, 18.04.-14.11.1999, 45♂ 1♀.

Haplodrassus moderatus (Kulczyński, 1897)

Species often recorded from peatbogs and fenlands. Two records from Lithuania are known, one from Kertušas peatbog overgrown with pines, the other from a sage swamp by Ežeraitis lake, near village Puvočiai (54°07'N, 24°19'E).

Material: Kertušas, 16.05.–30.05.1999, 1♀; Puvočiai, 19.06.–02.07.1999, 1♂.

Zelotes aeneus (Simon, 1878)

Known only from two localities in Lithuania. One is a former peat exploitation site planted with birch in Palios peatland (54°35'N, 23°42'E). The litter layer consisted mainly of undecomposed birch leaves. The species was caught 5-7 years after the end of commercial exploitation of the peatland. The other locality was a two year old area of open gravel ground in the central part of Vilnius (54°42'N, 25°18'E). For this species a preference for dry, ruderal habitats with areas of open ground can be noticed.

Material: Palios peatland, 29.07.– 12.08.1999, 8♂ 3♀; Vilnius, 23.07.–15.08.2000, 7♂ 1♀; 15.08.-07.09.2000, 2♀.

Agroeca dentigera Kulczyński, 1913

Locally occurring in peatbogs. Main part of individuals was collected in autumn (September - November).

Material: Laukėnai, 16.05.-30.05.1999, 1♀; 18.09.-17.10.1999, 1♂; 17.10.-14.11.1999, 3♂; Baloša, 15.04.-20.04.1999, 1♂; 02.06.-15.06.1999, 1♀; 28.09.-03.11.1999, 5♂.

Agroeca lusatica (L. Koch, 1875)

Species found in dry, ruderal and agrocultural habitats with sparse vegetation and areas of open ground. One locality was a slope of old sand and gravel quarry in the village Verkšionys (54°38'N, 24°55'E). The other locality was the undersized orchard of *Ribes* sp. in the Botanical Garden of Vilnius University in Kairėnai (54°44'N, 25°24'E).

Material: Verkšionys, 30.04.-22.05.1993, 1♂; 22.05.-07.06.1993, 1♀; Kairėnai, 26.06.-17.07. 1998, 1♂ 2♀.

Phrurolithus minimus C.L. Koch, 1839
The sparse records come from peatbogs. Due to insufficient investigations in other habitats in Lithuania it is difficult to state whether peatbogs are a typical habitat for this species in the region. Our data markedly contradict data about the habitat preferences of this species in Central Europe (Hänggi et al., 1995).

Material: Čepkeliai, 11.06.-06.07.1999, 2♂; 23.07.-23.08.1999, 1♀; Baloša, 20.05.-02.06.1999, 1♂; 12.08.-08.09.1999, 1♀.

Scotina palliardi (L. Koch, 1881)
Species found in all investigated peatbogs. Main activity in the end of April and May. This species often becomes dominant in spider communities in various peatbog habitats.

Material (total catches): Čepkeliai, 15.04.-12.11.1999, 135♂ 19♀; Laukėnai, 18.04.-14.11.1999, 147♂ 45♀; Baloša, 15.04.-03.11.1999, 2♂ 2♀; Kertušas, 18.04.-14.11.1999, 30♂ 8♀.

DISCUSSION

The three Baltic States present a narrow region bordering the eastern coast of the Baltic Sea. Due to insufficient arachnological data from some countries (especially Latvia and Lithuania), reliable patterns of the distribution of spider species in the region could not be distinguished. *Berlandina cinerea, Haplodrassus dalmatensis* and *Zelotes aeneus* are considered to be southern species missing in northern regions (Latvia, Estonia, Finland). Only older records of *Berlandina cinerea* in Finland (Koponen, pers. comm.) contradict this statement in the case of *B. cinerea*. In comparison with Central Europe, the Gnaphosidae and Liocranidae fauna in the Baltic region is very similar due to the occur-

rence of most common European species belonging to these families. 48 Gnaphosidae and 11 Liocranidae species were registered in the Baltic States until the present time. The records of *Gnaphosa sticta* and *Zelotes apricorum* are doubtful. We suggest they should be excluded from the regional species list, as well as *Micaria formicaria* and *Scotophaeus scutulatus* (due to the old records). For two species, *Drassodes hypocrita* and *Drassodes villosus*, mapped by Grimm (1985), no localities are known in this region. Following this, the current checklist of the Baltic States includes 42 Gnaphosidae and 11 Liocranidae species. Excluding some possible misidentifications and old records, 29 Gnaphosidae and 9 Liocranidae species are known in Lithuania, 31/10 in Estonia and 32/3 in Latvia. Only a few species known in Central Europe could be expected to be present in this region (*Poecilochroa conspicua, Apostenus fuscus, Liocranum rupicola*). The last two species are also known from Sweden (Kronestedt, pers. comm.). Also the possibility of the occurrence of some southern species cannot be rejected. Some species have not yet been found in one of the Baltic countries. It might be expected that *Gnaphosa microps* and *Zelotes electus* will be found in Latvia and *Drassodes lapidosus* in Lithuania. Elucidation of the status of *Gnaphosa lugubris* and *Gnaphosa nigerrima* in Latvia and Estonia requires revision of the whole material, because *G. lugubris* could be confused with *G. nigerrima*.

ACKNOWLEDGEMENTS
VR is greatly indebted to the Organizing Committee of 19th European Colloquium of Arachnology for granting his participation in this Colloquium, and to J. Dunlop for his help with the manuscript.

REFERENCES

Grimm, U. 1985. Die Gnaphosidae Mitteleuropas. *Abhandlungen des Naturwissenschaftlichen Vereins in Hamburg* 26, 1-318.

Grimm, U. 1986. Die Clubionidae Mitteleuropas Corinninae und Liocraninae *Abhand-*

lungen des Naturwissenschaftlichen Vereins in Hamburg 27, 1-91.

Hänggi, A., Stöckli, E. & Nentwig, W. 1995. *Habitats of Central European spiders. Characterisation of the habitats of the most abundant spider species of Central Europe and associated species.* Miscellanea Faunistica Helvetiae 4. Centre suisse de cartographie de la faune, Neuchatel.

Koponen, S., Rėlys, V. & Dapkus, D. 2001. Changes in structure of ground-living spider (Araneae) communities on peatbogs along a transect from Lithuania to Lapland. *Norwegian Journal of Entomology* 48, 167-174.

Platnick, N.I. 1993. *Advances in Spider taxonomy 1988-1991. With synonymies and transfers 1940-1980.* Entomological Society and American Museum of Natural History Press, New York.

Pupiska, F. 1939. Clubionidae, Drassidae I Dysderidae (Arachnida) okolic Wilna. *PraceTowarzystwa przyjaciol nauk w Wilne* 13, 163-195.

Rėlys, V. 1996. Ergänzungen zur Kenntnis der litauischen Spinnenfauna I. *Revue suisse de Zoologie* Vol. hors serie II, 555-560.

Rėlys V. 2000. Contribution to the knowledge of the spiders (Arachnida, Araneae) of Lithuania II. *Acta Zoologica Lithuanica* 10 (2), 57–63.

Rėlys, V. & Dapkus, D. (2001): Epigeic spider (Arachnida, Araneae) communities in ex-ploited peatbogs of Lithuania. *Norwegian Journal of Entomology*, 48, 153-160.

Rėlys, V. & Dapkus, D. 2002. Relations between epigeic spider (Arachnida, Araneae) communities in raised bog and surrounding pine forest habitat: a case study from Southern Lithuania. In: *European Arachnology 2000* (S. Toft & N. Scharff eds.), pp. 207-214. Aarhus University Press, Aarhus.

Spungis, V. & Rėlys, V. 1997. In memoriam Maris Sternbergs 1940–1996. *Arachnologische Mitteilungen* 14, 1-4.

Šternbergs, M. 1981. Materials on the spider (Aranei) fauna of Latvia. 5. Family Clubionidae. *Latvijas Entomologs* 24, 56-59. [In Latvian].

Šternbergs, M. 1990. Materials on the spider (Aranei) fauna of Latvia. 9. Family Gnaphosidae. *Latvijas Entomologs* 33, 27-30. [In Latvian].

Šternbergs, M. 1995. The spiders (Aranei) in the litter of *Fraxinetum dryopterioso* forest type in the 'Slitere' Nature Reserve. In: *Proceedings of 15th European Colloquium of Arachnology.* (V. Růžička eds.), pp. 169-171. Institute of Entomology, Česke Budějovice.

Tyshchenko, V. 1971. *Opredelitel paukov evropejskoj chasti SSSR.* Nauka, Leningrad.

Vilbaste, A. 1987. *Eesti amblikud.* Valgus, Tallinn.

European Arachnology 2000 (S. Toft & N. Scharff eds.), pp. 279-282.
© Aarhus University Press, Aarhus, 2002. ISBN 87 7934 001 6
(Proceedings of the 19th European Colloquium of Arachnology, Århus 17-22 July 2000)

Observations on the spider family Gnaphosidae (Araneae) in the nature reserve 'Oasis of Simeto' (Italy, Sicily)

FRANCESCA DI FRANCO

Dipartimento di Biologia Animale, Università di Catania, Via Androne 81, I-95124 Catania, Italy
(francesca_dfr@hotmail.com)

Abstract
New data on the Gnaphosidae fauna of the riparian nature reserve 'Oasis of Simeto' (Sicily - Italy) are presented. The specimens were collected by pitfall trapping over 13 months from May 1994 in different wetland environments, each characterised by a specific plant community. Among the species identified *Poecilochroa furcata* and *Poecilochroa senilis* are first records of the species in Italy, while *Leptodrassus albidus*, *Haplodrassus macellinus hebes* and *Zelotes reconditus* are first records for Sicily. Differences in both diversity and abundance of species and specimens are recorded for the environments investigated. Most species have a Mediterranean distribution.

Key words: Araneae, Gnaphosidae, faunistics, nature reserve 'Oasis of Simeto', Sicily, Italy

INTRODUCTION

The 'Oasis of Simeto' is a nature reserve located in the eastern part of Sicily near the mouth of the Simeto River, a few kilometres from Catania. It is one of the larger and more interesting riparian areas of Sicily and represents an important wintering area for migratory birds. It is also particularly interesting for scientific purposes as it is very rich in species that, due to their origin, distribution and ecology, provide significant evidence concerning the origin of the Sicilian fauna. In spite of the importance of this area the invertebrate fauna has not yet been well investigated; the only data available are on Coleoptera, Lepidoptera, Hymenoptera (Insecta), Isopoda (Crustacea) (Balletto & Toso 1982; Baroni-Urbani 1964; Caltabiano et al. 1984; Magistretti 1967; Vandel 1969) and on the orientation behaviour of some species of the spider genus *Arctosa* (Lycosidae) (Papi 1955ab, 1959; Papi & Tongiorgi 1963). New research on the invertebrate fauna of the nature reserve has been carried out and preliminary results on the spider fauna were reported by Di Franco & Lovetere (2000). Among the spiders living on the ground, the Gnaphosidae was the most abundant family in terms of both number of species and specimens. Data on this family is hereby provided.

MATERIALS AND METHODS

Specimens were collected with pitfall traps filled with acetic acid and 5% formalin. Five traps were placed in each of four different areas and were replaced every 20 days for 13 months, beginning in May 1994. The sampling areas were representative for the main vegetation types of the reserve, i.e. psammophilus and halophilus plants (Brullo et al. 1988). Two of these areas were about 2 km from the mouth of the Simeto River (Tamarisk and Quagmire areas) and the others were close to the sea (Saltwort area and Dunes). Sampling was also done at two other areas, situated near the sea, but they have been only partially investigated due to human disturbance. These environments included a pine forest and an adjacent area (Foredunes) characterised by the presence of plants typical of dunes like *Agropyrum junceum*, *Ammophila arenaria* and *Eryngium maritimum*.

The vegetation types of the four main areas sampled are listed below:

1. Quagmire area, plant community: *Juncetum-maritimo-acuti* Horvatic 1934, characterised by *Juncus acutus*, *Aster tripolium*, *Juncus maritimus*, *Carex extensa*, *Lotus preslii* and *Holoshoenus australis*, and halophilous plants such as *Inula crithmoides*, *Sarcocornia fruticosa* and *Limonium angustiofolium*.

2. Tamarisk area, plant community: *Nerio-Tamaricetea* Br.-Bl. & O. Obolòs 1957, characterised by *Tamarix gallica*, *Tamarix africana*, *Salix alba*, *Salix purpurea* and *Salix* cfr. *pedicellata*.

3. Saltwort area, plant community: *Salicornie-tum radicantis* BR.- Bl. 1993; the typical plants of this community are *Sarcocornia perennis* (=*Salicornia radicans*) and *Aleuropus lagopoides*.

4. Dunes, plant community: *Centaureo-Ononidetum ramosissimae* Br.-Bl & M. Frei 1937; the species that characterise this community are *Ononis ramosissima*, *Seseli tortuosum* var. *maritimum* and *Centaurea spherocephala*.

RESULTS

Faunistics

The identified specimens belong to 28 species and 12 genera (Table 1). It includes a male of *Poecilochroa furcata* that was captured outside the sampling area, but within the reserve. *P. furcata* and *Poecilochroa senilis* are first records of these species in Italy, and *Haplodrassus macellinus hebes*, *Leptodrassus albidus* and *Zelotes reconditus* are first records of these species in Sic-

Table 1. Total number of species and specimens found in the different areas studied. The table includes also a male of *Poecilochroa furcata* captured in another environment of the Reserve.

Species	Species distribution	Quagmire	Tamarisk	Saltwort	Dunes	Pine wood	Fore dunes
Berlandina plumalis (O. P.-Cambridge, 1872)	Holoarctic	1					
Drassodes lapidosus (Walckenaer, 1802)	Palearctic	3	1	1		2	
Drassodes persimilis Denis, 1937	S-euromaghrebian			3			
Drassyllus sp.n.	Endemic	5	4	1			
Gnaphosa alacris Simon, 1878	S-euromaghrebian	13	2	2	1		
Haplodrassus dalmatensis (L. Koch, 1866)	Palearctic	2			6		
Haplodrassus invalidus (O.P.-Cambridge, 1872)	Mediterranean	1	1	12			
Haplodrassus macellinus hebes (O.P.-Cambridge, 1874)	Not defined	4					
Haplodrassus severus (C.L. Koch, 1839)	S-euromaghrebian	14	1	1	3		
Leptodrassus albidus Simon, 1914	West Mediterranean		1		1		
Leptodrassus femineus (Simon, 1873)	W-medit.-atlantic	2		1	1		
Nomisia exornata (C.L. Koch, 1839)	Europe to central Asia	8		2	10	1	
Nomisia recepta (Pavesi, 1880)	West Mediterranean				8		
Poecilochroa furcata Simon, 1914							
Poecilochroa senilis (O.P.-Cambridge, 1872)	Mediterranean				1		
Scotophaeus blackwalli (Thorell, 1871)	Cosmopolitan					1	
Setaphis carmeli (O.P.-Cambridge, 1872)	Mediterranean	3	1		3		
Trachyzelotes barbatus (L. Koch, 1866)	Medit. to central Asia	18	14	17	2		
Trachyzelotes lyonnetii (Audouin, 1827)	Medit. to central Asia			334			
Trachyzelotes mutabilis (Simon, 1878)	Mediterranean	1	3	17			
Zelotes atrocaeruleus (Simon, 1878)	Paleartic	16	6	2			
Zelotes callidus (Simon, 1878)	North Mediterranean	5					
Zelotes denisi Marinaro, 1967	Sicily-Maghreb	1			5		
Zelotes labilis Simon, 1914	S-european			3			
Zelotes maccaricus Di Franco, 1998	Endemic				21	3	48
Zelotes nilicola (O.P. Cambridge, 1874)	Mediterranean	20		10	1	7	1
Zelotes reconditus Simon, 1914	Not defined	1	2	1			
Zelotes tenuis (L. Koch, 1866)	Mediteranean	11	52	13	1	2	
Number of species		19	12	16	13	7	2
Number of specimens		129	88	420	56	24	49

ily. *Drassodes persimilis, Leptodrassus femineus, Zelotes labilis* and *Zelotes maccaricus* are second records for Sicily (Di Franco 1986, 1993, 1997).

The distribution of the species collected depends on the habitat preferences of each species. The Quagmire area was the most diverse in terms of species (19 species) and ranged second in number of individuals found (129). Both *Berlandina plumalis, Zelotes callidus* and *Haplodrassus macellinus hebes* were only found in this area. The number of species found in the Saltwort area was also high (16 species) and this was also the area where most specimens were found (420). Most of this abundance was caused by high numbers of *Trachyzelotes lyonnetii* (334), a species only found in this habitat type. Adults of this species appear at the beginning of spring and reach a peak of abundance in May, whereafter their numbers decrease; in autumn and winter we have not captured individuals of this species. In spring 1995 individuals of *T. lyonnetii* were not captured as the weather was bad. They first appeared later, after the sampling period ended. Two other species were only captured in the Saltwort area: *D. persimilis* and *Z. labilis*. Two other species, *Trachyzelotes mutabilis* and *Haplodrassus invalidus*, were also quite abundant in the Saltwort area (17 specimens of each) and were only found in low numbers in the Quagmire area (1 specimen) and Tamarisk area (3 specimens).

The number of species (12) and specimens (88) was lower in the Tamarisk area and there were no species unique for this environment. *Zelotes tenuis* was more abundant in the Tamarisk area than in any other area. The number of species recorded from the Dune area was almost similar (13) to the number of species recorded from the Tamarisk area, but had fewer individuals (56 specimens). This could be due to the harsh environmental conditions. Interesting is the presence of *P. senilis*, captured only here, and *Z. maccaricus* that seems to be particularly linked to the dune environment. In the Pine forest it is interesting to note the unique presence of *Nomisia recepta*, a species particularly linked to wooded and shrubby habitats.

The data show that in each of the four environments there are different communities of Gnaphosidae. The communities of Quagmire, Saltwort and Tamarisk are more similar to each other whereas the community of the Dunes is quite different. The species that have broad habitat preferences have been captured in almost all the habitats, but they are more abundant where the ecological conditions are more suitable for them (Table 1). The species which have restricted habitat preferences have been found only in 1 or 2 areas, as *T. lyonnetii* that is a dominant species in the Saltwort area, and *Z. maccaricus* particularly linked to the Dune environment.

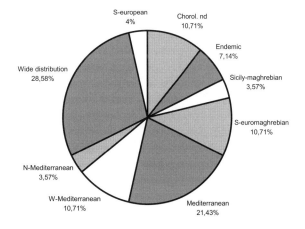

Fig. 1. Pie diagram showing species distribution: wide distribution, South-european, South-euromaghrebian, Mediterranean, West-Mediterranean, North-Mediterranean, Sicily-Maghrebian, endemic, chorology not defined.

Chorology

Most species of Gnaphosidae sampled in the nature reserve 'Oasis of Simeto' have a Mediterranean distribution (Table 1, Fig. 1). Among the species with a Mediterranean distribution there is a remarkable percentage of species with South-euromaghrebian, West Mediterranean, and Sicily-maghrebian distributions. The presence of these species in the Sicilian nature reserve is further evidence of the ancient links between Sicily and the west Mediterranean territories. At present the endemic species are *Zelotes maccaricus* and a possibly new species of *Drassyllus*.

ACKNOWLEDGEMENTS

The study was supported by 'Progetti di ricerca d'Ateneo 1999-2000, Conservazione della Biodiversità in ambienti mediterranei'. The author would like to acknowledge the support of Director Prof. A. Messina.

REFERENCES

Balletto, E. & Toso, G. 1982. Lepidotteri Ropaloceri dei litorali a duna dell'Italia meridionale. *Quad. 'Sruttura delle Zoocenosi terrestri' 3. Ambienti mediterranei i. Le coste sabbiose,* 153-158.

Baroni-Urbani, C. 1964. Studi sulla mirmecofauna d'Italia II. Formiche di Sicilia. *Atti Accademia Gioenia di Scienze Naturali Catania* ser. VI 16, 120-192.

Brullo, S., De Santis, C., Furnari, F., Longhitano, N. & Ronsisvalle, A. 1988. La vegetazione dell'Oasi della Foce del Simeto (Sicilia orientale). *Braun-Blanquetia (Camerino)* 2, 165-188.

Caltabiano, A.M., Caruso, S., Costa, G., Di Franco, F., Leonardi, M.E. & Petralia, A. 1984. Ricerche eco-etologiche sulla fauna del sistema costiero dell'Oasi di protezione faunistica della Foce del Simeto (Catania) I. Biologia comportamentale di *Scarites laevigatus* F. (Coleoptera, Carabidae). *Bollettino*

Accademia Gioenia di Scienze Naturali 17 (232), 25-41.

Di Franco, F. 1986. Gnaphosidae (Arachnida, Araneae) dell'Isola di Salina (Isole Eolie). *Animalia* 13 (1/3), 137-157.

Di Franco, F. 1993. New reports and remarks on Gnaphosidae (Arachnida, Araneae) of Sicily. *Bollettino Accademia Gioenia di Scienze Naturali* 26 (345), 85-92.

Di Franco, F. 1997. Gnaphosidae (Archnida, Araneae) della Riserva naturale orientata 'Oasi di Vendicari' (Siracusa). *Bollettino Accademia Gioenia di Scienze Naturali* 30 (342), 333-342.

Di Franco, F. & Lovetere, F. 2000. Preliminary results on the arachnofauna (Araneae) of the Nature Reserve 'Oasis of Simeto'. *Ekológia (Bratislava)* 19 Suppl. 4, 17-22.

Horvatic, S. 1963. Vegetatcijska karta otoka Paga s opcim pregledom vegetacizskih jedicina Hrvatskog primorja. *Acta Biologica (Zagreb)* 4, pages 10-15.

Magistretti, M. 1967. Coleotteri Cicindelidi e Carabidi della Sicilia. *Accademia Gioenia di Scienze Naturali Catania* ser. VI 19, 120-192.

Papi, F. 1955a. Astronomische Orientierung bei der Wolfspinne *Arctosa perita* (Latr.). *Zeitschrift für vergleichende Physiologie* 37, 230-233.

Papi, F. 1955b. Ricerche sull'orientamento astronomico di *Arctosa perita* (Latr.) (Araneae, Lycosidae). *Pubblicasioni Stazione Zoologica di Napoli* 27, 80-107.

Papi, F. 1959. Sull'orientamento astronomico in specie del gen. *Arctosa* (Araneae, Lycosidae). *Zeitschrift für vergleichende Physiologie* 41, 481-489.

Papi, F. & Tongiorgi P. 1963. Orientamento astronómico verso nord: una capacità innata dei ragni del gen. *Arctosa. Monitore Zoologico* 71, 485-490.

Vandel, A. 1969 Les Isopodes Terrestres de la Sicile. *Atti Accademia Gioenia di Scienze Naturali Catania* ser. VII 1, 1-59.

SAMPLING METHODS

European Arachnology 2000 (S. Toft & N. Scharff eds.), pp. 285-290.
© Aarhus University Press, Aarhus, 2002. ISBN 87 7934 001 6
(Proceedings of the 19th European Colloquium of Arachnology, Århus 17-22 July 2000)

Testing the efficiency of suction samplers (G-vacs) on spiders: the effect of increasing nozzle size and suction time

JAMES R. BELL[1,2], C. PHILIP WHEATER[1], REBECCA HENDERSON[1,3] & W. ROD CULLEN[1]

[1]*Environmental and Geographical Sciences, Manchester Metropolitan University, Chester Street, Manchester. M1 5GD, UK*

[2]*School of Life Sciences, University of Surrey Roehampton, Whitelands College, West Hill, London. SW15 3SN, UK (j.bell@roehampton.ac.uk)*

[3]*University College Northampton, Park Campus, Boughton Green Road, Northampton. NN2 7AL, UK*

Abstract

The efficiency of two G-vacs (Ryobi RSV 3100 and Flymo BVL 320) was tested to examine the effects of increasing the suction nozzle area and the sampling time whilst keeping total sampling area constant (0.49 m²). We tested an *a priori* hypothesis that increasing nozzle area was significant but that reducing sampling time was not, using a planned comparisons approach. We found that when the nozzle diameter was doubled in size, significantly fewer species and individuals of spiders and numbers of *Pachygnatha degeeri*, *Centromerita concinna* and *Lepthyphantes tenuis* were collected. However, the effect of increasing the suction time tenfold did not significantly increase numbers collected. This indicates that a significant proportion of the catch is collected in the first second of sampling. We conclude that in studies of short grasslands, small differences in suction time are unlikely to introduce confounding effects of under-sampling. Increasing the nozzle area may have serious and unwanted effects. G-vac users should be encouraged to give as much detail about their machines and their sampling method as possible and should see that reporting the nozzle size is paramount.

Key words: suction sampling, G-vac, D-vac, nozzle size, suction time, spiders

INTRODUCTION

Suction samplers have long been used in entomological studies, typically these have either been Dietrick (D-vac) or Burkhard machines (e.g. Dietrick 1961; Duffey 1974). These types of sampler are expensive, heavy and cumbersome and there is some suggestion that they have low suction speeds (Stewart & Wright 1995). Aware of the numerous limitations of these old style suction samplers, a different type of machine was required. Stewart & Wright (1995) describe a method of converting a garden 'blow & vac' machine to a modified D-vac, commonly referred to as a 'G-vac'. G-vacs are a quick and easy method of collecting invertebrates (Stewart & Wright 1995), although they cannot be used during wet weather or on grasslands with a heavy dew (Sunderland et al. 1995). G-vacs provide a good estimate of the population density (Dinter 1995) and are particularly useful in homogenous vegetation in which there are a number of small inactive species.

To date there have been many experiments

which have tested the efficiency of suction samplers against other methods such as sweep-nets (e.g. Churchill 1993; Samu & Sárospataki 1995), pitfall traps (e.g. Churchill 1993; Samu & Sárospataki 1995; Topping & Sunderland 1992), heat extraction (e.g. Curry & O'Neill 1979; Dinter 1995) and some papers have tested effects between D-vacs and G-vacs (e.g. Macleod et al. 1994; Stewart & Wright 1995). However, there has been very little research into different methods of using a G-vac to sample invertebrates (but see Samu et al. 1997). Here, we add to this limited knowledge and examine the effects of sampling time and nozzle size on the number of spiders caught.

MATERIALS AND METHODS
Study Site
The White Peak limestone grasslands cover an area of about 350 square kilometers rising to 500 metres above sea level. The White Peak is the southern section of the Peak District National Park in northern England, so named because of the white limestone geology which sometimes becomes exposed. The commonest National Vegetation Classification (NVC) stand type is the CG2d (*Festuca ovina* – *Avenula pratensis* grassland *Dicranium scoparium* subcommunity) which occurs on south facing slopes with shallow soils (Rodwell 1992). High Dale (grid reference: SK 155 719), in the central northern part of White Peak, was used for the experiment as it displayed large areas of CG2d. In this way, we were able to control the habitat type and assume stand type homogeneity.

Methods
High Dale was sampled on the 29.9.1998 using two G-vac machines. A Ryobi RSV 3100 G-vac (engine capacity 31 cm^3) with a nozzle diameter of 13 cm was used alongside a Flymo BVL 320 (engine capacity 32 cm^3) which had a nozzle diameter of 25 cm. Collection nets to catch the invertebrates were made from fine mesh nylon netting which fitted inside both suction nozzles. This net overlapped the external flange on the Ryobi model so that it could be secured

with elasticated cord to prevent it being sucked through the machine. However, the net for the Flymo was secured between the main nozzle and the detachable rim, which is the standard method of attachment on D-vac machines.

Three treatments were used in this experiment: a one-second sample with a small nozzle; a ten-second sample with a small nozzle; and a ten-second sample with a large nozzle. Ten replicates were allocated to each of the three treatments. To test the effect of sampling time, only the Ryobi was used: one-second samples were compared with ten-second samples. To test the effect of nozzle area, ten-second samples taken with the Ryobi were compared with ten-second samples taken with the Flymo.

A standard area of 0.49 m^2 was sampled in all three treatments. To achieve this, the number of sucks, which constituted a sample, differed between the G-vacs because of the different diameters of the two nozzles. Ten sucks constituted one sample for the Flymo, but 37 were required to make one sample for the Ryobi. Samples were taken alternately across the daleside in which samples were distributed evenly along a 50 m transect (i.e. treatments were not clumped in one part of the dale, but spread across the dale, one after the other). An invertebrate sample was taken by placing the nozzle on the ground at 1 m intervals for a specified time and frequency, depending on the G-vac and question under investigation. Between samples the G-vac ran on idle, whilst during samples (i.e the put-downs), the G-vac ran on full throttle. Once the sample was finished, the machine was then turned upside-down and the contents of the net was then emptied into an undiluted methanol solution.

Statistical Methods
All data were transformed using log(x+1) and then tested using Kolmogorov-Smirnov to see if the data differed significantly from the expected normal distribution (Sokal & Rohlf 1995). The Kolmogorov-Smirnov test is often conservative, but is a suitable test for detecting

Table 1. Log(x+1)-transformed numbers (means ± SE) of species and individuals caught with G-vacs depending on nozzle area (Flymo or Ryobi) and suction time, and One Way ANOVA with Contrast Analysis for Planned Comparisons.

	Effect of nozzle area				Effect of sampling time			
	Large nozzle	Small nozzle	$F_{1,27}$	P	1 second	10 seconds	$F_{1,27}$	P
Community effects								
Number of species	0.765±0.046	1.010±0.034	21.31	<0.0001	0.973±0.031	1.010±0.034	0.49	0.4895
Number of individuals	1.525±0.051	2.044±0.035	75.35	<0.0001	1.953±0.039	2.044±0.035	2.32	0.1386
Species effects								
Pachygnatha degeeri	0.258±0.091	0.928±0.114	20.92	<0.0001	0.981±0.105	0.928±0.114	0.13	0.7206
Centromerita concinna	0.614±0.074	1.258±0.063	47.78	<0.0001	1.191±0.060	1.258±0.063	0.58	0.4782
Lepthyphantes tenuis	0.576±0.088	0.951±0.036	19.59	<0.0001	0.884±0.042	0.951±0.036	0.62	0.4366

dispersion, skewness and location (Sokal & Rohlf 1995). All transformed data were found to be normally distributed.

We established an a priori hypothesis within the experimental design, that: increasing nozzle area decreases the number of spiders caught in the samples, but increasing the sampling time has no effect. These were tested as planned comparisons using a one-way ANOVA with contrast analysis. Planned comparisons necessitate the use of contrast over post hoc analysis because of the principle that an a priori hypothesis is being tested. Contrast analysis expresses the difference, if any, in terms of treatment effects by coding the data using +1, 0, -1 integer values to extract only the desired comparisons (Scheiner & Gurevitch 1993). In this experiment, the results are equivalent to separate one-way ANOVAs between treatments, the difference being that they are calculated within a larger ANOVA design. This approach does not require a Bonferroni correction for inequality (Sokal & Rohlf 1995).

We first tested the effect of nozzle size and suction time on the total number of individuals and the total number of species collected, and referred to later as 'community effects'. Three species occurred in large enough numbers to test separately: *Centromerita concinna* (Thorell);

Lepthyphantes tenuis (Blackwall), both species from the Linyphiidae, and *Pachygnatha degeeri* (Sundevall) (Tetragnathidae). The contrast analyses on each of the three spider species are referred to later as 'species effects'.

RESULTS

The treatments gave consistent results for both types of effects tested. When the nozzle diameter was doubled in size, the Flymo sample demonstrated that both the number of species and number of individuals were collected in significantly fewer numbers (Table 1). This was also true for the species effects, despite their differences in body size and overall general appearance (*Centromerita concinna* (Thorell) and *Lepthyphantes tenuis* (Blackwall) are small linyphiid spiders between 2-3 mm body length; *Pachygnatha degeeri* Sundevall is a medium size, bulky spider between 2.5-4 mm body length). When the size of spider increases to over 4 mm, a relatively large spider such as *Pardosa pullata* (Clerck) (4-6 mm body length) was not sampled at all by the Flymo (the large nozzle) but equal numbers (10) were recorded in the two Ryobi (the small nozzle) samples. The effect of increasing the sampling time tenfold did not significantly influence any of the community or species effects measured. We reject the null hypothesis and accept the alternative hypothesis (H_A) that increasing nozzle area decreases the number of spiders caught in the samples; but increasing the sampling time has no effect (H_0 not rejected).

DISCUSSION

There was no attempt made for the collection to reflect the total number of spiders that could be gathered within the sampled area, as this would require a much more intensive effort. Although G-vacs collect a large number of specimens, they should not be considered to be an absolute method (Samu et al. 1997). Some spiders will undoubtedly be missed by the G-vac because they are inaccessible (e.g. stone-dwellers and clubionids which may be hiding within silken cells) and others because they are too mobile and tend to escape when disturbed (e.g. lycosid spiders). Another source of error arises with dense vegetation, which can interrupt airflow and cause a filter to develop, under which predators can hide and avoid capture (Sunderland et al. 1995). However, whilst recognising these constraints, the species collected in this experiment were representative of CG2d grasslands in the White Peak (Bell 1999).

The effect of suction time

Table 1 indicates that there were no significant community or species effects between the one-second and ten-second samples. This would indicate that the majority of the animals were collected within the first second of sampling, with few individuals, if any, added when the suction time was increased tenfold. Macleod et al. (1994) found similar results in their study, suggesting that a fivefold increase in sampling time does little to enhance the catch. These results do not imply that all G-vac sampling henceforth should rely on a one-second sample to collect spiders, as suction time is dependent on habitat: the longer the vegetation and the more complex the structure of the sward, the more likely it is that a longer suction time will be required to collect a fauna representative of the habitat under investigation.

One problem with G-vacs is that suction

time errors can be made inadvertently, caused by lack of concentration by the user or difficult terrain. We have shown that if sampling time errors are made (i.e. ~10% error), then these are unlikely to have any statistical ramifications (i.e cause type I or II errors) when generating estimates of the population.

The effect of nozzle size

Significantly fewer individuals and species were collected by the larger nozzled Flymo when compared to the Ryobi (Table 1). When the engine capacity was not allowed to vary, but the nozzle size was doubled, there was effectively a reduction in suction power. Under a lower suction power, the Flymo was unable to dislodge spiders from their webs and retreats. Even if the Flymo was successful in removing animals, then it was clearly unable to retain them in the collection net until the sampling was completed.

Although the Flymo collected significantly fewer individuals, the Ryobi may have over-sampled the area because of an increased edge effect. Samu et al. (1997) established that edge effects are caused by differences in the diameter of the nozzle: the smaller the size of the nozzle, the greater the edge effect. The Ryobi G-vac may be forcing a significant result by over-sampling due to the smaller nozzle area (Samu et al. 1997). This phenomenon is not testable here because invertebrate quadrat sampling, which would verify the extent of edge effect, was not taken. However, it is probably a combination of edge effect (Samu et al. 1997) and a change in suction power (Macleod et al. 1994) which significantly increased the catch in the two Ryobi samples. Lack of suction power may account for the absence of *P. pullata* from any of the Flymo samples, but edge effect may, in part, be contributing to the increased abundances of *C. conccina*, *L. tenuis* and *P. degeeri*.

To test the effect of increasing the nozzle area, two different machines were used. Small variation in the fan shape and size, and in engine capacity did occur. However, at least for engine capacity, the Flymo was the larger of the two machines and would, if any bias occurred, be in favour of this machine and not the Ryobi. Thus, it is unlikely that a type I error has occurred in this experiment.

Recommendations

Suction samplers are one of the most useful field methods for collecting spiders (e.g. Dinter 1995; Samu et al. 1997) as they are cheap to purchase, easy to use and quick to produce results which are worthy of statistical analysis. Sampling designs which use either an enclosure or other methods, such as transect surveys, are an efficient way of collecting spiders (Samu et al. 1997). However, in light of the results presented here, enlargement of the nozzle should be avoided. To establish the correct suction time, it is possible to plot gradual increases in suction time against numbers of animals collected (e.g. see Macleod et al. 1994): either choose the suction time at the curve's asymptote or the point at which an acceptable proportion of the fauna has been collected (e.g. 75%). As a broad generalisation, a ten-second suction time for each put-down should be sufficient for most transect surveys. Quadrat surveys are the recommended approach to avoid oversampling caused by the edge effect (Samu et al. 1997), but these require much longer suction times to collect all the animals (e.g. 30 minutes, G.J. Bergthaler pers. comm.).

It would be desirable to report the make of the G-vac, its engine size (cm³), nozzle diameter (cm), the sample area (m²), suction time (seconds), and the design of the experiment (transect/quadrat/random) in all ecological investigations. Additionally, F. Samu (pers. comm.) suggests that the net design should not interfere so as to restrict airflow inside the nozzle, especially when large amounts of soil and detritus are prevalent in the samples. The net should be emptied frequently if a blockage is developing in the nozzle. We recommend that the design of the net should be tapered or a blunt spear shape, rather than conical, and longer (>30 cm) rather than short.

ACKNOWLEDGEMENTS

We would like to thank the Ted Locket Memorial Fund for supporting this research. Mr. S. Allen allowed us access to High Dale and English Nature were helpful in more ways than can be described here.

REFERENCES

Bell, J.R. 1999. Habitat use, community structure and biogeography of spiders (Araneae) in semi-natural and disturbed limestone grassland. PhD thesis, Manchester Metropolitan University.

Churchill, T.B. 1993. Effects of sampling method on composition of a Tasmanian coastal heathland spider assemblage. *Memoirs of the Queensland Museum* 33, 475-482.

Curry, J.P. & O'Neill, N. 1979. A comparative study of the arthropod communities of various swards using the D-vac suction sampling technique. *Proceedings of the Royal Irish Academy* (B) 79, 247-257.

Dietrick, E.J. 1961. An improved backpack motor fan for suction sampling of insect populations. *Journal of Economic Entomology* 54, 394-395.

Dinter, A. 1995. Estimation of epigaeic spider population densities using intensive D-vac sampling technique and comparison with pitfall trap catches in winter wheat. In: *Arthropod natural enemies in arable land. I. Density, spatial heterogeneity and dispersal* (S. Toft & W. Riedel eds.), pp. 33-45. Aarhus University Press, Aarhus.

Duffey, E. 1974. Comparative sampling methods for grassland spiders. *Bulletin of the British Arachnological Society* 3, 34-37.

Macleod, A., Wratten, S.D. & Harwood, R.W.J. 1994. The efficiency of a new lightweight suction sampler for sampling aphids and their predators in arable land. *Annals of Applied Biology* 124, 11-17

Rodwell, J.S. (Ed.). 1992. *British plant communities* Vol. 3. Cambridge University Press, Cambridge.

Samu, F. & Sárospataki, M. 1995. Design and use of a hand-hold suction sampler, and its comparison with sweep net and pitfall trap sampling. *Folia Entomologia Hungarica* LVI, 195-203.

Samu, F., Németh, J. & Kiss, B. 1997. Assessment of the efficiency of a hand-held suction device for sampling spiders: improved density estimation or oversampling. *Annals of Applied Biology* 130, 371-378

Scheiner, S.M. & Gurevitch, J. 1993. *Design and analysis of ecological experiments*. Chapman & Hall, London.

Sokal, R.R. & Rohlf, F.J. 1995. *Biometry*. Freeman, New York.

Stewart, A.J.A. & Wright, A.F. 1995. An inexpensive suction apparatus for sampling arthropods in grassland. *Ecological Entomology* 20, 98-102.

Sunderland, K.D., De Snoo, G.R., Dinter, A., Hance, T., Helenius, J., Jepson, P., Kromp, B., Jys, J.A., Samu, F., Sotherton, N.W., Toft, S. & Ulber, B. 1995. Density estimation for invertebrate predators in agroecosystems. In: *Arthropod natural enemies in arable land. I. Density, spatial heterogeneity and dispersal* (S. Toft & W. Riedel eds.), pp. 133-162. Aarhus University Press, Aarhus.

Topping, C.J. & Sunderland, K.D. 1992. Limitations to the use of pitfall traps in ecological studies exemplified by a study of spiders in a field of winter wheat. *Journal of Applied Ecology* 29, 485-491.

European Arachnology 2000 (S. Toft & N. Scharff eds.), pp. 291-297.
© Aarhus University Press, Aarhus, 2002. ISBN 87 7934 001 6
(Proceedings of the 19th European Colloquium of Arachnology, Århus 17-22 July 2000)

Suction sampling in alpine habitats: experiences and suggestions

GERNOT J. BERGTHALER[1] & VYGANDAS RĖLYS[2]

[1]*Bessarabierstr. 72/3, A-5020 Salzburg, Austria* (gjbergthaler@aon.at)
[2]*Department of Zoology, University of Vilnius, Ciurlionio St. 21/27, LT-2009 Vilnius, Lithuania* (vygandas.relys@gf.vu.lt)

Abstract
Experiences regarding the first use of a cheap, light and comfortable suction sampler for density estimations of spiders, harvestmen, pseudoscorpions and carabid beetles in steep slopes of alpine habitats (Salzburg, Austria) are shared and discussed. Data from suction sampling and pitfall trapping gained under the same measured period are compared. Suggestions for future usage of this method are provided. Suction sampling in steep slope habitats involves two people, is time-consuming and thus makes the study expensive. Nevertheless, this is probably the most effective method if quantitative data on arachnids from alpine habitats is to be obtained.

Key words: Arachnida, Araneae, Opiliones, Pseudoscorpiones, Alps, density estimations, methods

INTRODUCTION

There are several methods for estimating the number of epigeic invertebrates per unit area: hand searching, use of fenced traps or emergence traps, heat extraction of soil samples, or sampling with a suction device. During the last decades data obtained by using different types of suction samplers have been published and questions about the efficiency of these samplers have been discussed (Kauri et al. 1969; Solhøy 1972; Duffey 1980; Hand 1986; Sunderland & Topping 1995; Sunderland et al. 1995). The Dietrick vacuum insect sampler or D-vac (Dietrick 1961) and related designs, e.g. the Burkhard (Duffey 1974) or the Thornhill vacuum sampler (Thornhill 1978), have mostly been replaced by the leaf-blowers or leaf-gathering suction devices for absolute density estimations (De Barro 1991; Macleod et al. 1994; Samu & Sárospataki 1995; Samu et al. 1997).

Since the latter devices have never been used in steep alpine habitats, we decided to test them under such conditions. The test was performed parallel to a pitfall trap inventory of arachnid and carabid communities. This report about our first experiences includes suggestions for future suction sampler studies in the steep slopes of alpine habitats.

STUDY AREA

Five different sites within a mosaic of alpine habitats along the Grossglockner Hochalpenstrasse were chosen. The panoramic road lying between the Austrian counties of Salzburg and Carinthia crosses the Hohe Tauern National Park (1800 km² in the Eastern Alps) comprising crystalline rocks with a high mica and a low base mineral content. Sites I, II, and III as well as IV and V were situated close to each other between 1960 m and 2280 m a.s.l.; however,

Fig. 1. The five study sites along the Grossglockner Hochalpenstrasse. I, II, and III are some hundred meters away from IV and V.

they are separated by the road and a mountain stream respectively (Fig. 1):

-Site I: site poor in plant species (*Rumex alpinus* dominates) within a bend of the road; nitrogenous soil with a thick litter accumulation; 1960 m a.s.l.; not sloping.

-Site II: peripheral area of a cattle pasture near a mountain stream; a species rich *Nardetum* with few dwarf shrubs; almost no litter; 2020 m a.s.l.; ≤ 45° slope facing WNW.

-Site III: grassy site (*Anthoxanthum alpinum, Deschampsia cespitosa, Juncus trifidus, Nardus stricta, Poa alpina*) interspersed with dwarf shrub species *Vaccinium myrtillus, V. gaul-*

therioides, and *Calluna vulgaris;* near a mountain stream; low litter accumulation; 2020 m a.s.l.; ≤ 45° slope facing WNW.

-Site IV: artificial embankment below the road with herbaceous vegetation dominated by *Alchemilla vulgaris* agg., *Geranium sylvaticum, Pimpinella major, Silene vulgaris* and some grasses (*Agropyron repens, Festuca rubra, Phleum pratense*); low litter accumulation; 2200 m a.s.l.; ≥ 45° slope facing SW.

-Site V: grass-dominated site (*Agropyron repens, Carex sempervirens, Deschampsia cespitosa, Festuca rubra, Juncus jackquinii, Poa alpina*) some meters above a concrete embankment wall

A.

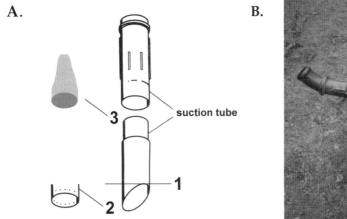

B.

Fig. 2. The Partner 32 GBI suction sampler: assembly **(A)** and usage **(B)**. 1: slanting end of nozzle cut off (see 2); 2: holes (5 mm diameter) drilled around nozzle end; 3: collecting net, to be fitted between the two parts of the suction tube.

along the road; low litter accumulation on a stony underground; 2280 m a.s.l.; ≥ 45° slope facing SW.

MATERIAL AND METHODS

Suction sampling

The suction sampler used was a petrol driven Partner 32 GBI Blower: ca. 6 kg, 32 cm^3 air cooled engine operating at 7600 rpm and 86 dB maximum, nozzle width 0.01 m^2. The distal part of the suction tube needed modification (Fig. 2): The slanting end of the nozzle was cut off so that the sampler can be used in a perpendicular position (1 on Fig. 2A). Several holes (ca. 5 mm diameter) were drilled around the nozzle's end (2) to allow a continuous air-stream also when the suction-tube is firmly pressed down to the ground. A collecting net (3) can easily be fitted between the two parts of the suction tube. A metal cylinder enclosure (Fig. 2B) was made (ca. 7 kg solid steel which allowed us to press it some cm into the soil). It was dimensioned with an area of 0.1 m^2 and 0.5 m height.

The upper part of the vegetation was vacuumed first, then cut, and the plants were searched for remaining specimens. Afterwards, the sampling was continued by repeatedly pressing down the nozzle, so that the area within the cylinder was completely covered. Depending on the vegetation and soil type the collecting net had to be emptied after varying numbers of sub-samples to maintain the suction power of up to 66 m/s. We emptied the collecting net by lowering the suction power to a minimum using the throttle lever while the nozzle was held over a plastic bucket. The material was transferred into a glass with ethanol. We finished the sampling procedure within one enclosure when no more specimens could be found either in the sub-sample material or on the soil surface. Five enclosures were sampled in each of the five sites on 07.08.1998. The sampling duration for one enclosure was up to 20 minutes on average.

Pitfall trapping

Between 13 July and 8 September 1998 five pitfall traps (7.5 cm Ø, 0.25 l) with 4% formalin solution (and detergent) were placed in the centre of each of the habitat sites either in a row (sites I-III) or within a 6 x 6 m square (IV and V). The traps were placed about 3 m apart, protected by aluminium covers and emptied every two weeks (28 July, 11 Aug., 24 Aug., and 8 Sept.).

RESULTS

Time consumption

Starting with collecting in the field and finishing with a simple species list, the time needed for one suction sampling and one pitfall trapping period (28.07. - 11.08.98), was approximately the same. Suction sampling had to be done by two persons, because of the steepness of the sites: 12.6 h (field work) + 6 h (identification) = 18.6 h for pitfall traps vs. 16.9 h (field work) + 2.2 h (identification) = 19.1 h for suction sampling. Thus our suction sampling effort resulted in 0.148 m^2/h covered. The difference of 0.5 h less for gaining pitfall trap results can be subtracted from the time needed for an extra travel to the study site, since pitfall traps have to be installed and emptied.

Suction sampling

A total of 177 arachnids and 1 carabid were captured. Of the arachnids 20.9% were adult and belonged to 10 spider, 2 harvestman, and 1 pseudoscorpion species (Table 1). The density of spiders was highest with a median of 6 specimens/m^2 in III, and lowest with a median of 2 specimens/m^2 in II (Fig. 3). The density of adult spiders was highest with a median of 2 specimens/m^2 in V. Harvestman density was highest in the herbaceous sites I and IV and lowest in the grassy sites, especially in II, where not even one specimen was sampled.

Suction sampling precision

The coefficient of variation (CV = SD/mean) is a dimensionless measure of sampling variability that allows comparison between sites and years. For adult spiders it was 0.32-0.61, for adult harvestmen 0.66-1.00. It was within the same

Tab. 1. Species recorded at five sites in 25 suction samples (0.1 m² each) on 07.08.1998 as well as with 25 pitfall traps (28.07. - 11.08.1998). Species captured only with pitfall traps are not listed. The number of juveniles are given in brackets.

FAMILY/Species	Pitfall	Vac
CLUBIONIDAE		
Clubiona sp.	-	(1)
LINYPHIIDAE		
Gen. sp.	(3)	(11)
ERIGONINAE		
Ceratinella brevipes (WESTRING, 1851)	1	3
Diplocephalus latifrons O.P.-CAMBRIDGE, 1863	2	1
Erigonella subelevata (L. KOCH, 1869)		4
Pelecopsis radicicola (L. KOCH, 1872)	11	5
Walckenaeria alticeps (DENIS, 1952)		1
Gen. sp.	(4)	(9)
LINYPHIINAE		
Bolyphantes luteolus (BLACKWALL, 1833)		(5)
Centromerus pabulator (O.P.-CAMBRIDGE, 1875)	19	6
Meioneta rurestris (C. L. KOCH, 1836)	20	6
Gen. sp.	(34)	(31)
LYCOSIDAE		
Pardosa oreophila SIMON, 1937	224 (33)	4 (78)
NEMASTOMATIDAE		
Nemastoma triste (C.L. KOCH, 1835)	57	2
PHALANGIIDAE		
Mitopus morio (FABRICIUS, 1799)	139 (223)	4 (5)
NEOBISIIDAE		
Neobisium noricum BEIER, 1939		1
CARABIDAE		
Calathus micropterus (DUFTSCHMID, 1812)	3	1
Total adults (juveniles)	476 (297)	38 (140)

range for total spiders, but the minimum with 0.41 was lower for total harvestmen.

Pitfall trapping

In the period from 28.07. - 11.08.98, 941 arachnids and 42 carabids were captured with 25 pitfall traps. Among 412 spiders (79.9% adults), 529 harvestmen (52.7% adults), and 42 carabids were 24 spider, 5 harvestman, and 14 carabid species. In Table 2 the number of specimens is given only for species also captured by suction sampling.

Relative effectiveness

A total of 28 spider, 5 harvestmen and 1 pseudoscorpion species were captured by both methods. In the suction samples 4 spider and 1 pseudoscorpion species were exclusively taken.

Thus more specimens and species were captured with pitfall traps. In the suction samples the number of juveniles was higher than the number of adults, whereas in pitfall traps it was vice versa. More juvenile and female erigonines were captured by suction sampling, whereas linyphiids, lycosids, harvestmen, and carabids were better represented in pitfall traps (Table 1).

DISCUSSION

Experiences

In agreement with previous studies these results from alpine sites demonstrate that pitfall traps result in a better return of species and specimens per unit effort. But pitfall trapping cannot be used as a substitute to give an index of abundance because the relationship between pitfall catch and density is unreliable (Sunderland & Topping 1995). For studies where the age structure of a population is needed or where at least relative density estimates are important, techniques such as the suction sampling are probably preferable. After Merrett & Snazell (1983) many other large-scale comparisons of the results of pitfall trapping and suction sampling have been made, especially regarding farmland spiders. A comparison not only provides information about the collecting methods, but also indicates differences in the ecology or behaviour of species which render them more susceptible to capture by one method or the other, as reported for example by Flatz (1986) from high-mountainous sites. Method-specific species were regularly found with both sampling methods. Thus both techniques contribute to assessing species assemblages in lower vegetation strata as reported by Standen (2000).

The efficiency of the sampling method, the area sampled, and the amount of samples per site are factors which influence the accuracy of density estimations (Meyer 1981). The efficiency itself is dependent on the duration of suctioning and on the vegetation height (Henderson & Whitaker 1977). Since these factors vary in the previously published studies on density estimations in alpine habitats, comparisons are difficult or impossible. Sometimes even a description of the method or vegetation type is missing or important details are lacking. Hence, the results gained by soil extractions in a Kempson apparatus (Meyer 1980, 1981) are more accurate. The latter reference mentions densities of 7-78 Araneae/m^2, 3-5 Opiliones/m^2, 52 Pseudoscorpiones/m^2, and 4-26 carabids/m^2 (adult specimens only) for 6 cm deep soil samples taken at the beginning of August within different sites (1850 - 2300 m a.s.l) in the Hohe Tauern National Park. A density of 35 adult spider specimens/m^2 is stated by Puntscher (1980) for a site in the Ötztaler Alps (2650 m a.s. l.) sampled with a Burkhard Univac at the beginning of August. The comparison with our data reveals an enormous difference in the carabid and pseudoscorpion density which is lower in our study, since we only sampled the soil surface.

The density estimation method used here was less labour-intensive than the combination of methods used by Sunderland et al. (1987) in winter wheat: 0.15 m^2/h vs. 0.1 m^2/h covered. The combination of suction sampling and hand-searching by Topping & Sunderland (1994) in a cereal crop was even much less labour-intensive (1.66 m^2/h covered). Thus the

suction sampling presented here seems to be a preferable method also if time is limited. However, the suggestions below should be taken into account in future studies.

Suggestions

This cheaper, lighter and more comfortable alternative to the former generation of suction samplers and to less absolute, uncontrolled, time consuming or more destructive methods can be recommended for density estimations of arachnids also in steep alpine habitats. As a result of the experiences made by previous users (De Barro 1991; Mcleod et al. 1994; Samu & Sarospataki 1995; Samu et al. 1997) we tried to increase the efficiency. Thus we excluded oversampling with the use of an enclosure and we made varying numbers of sub-samples until no remaining specimen could be found either in the sub-sample material or on the soil surface.

Nevertheless, it was far from being an 'absolute' density estimation, because of inaccessibility of some spiders, harvestmen and carabids which have retreats in the stony underground.

Our sampling was restricted to one date and five replicates per site. In the future, sampling should be done more extensively. Although the coefficients of variation are relatively low

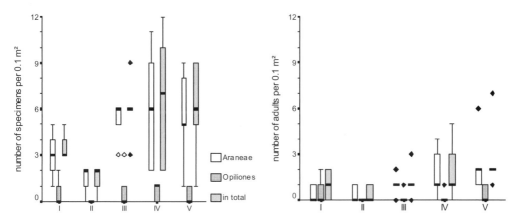

Fig. 3. Box plots relating to the 25 suction samples at the five sites (I-V) on 07.08.1998. They describe the first quartile, the median, the third quartile, the range, as well as the outliers (◊) and extremes (w) for the specimens: (A) all specimens (Araneae, Opiliones, in total) excl. pulli of lycosids and (B) adults (Araneae, Opiliones, in total) per 0.1 m^2

(Sunderland & Topping 1995), more replicates should be made on at least three dates during the short season in the alpine region from June to September. Since suddenly changing weather conditions in the Alps limit the time for optimal sampling in dry vegetation, i.e. there are often rain showers in the afternoon, the dew in the morning and intensive snowfall periods during summer, respectively, these factors have to be taken into account when scheduling such a study in the alpine region. Hence, we suggest a lower number of sites than presented here.

Using a wider cylinder would be difficult on the stony and uneven ground of alpine slopes. Of course it would also be more time-consuming, much heavier and more inconvenient to carry and handle. The enclosure's height of 50 cm is appropriate for alpine areas. In any case, it should be of solid steel, so that it can be rammed down into the stony soil.

ACKNOWLEDGEMENTS

Thanks go to Balázs Kiss and Ferenc Samu (Budapest) for demonstrating the suction sampler and sharing experiences. We thank Konrad Thaler (Innsbruck) and Volker Mahnert (Geneva) for their valuable help in identifying some arachnids, and Erich Traugott (Salzburg) for providing us with the carabid data. Thanks also go to Gerald Moser (Salzburg) for the botanical excursion to the sites. Last but not least, we are indebted to Søren Toft (Århus) and two anonymous referees who provided very helpful comments on a previous version. This study was part of an invertebrate inventory project financed by the Glockner-Öko-Fonds 1998 of the Großglockner Hochalpenstraßen AG.

REFERENCES

De Barro, P.J. 1991. A cheap lightweight efficient vacuum sampler. *Journal of the Australian Entomological Society* 30, 207-208.

Dietrick, E.J. 1961. An improved backpack motor fan for suction sampling of insect populations. *Journal of Economic Entomology* 54, 394-395.

Duffey, E. 1974. Comparative sampling methods for grassland spiders. *Bulletin of the British Arachnological Society* 3 (2), 34-37.

Duffey, E. 1980. The efficiency of the Dietrick vacuum sampler (D-vac) for invertebrate population studies in different types of grassland. *Bulletin of Ecology* 11 (3), 421-431.

Flatz, U. 1986. Zur Biologie und Ökologie epigäischer Wiesenspinnen des Innsbrucker Mittelgebirges (Nordtirol, Österreich). In: *Actas X Congreso Internacional de Aracnologia, Jaca (España)* (J.A Barrientos ed.) I, pp. 225-230.

Hand, S.C. 1986. The capture efficiency of the Dietrick vacuum insect net for aphids on grasses and cereals. *Annals of Applied Biology* 108, 233-241.

Henderson, I.F. & Whitaker, T.M. 1977. The efficiency of an insect suction sampler in grassland. *Ecological Entomology* 2, 57-60.

Kauri, H., Moldung, T.J. & Solhøy, T. 1969. Turnbull and Nicholls' 'quick trap' for acquiring standing crop of evertebrates in high mountain grassland communities. *Norsk Entomologisk Tidsskrift* 16, 133-136.

Macleod, A., Wratten, S.D. & Harwood, R.W.J. 1994. The efficiency of a new lightweight suction sampler for sampling aphids and their predators in arable land. *Annals of Applied Biology* 124, 11-17.

Merrett, P. & Snazell, R. 1983. A comparison of pitfall trapping and vacuum sampling for assessing spider faunas on heathland at Ashdown Forest, south-east England. *Bulletin of the British Arachnological Society* 6, 1-13.

Meyer, E. 1980. Ökologische Untersuchungen an Wirbellosen des zentralalpinen Hochgebirges (Obergurgl, Tirol). VI. Aktivitätsdichte, Abundanz und Biomasse der Makrofauna. Abundanz und Biomasse von Invertebraten in zentralalpinen Böden (Hohe Tauern, Österreich). *Veröffent-lichungen der Universität Innsbruck* 125, *Alpin-Biologische Studien* XIII. Innsbruck University, Innsbruck.

Meyer, E. 1981. Abundanz und Biomasse von Invertebraten in zentralalpinen Böden

(Hohe Tauern, Österreich). In: *Veröf-fentlichungen des Österreichischen MaB-Hochgebirgsprojektes Hohe Tauern 4: Boden-biologische Untersuchungen in den Hohen Tau-ern 1974-78*, pp. 153-178. Universitätsverlag Wagner, Innsbruck.

Puntscher, S. 1980. Ökologische Untersuchun-gen an Wirbellosen des zentralalpinen Hochgebirges (Obergurgl, Tirol). Verteilung und Jahresrhythmik von Spinnen. *Veröf-fentlichungen der Universität Innsbruck* 129, Innsbruck University, Innsbruck.

Samu, F. & Sarospataki, M. 1995: Design and use of a hand-hold suction sampler, and its comparison with sweep net and pitfall trap sampling. *Folia Entomologica Hungarica* 56, 195-203.

Samu, F., Németh, J.& Kiss, B. 1997: Assess-ment of the efficiency of a hand-held suc-tion device for sampling spiders: improved density estimation or oversampling? *Annals of Applied Biology* 130, 371-378.

Solhöy, T. 1972. Quantitative invertebrate stud-ies in mountain communities at Hardanger-vidda, South Norway. I. *Norsk Entomologisk Tidsskrift* 19, 99-108.

Standen, V. 2000. The adequacy of collecting techniques for estimating species richness of grassland invertebrates. *Journal of Applied Ecology* 37, 884-893.

Sunderland, K.D. & Topping, C.J. 1995. Esti-mating population densities of spiders in cereals. In: *Arthropod natural enemies in ar-able land. I. Density, spatial heterogeneity and dispersal.* (S. Toft & W. Riedel eds.), pp. 13-22. Aarhus University Press, Aarhus.

Sunderland, K.D., Hawkes, C., Stevenson, J.H., McBride, T., Smart, L.E., Sopp, P.I., Powell, W., Chambers, R.J. & Carter, O.C.R. 1987. Accurate estimation of invertebrate density in cereals. *Bulletin SROP/WPRS* 10, 71-81.

Sunderland, K.D., De Snoo, G.R., Dinter, A., Hance, T., Helenius, J., Jepson, P., Kromp, B., Lys, J.-A., Samu, F., Sotherton, N.W., Toft, S. & Ulber, B. 1995. Density estimation of arthropod predators in agroecosystems. In: *Arthropod natural enemies in arable land. I. Density, spatial heterogeneity and dispersal.* (S. Toft & W. Riedel eds.), pp. 133-162. Aarhus University Press, Aarhus.

Thornhill, E.W. 1978. A motorised insect sam-pler. *Pest Articles and News Summaries* 24, 205-207.

Topping, C.J. & Sunderland, K.D. 1994. Meth-ods for quantifying spider density and mi-gration in cereal crops. *Bulletin of the British Arachnological Society* 9, 209-213.

European Arachnology 2000 (S. Toft & N. Scharff eds.), pp. 299-300.
© Aarhus University Press, Aarhus, 2002. ISBN 87 7934 001 6
(Proceedings of the 19th European Colloquium of Arachnology, Århus 17-22 July 2000)

An improved version of the 'aspirator gun' – a device for collecting arthropods

FERENC TÓTH

Szent István University, Faculty of Agricultural and Environmental Sciences, Department of Plant Protection, H-2103 Gödöllő, Páter K. u. 1., Hungary (tothf@fau.gau.hu)

Abstract

In this paper I discuss improvements to a one-handed mechanical device known as an 'aspirator gun'. The main structural change is that while the previous version of the aspirator gun contains one pump and two check valves, the new device contains two pumps (a pair of bellows), four check valves and a direction switch. As a result of this modification, the airflow can now be either long and controlled (e.g. to vacuum many ants or fruit flies) or short, pre-set and triggered (e.g. to capture fast-moving insects or spiders). The direction of the airflow is also alterable, so that the captured insects can be blown out.

Key words: aspirator gun, arthropod collecting, pooter

INTRODUCTION

Lung-operated aspirators (pooters) provide a continuously controlled long airflow, but their use can cause discomfort and can be dangerous (inhalation of fungal spores, pollen, dust, excrements or hairs of the arthropods, etc). There are several types of gun-like aspirators. Motorised aspirators (Wade and Wade 1993) require additional energy sources, e.g. batteries. Mechanical aspirators (Schuman 1976; Zoz 1987; Winnicki 1988; Fahringer 1989; Brandstetter 1993; Ott 1994; Tóth 1999, 2000) provide only short and pre-set airflow. Most of these machines require both hands when the machine is made ready for shooting. Mechanical aspirators operating with a piston pump (Schuman 1976; Zoz 1984; Winnicki 1988; Ott 1994) and/or a stored vacuum (Schuman 1976; Zoz 1987; Fahringer 1989) are sensitive to dust.

DESCRIPTION

The new device (Fig. 1) consists of a catching tube (1), a flap-door (2), a transparent collection chamber (3) with a sieve (4), a direction switch (5), a flexible suction tube (6), a flexible blow-tube (7), a pair of bellows (8) with two check valves (9) on both, three springs (10), a pump arm (11), and a trigger (12). Pressing or releasing the pump arm makes one of the bellows expand and the other compress. Repeated movements of the pump arm generate a continuous airflow in the catching tube and the collection chamber. Accelerating or decelerating of the movements regulates the air speed. The position of the direction switch determines whether the air moves into or out of the catching tube and the collection chamber. When the direction switch is in the 'suck' position, the suction tube is connected with the collection chamber and the blow tube directly with the atmosphere. The vacuum draws the insect into the collection chamber through the catching tube and the sieve stops it. The 'blow' position creates the opposite effect: the captured insect

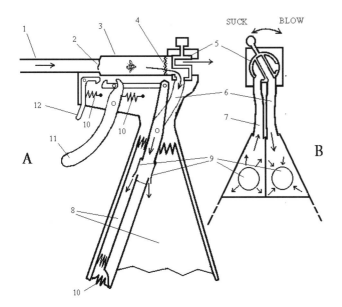

SUCK   BLOW

A   B

Fig. 1.
Improved aspirator gun.
(A) Through section;
(B) Cross-section.
1, catching tube;
2, flap-door;
3, transparent collection
   chamber;
4, sieve;
5, direction switch;
6, flexible suction tube;
7, flexible blow-tube;
8, a pair of bellows;
9, check valves;
10, springs;
11, pump arm;
12, trigger.
Arrows inside the device
indicate the direction of the
airflow.

is ejected from the collection chamber through the catching tube. Airflow can be very quick and short as well: a firm pumping movement fixes the bellows; with bellows fixed triggering is made possible. This is essential: collecting fast moving arthropods requires high aiming precision.

ADVANTAGES AND DISADVANTAGES

The improved aspirator gun is non-motorised and one-handed. In contrast to the piston pumps the bellows are not sensitive to dust. The airflow can be either long and controlled (e.g. to vacuum many ants or fruit flies), or short, pre-set and triggered (e.g. to catch fast-moving arthropods). The direction of the air-flow is alterable, so that the captured animals can be blown out. Inevitably, when these im-provements were made to the aspirator gun, the structure became more complicated: as a result, this has increased the potential manufac-turing time and cost.

ACKNOWLEDGEMENTS

I would like to express my thanks to the organ-isers of the Colloquium for a grant to attend the meeting.

REFERENCES

Brandstetter, H.P. 1993. Verfahren und Vor-richtung zum Fangen von Insecten. Pat. Id: DE 4327150 A1.

Fahringer, S.A. 1989. Insect capturing device. Pat. Id: US 4817330 A.

Ott, G. 1994. Suction insect trap apparatus. Pat. Id: US 5367821 A.

Schuman, M. 1976. Manually operated suction device for capturing small objects. Pat. Id: US 3965608 A.

Tóth, F. 1999. Sniffing pistol for gathering of small animals especially arthropods and fries. Applic. Id: P 9901059.

Tóth, F. 2000. Aspirator gun - a new device for sampling spiders and insects. *Ekológia (Bratislava)* 19 Suppl. 3, 279-280.

Wade, B.R. & Wade, T.L. 1993. Pest collection-disposable device. Pat. Id: US 5175960 A.

Winnicki, J. 1988. Flying insect exterminator. Pat. Id: US 4733495 A.

Zoz, J.R. 1984. Apparatus for sucking isolated objects, for example insects, flies, wasps, etc. Pat. Id: WO 8400280 A1.

TAXONOMY

SYSTEMATICS

EVOLUTION

European Arachnology 2000 (S. Toft & N. Scharff eds.), pp. 303-314.
© Aarhus University Press, Aarhus, 2002. ISBN 87 7934 001 6
(Proceedings of the 19th European Colloquium of Arachnology, Århus 17-22 July 2000)

Why no subspecies in spiders?

OTTO KRAUS

Zoologisches Institut und Zoologisches Museum, Universität Hamburg, Martin-Luther-King-Platz 3, D-20146 Hamburg, Germany (otto.kraus@zoologie.uni-hamburg.de)

Abstract

The several tens of thousands of spider species that have been named so far have almost exclusively proved to be morphospecies; some of them have been tested and proved to be biospecies, but subspecies have only been recognized in a few exceptional instances. This phenomenon cannot be explained by regular speciation mechanisms alone: obviously, these are no different from those ocurring in other groups of the animal kingdom. However, the species groups that have been sufficiently investigated in spiders indicate that separation fairly quickly produces superspecies composed of allospecies ('semispecies'). The origin of allospecies patterns can be explained by a combination of the regular effects of separation with functional needs to safeguard sperm transfer between sexual partners. Selection pressure towards optimal co-adaptation between male and female copulatory organs may shorten transitory phases at a subspecies level. This could explain the origin of superspecies composed of similar but clearly distinguishable biospecies (allospecies); primarily, they are unable to coexist sympatrically or even syntopically.

Key words: Arachnida, Araneae, speciation, genital coadaptation, superspecies, allospecies

INTRODUCTION

It was Ted Locket who raised the critical question of whether some linyphiid spiders from remote localities in Europe really represented true species, as there were only slight morphological character differences on which to base this decision. This was almost 25 years ago on the occasion of the 7th International Congress of Arachnology held in 1977 in Exeter. His preliminary remarks did not form part of the congress volume (see Merrett 1978), but the problem was of course not forgotten by systematicists, who were interested in the general aspects of speciation and evolutionary biology. However, I am not aware of any sound discussion later. Not even a hypothesis has been published. So far, spider taxonomists have stamped tens of thousands of morphospecies, but subspecies have

only been named in exceptional instances. It may be questionable as to whether such nominal taxa really form subspecies, i.e. subunits of polytypic species. The aim of the present contribution is to stimulate discussion by investigating this comparatively unusual situation in more detail. Of course, tradition has played and is continuing to play a predominant role, i.e. the stabilizing effect of long-established taxonomic practice. However, the real biological situation may not be reflected by practicing conventional procedures of this kind. There is accordingly a real need to try to explain why there is practically no problem in distinguishing morphospecies from morphospecies without any intermediate forms. This aspect is directly linked to modes of speciation in spiders.

GENERAL REMARKS ON SPECIES AND SPECIATION

The investigation and subsequent attempts to explain this situation will first need some general remarks.

Allopatric versus sympatric speciation

Allopatric speciation is the predominant mode of speciation in animals. Separation by barriers causes interruption of genetic exchange between groups of populations; peripheral isolates especially play a major role. Accordingly, interruption of the coherent function of gene flow permits divergence (see Mayr 1942, 1963). Sympatric speciation, i.e. speciation without geographic isolation, cannot be definitively excluded, but there is no reason to believe that mechanisms of this kind could be of major importance.

Until now, no sound argument has been put forward to contradict this general view.

Polytypic and monotypic species

In most sufficiently studied major groups of the animal kingdom, many biospecies[1] comprise two or more subspecies. As defined by Mayr (1963), subspecies form an 'aggregate of local populations of a species inhabiting a geographic subdivision of the range of the species and differing taxonomically from other populations of the species'. Numerous examples are well-documented from mammals to birds and from lizards to salamanders, etc. This is also true for invertebrates such as carabid beetles (Mossakowski & Weber 1976), and pulmonate gastropods (Knipper 1939; Mayr & Rosen 1956). See Mayr (1963) for details and further documentation.

According to this definition, the subdivision of polytypic species into subspecies is based on typology. This explains why, in a given species, different numbers of subspecies have occasion-

ally been distinguished by different authors. Polytypic species seem to occur almost universally (Rensch 1929).

The counterpart - monotypic species - seems to be exceptional. This term designates species without subspecies. They may be geographically widely distributed. I refer to the holarctic distribution of the araneid *Araneus diadematus* Clerck, 1757 (Fig. 1) and similar distribution patterns of various species of the Erigonidae. Could it be that species of this kind consist of just one single panmictic population? If so, they may be barred from geographic speciation. The same seems to be true for freshwater Bryozoa and various species of the Tardigrada, e.g., *Macrobiotus hufelandi* C.A.S. Schultze, 1834 but there is evidence now that this tardigrad species is composite (Bertolani & Rebecchi 1993). These and other cases have in common the trait that the species concerned share extraordinary dispersal abilities. I refer to ballooning in spiders, to blepharoblasts in freshwater bryozoans and to the dessicated cryptobiotic states in tardigrades.

ANALYSIS

The situation in spiders is heterogeneous. There are sedentary species, and in the other extreme, others are world champions in ballooning like the Tetragnathidae in the Northern Hemisphere. Species may have extremely wide holarctic or pantropic distribution patterns, such as *Neoscona nauticus* (L. Koch, 1875). Others are confined to relatively small areas (see, for example, Thaler's work on alpine species (1994; Thaler et al. 1994)).

Almost all these species have in common the fact that they are distinct morphospecies. There is practically no variation. This is especially true for the taxonomically decisive genital structures. Examples of the narrow limits for variation of such details are presented in Figs. 9-10; they are partly correlated with the occurrence of separated populations (see Kraus & Kraus 1988). Intermediate populations or at least single intermediate specimens between different morphospecies are almost unknown in nature[2] — with the exception of extremely rare teratological individuals. There is usually no difficulty in

---

[1]The present author continues to accept the biospecies concept sensu Mayr (1963) as it is biological. What is called the phylogenetic species concept does not seem to be applicable in the present discussion: it is based on individuals, not classes, and sets of characters (Goldstein & DeSalle 2000), i.e. close to typology.

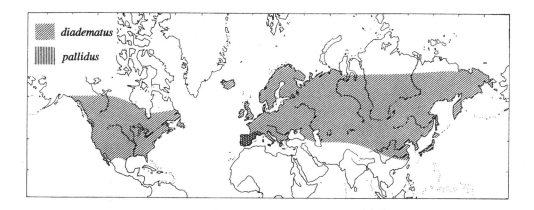

Fig. 1. Holarctic distribution of *Araneus diadematus* and limited distribution of its presumed sister species, *A. pallidus* (from Grasshoff 1968).

assigning even single specimens to morphospecies A, B, or C.

It is highly improbable that the modes of speciation in spiders could be principally different from those in other animals or at least in other groups of terrestrial arthropods. For better understanding, three aspects should be considered:

1. Arguments in favor of regular allopatric speciation mechanisms

Various cases are wellknown enough to provide information on speciation events caused by geographical separation. Some selected examples are:

Mesothelae
Species differentiation in island representatives

of the genus *Heptathela* and closely related forms (*Ryuthela*) studied by Haupt (1983) indicates a correlation with their separation on different islands (Fig. 2). With respect to only slight differences between material of different origins, Haupt divided *H. kimurai* (Kishida, 1920) into the nominate and three further subspecies, and

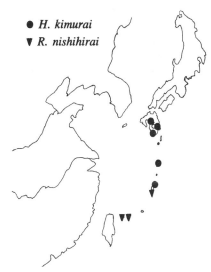

Fig. 2. Distribution pattern of island representatives of the genera *Heptathela* and *Ryuthela* on southern Japanese islands. Further differentiation into presumed subspecies (*H. k. kimurai, k. amamiensis, k. higoensis, k. yanbaruensis* and *R. n. nishirai, n. ishigakiensis*) not indicated (from Haupt 1983).

[2]Hybrids between *Tegenaria gigantea* Chamberlin & Ivie, 1935 and *T. saeva* Blackwall, 1844 were found in Yorkshire, England, where both species occur sympatrically (Oxford & Smith 1987; Oxford & Plowman 1991). Only 3% of the sampled male specimens were identified as intermediates. They occur occasionally; no hybrid populations were found. Hence, the specific status of the two *Tegenaria* species is not invalidated. There is evidence that two closely related synanthropic species are expanding their ranges, perhaps caused by human interference (see Mayr 1963). This case does not seem to be of any relevance for the present discussion.

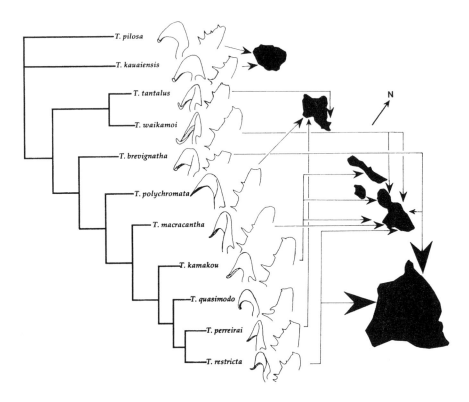

Fig. 3. Phylogenetic biogeography of a clade of Hawaiian *Tetragnatha* species (from Gillespie 1993); (distribution of the widely distributed species *T. quasimodo* [Hawaii, Maui, Molokai, Lanai, Oahu, Kaui] not indicated).

R. nishihirai Haupt, 1979 [= *Heptathela* ?] into the nominate and one subspecies. Because of the discontinuous distribution of the species, especially those occurring on islands, it is hard to decide whether these phena form part of two polytypic species or whether differentiation into biospecies has already been achieved.

Anyway, the *Heptathela* case is a convincing case of allopatric differentiation.

Colonization and differentiation in the Hawaiian Archipelago

In her extensive work on Hawaiian spiders, Gillespie (e.g. 1993; see also Roderick & Gillespie 1998) investigated the radiation of certain spider groups, especially tetragnathids. She discovered a fascinating example of differentiation and speciation (Fig. 3). Among others, there is one clade of cursorial hunters, with the typical

web-building behavior totally reduced. Gillespie concluded that multiple founder events occurred and that speciation required strict geographic isolation; ecological (more than sexual) shifts appear to play a role in initiating divergence. In principle, but not exclusively, islands were primarily colonized in the sequence of their age, i.e., from north to south. Multiple invasions may have happened, and present distribution patterns indicate cross-colonizations. This development parallels the diversification of the Hawaiian Drosophilidae into more than 500 species, approximately 98% of them endemic (for a review see White 1978).

The conclusion is that even tetragnathids can be subject to complex allopatric speciation events. Their diversity on the Hawaiian islands apparently forms a typical case of archipelago speciation (Mayr 1963); together with various other

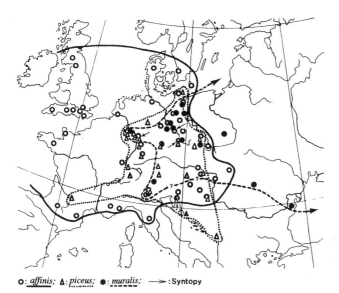

Fig. 4. Distribution of *Atypus* species in the western Palaearctic Region (from Kraus & Baur 1974).

o: *affinis;* ▲: *piceus;* ●: *muralis;* ——►: Syntopy

examples, it is more spectacular than the well-known radiation of Darwin's finches in the Galapagos archipelago (see also Wagner & Funk 1995).

Atypus species in the western Palaearctic Region (Atypidae).
Three *Atypus* species occur in the western Palaearctic Region (Kraus & Baur 1974)[3]. Their differentiation is obviously correlated with Pleistocene separation into habitats in Western Europe, southern Siberia, and a southeastern refugium in the Balkan region (Fig. 4). Corresponding to these refugia, the recolonization of Central Europe apparently occurred from three directions [for principles of reasoning see de Lattin 1967]. There is no reason to doubt that the three species originated by allopatric speciation; they prove to be biospecies as localities are known where they coexist sympatrically; hybrids were never found.

Heriaeus species in the western Palaearctic Region (Thomisidae).
Loerbroks (1983) revised the species of the

thomisid genus *Heriaeus* of the Western Palaearctic. He distinguished three species groups and mapped the distribution of 11 species of the *H. hirtus* group (Fig. 5). This distribution pattern may be explained by referring to various well-known glacial refugia. They are currently designated as Adriato-mediterranean (plus tyrrhenian), ponto- mediterranean, Syrian and Mauretanien faunal elements (see de Lattin 1967). They recolonized Central Europe. This is another example of regular allopatric speciation.

One could continue, but this would lead to redundancies. The conclusion is that there is no evidence at all for assuming special modes of speciation in spiders. But this kind of review does not provide an answer to the central question: why were all of the tens of thousands of spider species described almost exclusively as monotypic units?

2. Extremely wide distributional areas versus patterns of local species
As already mentioned, there is a wide spectrum between holarctic and also pantropic distributional patterns, on the one hand, and considerably small areas of occurrence on the other.

---

[3]There is evidence that *Atypus muralis* Bertkau, 1890 is not different from *A. karschi* Dönitz, 1887 (see Kraus & Baur 1974).

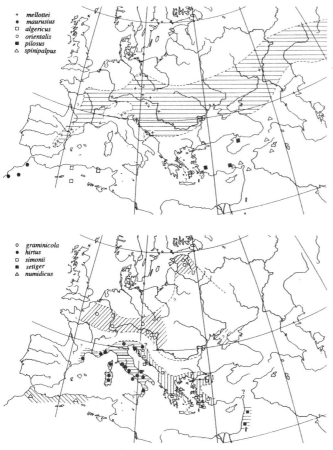

Fig. 5. Distribution of species of the *Heriaeus hirtus* group in the western Palaearctic Region (from Loerbroks 1983).

Geographically wide distributional areas

It is difficult to explain why certain species were extremely successful in extending their range, cosmopolitan areas included. The impression is that they form one single panmictic population in certain instances. This can be assumed, e.g. in the case of *Araneus diadematus*, as ballooning spiderlings may rotate eastwards because of the prevailing wind direction in the Northern Hemisphere. Grasshoff (1968) stated that specimens of *A. diadematus* from Japan or from North America cannot be distinguished from their European conspecifics. Various ecophenotypes, color variants included, occur repeatedly at different localities. However, there is no discernible difference at all in the complicated genital structures. Uhl et al. (1992) made a similar observation in their study on the North American species *Tet-*

ragnatha shoshone Levi, 1981, then newly discovered in Europe. As the differences on both sides of the Atlantic proved to be gradual, not even subspecific rank was assigned.

Geographically limited distributional areas

Some presently limited distributional areas may be relics of formerly wider distributions, but it seems to be much more probable that most of them are the result of speciation events caused by complex separations. Glaciations in the European Alps form a good example (for review see de Lattin 1967). There was a complicated pattern of relatively small 'Nunatakker' and also of major refugia ('massifs de refuge'). This kind of fragmentation of previously coherent distributions may well have induced allopatric differentiations, including speciation. Attempts to link separations in the past (in refugia) of this kind with geographically limited recent distribution patterns of spiders may — but must not — be speculative.

Despite the fact that no reliable method is known for measuring dispersal ability, spiders should not be underestimated. Wunderlich, for instance, collected thousands of individuals of spider species in a limited area in Berlin. Among others, he also found one single male specimen of the hitherto unknown species *Entelecara berolinensis* (Wunderlich, 1969). As in birds, it could well be that this was nothing other than an accidential occurrence - as a result of dispersal ability and different from the existence of an established reproducing population. The same is true for the discovery of one single female of *Araneus grossus* (C.L. Koch, 1847) in south-western Germany. The species was reported by Wiehle (1963) as 'new' for Germany; but until now, no further specimen has been found. This demonstrates that occurrence due to dispersal ability should be distinguished from the existence of firmly established, reproducing populations.

The conclusion is that the extension of distributional areas does not depend on dispersal ability alone. Abiotic factors (such as climatic conditions), narrow specialization in biological properties, the ability to form an ecological niche within the framework of complex interdependencies (including competition with other, already well-established species) may effectively prevent the extension of geographically limited distributions. Situations cannot be excluded in which closely related species occur allopatrically, as they lack sufficient mutual differentiation. This may explain why allopatry in species with limited distributions could be obligatory.

Similar situations of mutual exclusion have already been found in other animal groups, including even mammals. As an example, I refer to species of the coccinellid beetle genus *Chilocorus* with obligatory allopatric species complexes, both in North America and in the Palaearctic (see White 1978 for review and other cases). There are differences in the karyotypes, but this aspect does not seem to have been considered so far in allopatric spider species complexes. Future work in this direction could be promising.

3. No transition zones, not even hybrid belts ?

The most crucial problem mentioned at the very beginning of this contribution remains: Until now, there has been no reliable information on potential transition zones or at least hybrid belts between allopatric phena traditionally classified as morphospecies. The probability is high that intermediates of this kind do not exist at all (but see footnote 2). Genital differences may be minute, but they are always distinct in such allopatric forms, including presumably isolated populations. This is true for Mediterranean species of the genus *Amaurobius*, for representatives of the linyphiid genera *Lepthyphantes* (see Thaler 1994) and *Agyneta* (see Tanasevitch 1999), but also for *Theridion* (Theridiidae) and *Acantholycosa* (Lycosidae) morphospecies (Thaler, in litt.).

HYPOTHESIS

The present analysis demonstrates normality: Compared with other higher animal taxa, modes of speciation in spiders, as well as available biogeographical data, do not offer any exclusive features. However, there is no peculiarity discernible that could explain the traditional morphospecies by morphospecies situation in spiders. Are there still other, different aspects that should be taken into consideration?

Application of the superspecies concept
Superspecies patterns
I would like to refer to the superspecies concept first proposed by Mayr (1931). The definition given by Mayr & Ashlock (1991: 430) reads as follows: 'A monophyletic group of closely related and entirely or largely allopatric species that are too distinct to be included in a single species or that demonstrate their reproductive isolation in a zone of contact.' This corresponds exactly to the situation in very similar but allopatric morphospecies in spiders. The subunits of superspecies were called 'semispecies'. This term could be misleading, as one could conclude

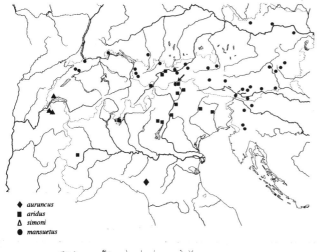

Fig. 6. Distribution of species of the *Lepthyphantes mansuetus* group [=*Mansuphantes* Saaristo & Tanasevitch, 1996] in the European Alps; arrow indicates syntopic occurrence (from Thaler 1994).

Fig. 7. Distribution of species of the *Lepthyphantes annulatus* group [=*Incestophantes* Tanasevitch, 1992] in the western Palaeartic Region (from Thaler et al. 1994).

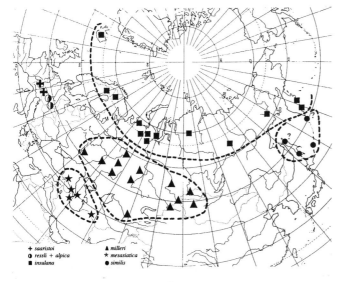

Fig. 8. Distribution of the Palaearctic species of the *Agyneta similis* group (from Tanasevitch 1999).

that full biospecies rank has not yet been achieved. Thus, Amadon (1967) replaced 'semispecies' by introducing the more appropriate term 'allospecies'.

It seems to be easy to group a lot of spider phena — currently classified as morphospecies — in complexes of superspecies. The impression is that polytypic species structures (with a certain number of more or less intergrading subspecies) are replaced by superspecies, with allospecies as subunits. Well-documented examples are already known.

One could refer to the already mentioned *Heriaeus hirtus* complex (Loerbroks 1983) in thomisids (Fig. 5). The linyphiid species *Lepthyphantes mansuetus* (Thorell, 1875) and its relatives (Fig. 6) were recently studied by Thaler (1994). Thaler et al. (1994) investigated the complex around *L. annulatus* (Kulczynski, 1882) (Fig. 7). Similar results were obtained by Tanasevitch (1999) who investigated *Agyneta* species (Fig. 8). Thaler and also Tanasevitch had already used the appropriate term 'superspecies'. That the subunits had already achieved full biospecies level was confirmed in at least one case: a locality is known where the otherwise allopatric species *L. mansuetus* and *L. aridus* (Thorell, 1875) were found to occur sympatrically; there was not even a single intermediate individual (see Fig. 6, arrow).

Why superspecies with allospecies as subunits?
Pure application of the superspecies concept is primarily a matter of correct terminology and should not be misunderstood as an *explanation* of the biological background. But genital structures and what was formerly called the 'lock-and-key principle' (Dufour 1844) may provide the key for a biological interpretation of the origin of allospecies patterns. Two aspects should be considered: a) It is generally supposed that complex genital structures in terrestrial arthropods are extremely sensitive in reflecting genetic differences (see e.g. Arnqvist 1997), and b) the need for an optimal co-adaptation between male and female copulatory organs. Both factors are interlinked. As already explained, there are many

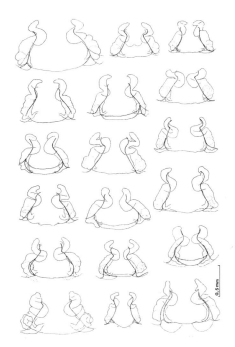

Fig. 9. Intraspecific variation of vulvae of *Stegodyphus dufouri* specimens from different localities in northern Africa, Aden, the Sudan, Mali and Niger (from Kraus & Kraus 1988).

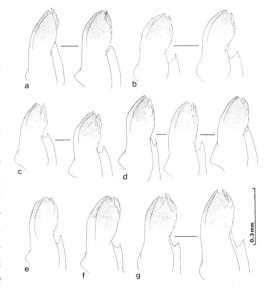

Fig. 10. Intraspecific variation of terminal lamellae of male bulbi of *Stegodyphus mimosarum* specimens from the Congo (a), the Serengeti (b), Durban (c), Natal (d), the Kruger Park (e), the Transkei coast (f), and Madagascar (g) (from Kraus & Kraus 1988).

Fig. 11. Hypothesis: Diagram illustrating the possible origin of allospecies, including the potential ability to coexist with similar species at later stages of evolutionary divergence.

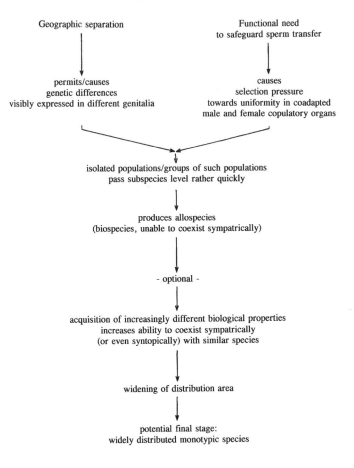

examples of comparatively slight but regionally invariable genital differences. This is not only true for species complexes in spiders, but has also been found in the Lepidoptera and in many other arthropod groups, carabid beetles included (for review, see de Lattin 1967). Optimal coadaptation has been extensively discussed by Eberhard (1985). He believes in selection by female choice (see Arnqvist 1997). However, in spiders female choice may be highly dangerous for males and could probably cause unbalanced sex ratios. Hence, any selective advantage of female choice seems to be doubtful; it was apparently not found to occur in nature. Furthermore, male bulbs are not innervated at all, and corresponding (tactile) receptors remain to be discovered in sclerotized female copulatory organs.

Instead, the function of complicated cou-

pling mechanisms simply seems to safeguard sperm transfer (Kraus 1968), in some instances by more than one copulating male (see Wiehle 1967). It is this requirement that apparently stabilizes details of copulatory organs. Loerbroks (1983) demonstrated that limited variability may be observed in the epigynes but not in the vulvae of certain thomisid spiders. Similar variation was observed by Kraus & Kraus (1988) in female specimens of African Eresidae, especially in *Stegodyphus dufouri* (Audouin, 1826) (Fig. 9), and in the male terminal lamella (Fig. 10) of *S. mimosarum* Pavesi, 1883. This kind of slight variability remains within narrow limits and does not seem to influence perfect function.

The hypothesis deduced is as follows: The main function of coupling mechanisms is to safeguard sperm transfer. Almost perfect homogeneity of the functioning components of both male and female genitalia is of high selective advantage. Less perfect mutual adaptation is regarded as counterproductive. Selection pressure favors uniformity and obliterates deviations. Two mechanisms may shorten transitory phases at a subspecies level and hence accelerate the formation of distinct allospecies:

—the expression of genetic differences in genital structures in geographically separated populations (or in separated groups of intercommunicating populations, respectively), and

—selection pressure towards optimal coadaptation of male and female copulatory organs within such units, i.e. the origin of obligatory allopatric biospecies (Fig. 11).

All progressive transitions — from allospecies to regular species — can be expected when the efficiency of primarily unsolved problems of coexistence becomes increasingly reduced by the acquisition of diverging biological properties.

An alternative hypothesis would be most welcome, but I am presently not aware of another conceivable model. One possible way of testing the hypothesis presented here would be to investigate whether there could be any correlation between the occurrence of allospecies in spiders and the presence of coadapted complex, instead of comparatively simple, genital structures.

PERSPECTIVES AND FUTURE QUESTIONS

1. Zones of contact between allospecies should be studied.

2. As chromosomal differences between allospecies have been observed in many other animal taxa, future studies should include appropriate analyses; in spiders, the present information is close to zero (for details see e.g. White 1978).

3. The problem of tens of thousands of independently described species without subspecies can be solved by applying the superspecies concept. Instead of just 'stamping' the morphospecies, this approach would lead to a better integrative understanding of evolutionary and especially of biogeographic interdependencies.

REFERENCES

Amadon, D. 1967. The superspecies concept. *Systematic Zoology* 15, 245-249.

Arnqvist, G. 1997. The evolution of animal genitalia: distinguishing between hypotheses by single species studies. *Biological Journal of the Linnean Society* 60, 365-379.

Bertolani, R.B. & Rebecchi, L. 1993. A revision of the *Macrobiotus hufelandi* group (Tardigrada, Macrobiotidae), with some observations on the taxonomic characters of eutardigrades. *Zoologica Scripta* 22 (2), 127-152.

Dufour, L. 1844. Anatomie générale des Diptères. *Annales des Sciences naturelles* 1, 24-264.

Eberhard, W.G. 1985. *Sexual selection and animal genitalia.* Harvard Univ. Press, Cambridge.

Gillespie, R.G. 1993. Biogeographic patterns of phylogeny in a clade of endemic Hawaiian spiders (Araneae, Tetragnathidae). *Memoirs of the Queensland Museum* 33 (2), 519-526.

Goldstein, P.Z. & DeSalle, R. 2000. Phylogenetic species, nested hierarchies, and character fixation. *Cladistics* 16 (4), 364-384.

Grasshoff, M. 1968. Morphologische Kriterien als Ausdruck von Artgrenzen bei Radnetzspinnen der Subfamilie Araneinae (Arachnida: Araneae: Araneidae). *Abhandlungen der senckenbergischen naturforschenden Gesellschaft* 516, 1-100.

Haupt, J. 1983. Vergleichende Morphologie der Genitalorgane und Phylogenie der liphistiomorphen Webspinnen (Araneae: Mesothelae), I. Revision der bisher bekannten Arten. *Zeitschrift für zoologische Systematik und Evolutionsforschung* 21 (4), 275-293.

Knipper, H. 1939. Systematische, anatomische und tiergeographische Studien an südosteuropäischen Heliciden (Moll., Pulm.). *Archiv für Naturgeschichte* (NF) 8 (3/4), 327-517.

Kraus, O. 1968. Isolationsmechanismen und Genitalstrukturen bei wirbellosen Tieren. *Zoologischer Anzeiger* 181 (1-2), 22-38.

Kraus, O. & Baur, H. 1974. Die Atypidae der West-Paläarktis. Systematik, Verbreitung und Biologie (Arach.: Araneae). *Abhandlungen und Verhandlungen des naturwissenschaftlichen Vereins in Hamburg* (NF)17, 85-116.

Kraus, O. & Kraus, M. 1988. The genus *Stegodyphus* (Arachnida, Araneae). Sibling species, species groups and parallel origin of social living. *Verhandlungen des naturwissenschaftlichen Vereins in Hamburg* (NF)30, 151-254.

Lattin, G. de 1967. *Grundriss der Zoogeographie*. G. Fischer, Jena.

Loerbroks, A. 1983. Revision der Krabbenspinnen-Gattung *Heriaeus* Simon (Arachnida: Araneae: Thomisidae). *Verhandlungen des naturwissenschaftlichen Vereins in Hamburg* (NF)26, 85-139.

Mayr, E. 1931. Notes on *Halcyon chloris* and some of its subspecies. *American Museum Novitates* 469, 1-10.

Mayr, E. 1942. *Systematics and the origin of species from the viewpoint of a zoologist.* 2nd Ed. Dover, New York.

Mayr, E. 1963. *Animal Species and Evolution.* Belknap Press at Harvard Univ. Press, Cambridge.

Mayr, E. & Ashlock, P.D. 1991. *Principles of Systematic Zoology.* 2nd Ed. Mc Graw Hill, New York.

Mayr, E. & Rosen, C.B. 1956. Geographic variation and hybridization in populations of Bahama snails (*Cerion*). *American Museum Novitates* 1806, 1-48.

Merret, P. (Ed.) 1978. *Arachnology. Seventh International Congress.* Symposia of the Zoological Society of London 42. Academic Press, London.

Mossakowski, D. & Weber, F. 1976. Chromosomale und morphometrische Divergenzen bei *Carabus lineatus* und *C. splendens* (Carabidae). Ein Vergleich sympatrischer und allopatrischer Populationen. *Zeitschrift für zoologische Systematik und Evolutionsforschung* 14, 280-291.

Oxford, G.S. & Smith, J.C. 1987. The distribution of *Tegenaria gigantea* Chamberlin & Ivie, 1935 and *T. saeva* Blackwall, 1844 (Araneae, Agelenidae) in Yorkshire. *Bulletin of the British Arachnological Society* 7(4), 123-127.

Oxford, G.S. & Plowman, A. 1991. Do large house spiders *Tegenaria gigantea* and *T. saeva* (Araneae, Agelenidae) hybridize in the wild? A multivariate approach. *Bulletin of the British Arachnological Society* 8(9), 293-296.

Rensch, B. 1929. *Das Prinzip geographischer Rassenkreise und das Problem der Artbildung.* Bornträger, Berlin.

Roderick, G.K. & Gillespie, R.G. 1998. Speciation and phylogeography of Hawaiian terrestrial arthropods. *Molecular Ecology* 7, 519-531.

Tanasevitch, A.V. 1999. On some palaearctic species of the spider genus *Agyneta* Hull, 1911, with description of four new species (Aranei: Linyphiidae). *Arthropoda Selecta* 8(3), 201-213.

Thaler, K. 1994. Vikariante Verbreitung im Artenkreis *Lepthyphantes mansuetus* in Europa und ihre Deutung (Araneae, Linyphiidae). *Entomologia Generalis* 18, 171-185.

Thaler, K., van Helsdingen, P. & Deltshev, V. 1994. Vikariante Verbreitung im Artenkomplex von *Lepthyphantes annulatus* in Europa und ihre Deutung (Araneae, Linyphiidae). *Zoologischer Anzeiger* 232 (3/4), 111-127.

Uhl, G., Sacher, P., Weiss, I. & Kraus, O. 1992. Europäische Vorkommen von *Tetragnatha shoshone* (Arachnida, Araneae, Tetragnathidae). *Verhandlungen des naturwissenschaftlichen Vereins in Hamburg* (NF)33, 247-261.

Wagner, W.L. & Funk, V.A. 1995. *Hawaiian biogeography.* Smithsonian Series in Comparative Evolutionary Biology.

White, M.J.D. 1978. *Modes of speciation.* Freeman, San Francisco.

Wiehle, H. 1963. Beiträge zur Kenntnis der deutschen Spinnenfauna, III. *Zoologische Jahrbücher (Systematik)* 90, 227-298.

Wiehle, H. 1967. Steckengebliebene Emboli in den Vulven von Spinnen (Arach., Araneae). *Senckenbergiana Biologica* 48 (3), 197-202.

Wunderlich, J. 1969. Zur Spinnenfauna Deutschlands, IX. Beschreibung seltener oder bisher unbekannter Arten (Arachnida: Araneae). *Senckenbergiana Biologica* 50 (5/6), 381-393.

European Arachnology 2000 (S. Toft & N. Scharff eds.), pp. 315-327.
© Aarhus University Press, Aarhus, 2002. ISBN 87 7934 001 6
(Proceedings of the 19th European Colloquium of Arachnology, Århus 17-22 July 2000)

Generic revision of some thomisids related to *Xysticus* C.L.Koch, 1835 and *Ozyptila* Simon, 1864

PEKKA T. LEHTINEN

Zoological Museum, Centre for Biodiversity, University of Turku, FIN-20014 Turku, Finland
(pekleh@utu.fi)

Abstract

The changes in the concepts and reciprocal relationships of *Xysticus* s.l., *Ozyptila* s.lat., and *Coriarachne* are reviewed. *Coriarachne, Bassaniana, Psammitis,*and *Spiracme* comprise a group of genera and/or sub-genera with slightly a variable male palpal structure and more variable epigyna. Different body shapes explain the traditional treatment of this group as several genera. *Bassaniana* is accepted here only as a species group of *Coriarachne*. *Proxysticus* Dalmas, 1922 is removed from *Xysticus* s.lat. and is found to be a junior synonym of *Bassaniodes* Pocock, 1903. This Old World genus has its centres of speciation in the Mediterranean region, Central Asia, and South Africa.

Key words: Thomisidae, generic revision, *Xysticus, Ozyptila, Coriarachne*

INTRODUCTION

This work forms a part of a worldwide generic revision of Thomisidae. The group discussed here is mostly Holarctic while most other groups of Thomisidae are mainly tropical.

As for most groups of Thomisidae, no modern classification of *Xysticus* s.lat., *Ozyptila* s.lat. and related groups has been generally accepted, although a lot of work has been done at specific level and many nomenclatorally less important species-groups have been suggested (Locket & Millidge 1951; Gertsch 1953; Schick 1965; Turnbull & al. 1965; Ono 1978 & 1988; Marusik & Logunov 1995; Wunderlich 1987, 1995; Jantscher 2002). The first discussions simply included the important statement that a part of the species have one or two tegular apophyses.

Some type species of genera in this group have never been studied by specialists after the original description. Sometimes the correct placing would have been difficult, as the samples of type specimens happen to be juveniles or represented only by one sex. Much effort has therefore been directed to possibilities to find adult topotypical material for such species. This has been quite successful for *Pycnaxis* and *Ocyllus*. Parallely, topotypical material of unknown sexes of *Philodamia* and *Demogenes* spp. have facilitated the work. Topotypical material of many other tropical species of this group has been collected by myself.

No specific revisions have been carried out and new synonyms are listed here mainly for supraspecific taxa and their type species.

MATERIAL AND METHODS

The material studied includes all type material of taxa discussed that is still preserved and available, checked on site or borrowed from different museums for the worldwide generic revision of Thomisidae. These museums are in Stockholm, Paris, Genoa, London, Berlin, New York, Hamburg, Los Baños and Calcutta. Other thomisid material in these museums has also been screened. All thomisid material collected by myself during numerous expeditions to temperate and tropical areas of the world has been carefully studied. Some unpublished material collected by my colleagues (Yuri Marusik and Dmitri Logunov, Russia, Cor Vink, New Zealand, and Seppo Koponen, Finland) has also been available for this study.

The material has been studied with traditional methods of taxonomy, with strong emphasis laid to the study of male palpi and surface structures of legs, modified hairs of carapace and abdomen.

Acronyms for museums:

IRRI = International Rice Research Institute, Los Baños, Philippines

MCSN = Museo Civico di Storia Naturale 'Giacomo Doria', Genoa, Italy

MNHM = Museum National d'Histoire Naturelle, Paris, France

MZT = Zoological Museum, University of Turku, Turku, Finland

NHRS = Naturhistoriska Riksmuseet, Stockholm, Sweden

ZSI = Zoological Survey of India, Calcutta, India Abbreviations for structural details: MOT median ocular triangle, RTA retrolateral tibial apophysis, VTA ventral tibial apophysis.

TAXONOMIC RESULTS

The concepts and limitation of *Ozyptila, Coriarachne,* and *Xysticus* (Figs 1-3) have been widely confused in the past (C.L. Koch, 1838, Thorell, 1872, Menge 1875, Simon, 1864 & 1875, Kroneberg, 1875, Keyserling, 1880, Ono, 1979 & 1988, etc.). It must be emphasized here that all species of *Ozyptila* s.lat. known to Menge (1875) were listed in *Coriarachne.* The status, diagnosis, and limitation of various species groups in *Xysticus* s.str., *Psammitis*, and *Ozyptila* have been widely discussed e.g., by Gertsch (1953), Dondale & Redner (1975) Marusik & Logunov (1991), Wunderlich (1995), Logunov & Marusik (1998), and Jantscher (2002), but these species groups are not further discussed here.

Suprageneric taxa

Although *Xysticus* and *Ozyptila* are well known

thomisid groups, the only suprageneric taxon so far based on either of them is the family Xysticidae proposed by Dahl (1907). His classification of crab spiders has not been accepted by later authors. Simon (1895) created the tribe Coriarachneae for two thomisids with flat bodies (*Coriarachne* Thorell, 1870 & *Tharpyna* L. Koch, 1874) and Roewer (1954) added *Firmicus* Simon, 1895, in which some species have flat bodies. The name based on *Coriarachne* has priority and must be used as the name of suprageneric groups including *Coriarachne*, even in the case that the limitation of the newly accepted group is much different from that of Simon. Most authors have simply placed *Xysticus* and *Ozyptila* to Thomisinae in the same group as *Diaea*. Ono (1988) resurrected the tribe Coriarachnini including *Bassaniana, Bassaniodes, Coriarachne, Narcaeus, Ocyllus, Oxyptila* and *Xysticus*. The general concept of Coriarachnini sensu Ono has been accepted here, although its final delimitation in regard to poorly known genera is waiting for a phylogenetic analysis. Ono (1988) included *Ocyllus* Thorell (1887), but careful comparison of the juvenile holotype of the type species in MCSN has revealed that *Ocyllus* is a synonym of *Cebrenninus* Simon, 1887, a genus never placed in Thomisinae. The other species of *Ocyllus* must be transferred to *Oxytate* L. Koch, 1878. A more detailed discussion of this genus with complex nomenclatoral problems is in preparation, but cannot be included in the limited space here. *Narcaeus* Thorell, 1890 was also included by Ono. I have checked the female holotype of the type species *N. picinus* Thorell, 1890 in MCSN and its inclusion in Coriarachnini s.lat. could be accepted only together with the closely related *Demogenes* Simon, 1895. If so, the concept of Coriarachnini must be greatly expanded, as some other thomisine groups (*Pycnaxis* Simon, 1895, *Lysiteles*-group, and the *Haplotmarus-Philodamia*-group) are at least as close to Coriarachnini (as accepted here) as the *Narcaeus* group. The two latter groups are not further discussed here.

The concept of *Xysticus* s.lat.

The generic name *Xysticus* C.L. Koch, 1835 has

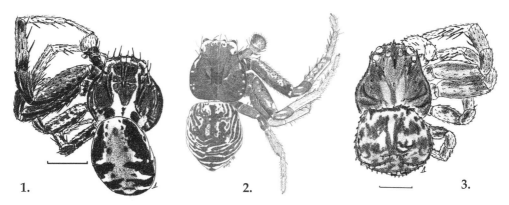

Fig. 1-3. Main genera of well-known Coriarachnini sensu Ono. **Fig. 1.** *Xysticus audax* (Finland, Korpoo), male dorsally. **Fig. 2.** *Coriarachne depressa* (Russia, Burjatia), female dorsally. **Fig. 3.** *Ozyptila nigrita* (Turkey, Yamanlar Mts.), juvenile female dorsally.

been a waste basket during the long history of thomisid taxonomy. In addition to the contemporaries of C.L. Koch , e.g., Walckenaer (1837), many authors of the 19th century knew only two thomisid genera, *Thomisus* and *Xysticus*. The author of the generic name *Xysticus* himself placed even some *Tmarus* to *Xysticus* (C.L. Koch 1838). Later on, many tropical species from Australia and Melanesia (e.g., L. Koch 1867; Karsch 1878; Roewer 1938), South East Asia (e.g., Simon 1909; Bristowe 1931: *Pycnaxis* spp.), South, Central, and North America (Taczanowski 1872; Emerton 1893: 'Synaema' sp.), and Africa (Berland 1922; Lawrence 1928, 1936, 1952; Caporiacco 1941; Jézéquel 1966) were originally described as *Xysticus*. Even some green thomisids of Europe were placed in *Xysticus*, e.g., by Simon (1864: *Thomisus*) and Herman (1879: *Runcinia*). The Australian species were later transferred to *Diaea* s.lat., the first ones by the original author (L. Koch 1874). These species actually belong to several undescribed genera, but all of them are here left outside Coriarachnini sensu Ono. Most 'Xysticus' from tropical Africa are still unrevised.

A concept of *Xysticus*, including both apophysate and nonapophysate species was generally used for almost one hundred years. This concept of *Xysticus* s.lat. is no more generally used in strictly taxonomic papers, although it is still common in faunistic and other non-taxonomic papers. First, several species groups

were proposed (Locket & Millidge 1951; Gertsch 1953; Schick 1965; Turnbull et al. 1965). The splitting of *Xysticus* s.lat. to named taxa has been proposed by many authors (e.g., Menge 1875; Ono 1978; Wunderlich 1987, 1995), and it is quite justified, although many species outside Europe and North America have never properly been placed to some of these genera. Most of the taxonomic papers dealing with a single genus of Thomisidae are discussing *Xysticus* s.lat. Up till now no real revision has been done and generic names other than *Xysticus* have been rather sparingly used.

Menge (1875) erected the genera *Psammitis* (type species *Thomisus sabulosus* Hahn, 1832) and *Spiracme* (type species *S. striata* Menge, 1875 = *Xysticus striatipes* L. Koch, 1870), both from Central Europe. Both names have been used for *Xysticus* spp. without tegular apophyses and sometimes the latter has been regarded as a subgenus of the former (Gertsch 1953; Ono 1978). *Proxysticus* was established by Dalmas (1922). Bonnet (1958) accepted the genus *Psammitis*, but included with the two original ones only one additional species, *P. doriai* Dalmas, 1922 (from Italy). Many westpalaearctic species of non-apophysate *Xysticus* were transferred by Wunderlich (1987, 1992, 1995) out of *Xysticus*. He used first the name *Proxysticus* for all non-apophysate species (Wunderlich, 1987) but later (Wunderlich, 1992, 1995) he used *Psammitis* in the same meaning (including also many species of *Proxysticus* auct. (now *Bassaniodes*). Jézéquel (1964) used the generic name *Proxysticus* for a species from tropical Africa.

Character evolution in the relatives of *Xysticus*

Tegular apophyses in *Xysticus* s.lat. and *Ozyptila* s.lat. as well as tutacular structures and clavate or spatulate dorsal and leg setae are characteristic for many well-known Coriarachnini sensu Ono, but none of them seems to be a synapomorphy.

The presence of one or two tegular apophyses in *Xysticus* s.lat. (Fig. 4) and *Ozyptila* s.lat. (Fig. 5) is problematic. Most of the species groups have no tegular apophysis at all. *Xysticus* and *Ozyptila* are mainly characterized by non-genitalic characters, including size, type of dorsal setae, shape of MOT and colour patterns.

A membranous apophysis has been found in the subdistal part of tegulum at least in one undescribed tropical genus (aff. *Demogenes*: Fig. 6). The structure and place of this apophysis looks like the median apophysis of some other spider families, where such an apophysis is present in some genera and is lacking in others (e.g., Hahniidae)

Various modifications of the cymbial margin connected to the distal part of embolus are often called tutacular grooves or simply as tutaculum. There are several different types and sites for these tutacular structures. The tutaculum of many *Xysticus* s.str. (Fig. 7) and *Psammitis* (Fig. 8) is a lateral process of cymbium, while quite dissimilar and probably non-homologous types of tutacular structures are present at least in some Misumenini (distal) and Hedanini (basal).

Clavate or spatulate dorsal setae have certainly been developed in many thomisid groups that live in the ground. However, the ground-living tropical genera *Demogenes*, *Narcaeus* (Fig. 9), *Pycnaxis* and '*Phrynarachne*' *clavigera* (Fig. 10) also share a simple palpal pattern and general body shape with *Ozyptila*, in which many of these species were originally described.

Confusions within the true relatives of *Xysticus*

Almost all theoretically possible misplacings have been done between *Xysticus* s.lat., *Ozyptila* s.lat. and *Coriarachne* s.lat. The space here does not allow a more detailed discussion of them. The reason is very simple: some authors have regarded the size and habits as the differential character (e.g., Jocqué 1993), some have used the type of dorsal hairs (e.g., Crome 1962), and some the eye pattern (e.g., Locket & Millidge 1951). In this situation it is not surprising that the large relatives of *Ozyptila*, *Bassaniodes* spp. have quite recently been regularly included in *Xysticus*, in spite of epigynal hood and also dorsal setae typical of *Ozyptila*.

A misplacing of other relatives of Coriarachnini in the present sense, especially the partly tropical *Pycnaxis* and *Lysiteles* has also taken place many times. An eight-eyed species *Demogenes* s.lat. has been placed in *Ozyptila*.

Xysticus C.L. Koch, 1835
    Araneus Clerck, 1757: 136, in part
    Aranea Linné, 1758: 623, in part; Schrank, 1803: 235, in part
    Thomisus Audouin, 1826, in part: 398; Hahn, 1831: 1, in part, et auct. seq. until Lebert, 1877: 260 & 268
    Thomisus (Pachyptile) Simon, 1864: 433, in part (*luctans* only)
    Coriarachne Menge, 1875: 424 (*fusca*, misidentified)
    Oxyptila Emerton, 1893: 366, in part; Sørensen, 1898: 230, et auct seq. until Caporiacco, 1935: 188 (*xysticiformis*)
    Synema Keyserling, 1880: 64; Keyserling, 1884: 667; Dahl, 1907: 379 et auct. seq. until Kaston, 1948: 417
    Xysticus (Pellysticus) Schick, 1965: 146
    Xysticus (Lassysticus) Schick, 1965: 161

Type species by original designation *Aranea audax* Schrank, 1803 from Central Europe. *Xysticus* s.str. is the most speciose group of *Xysticus* s.lat., but more restricted to the temperate regions than both *Psammitis* and especially *Bassaniodes*. *Pachyptile* Simon, 1864 is a senior synonym of

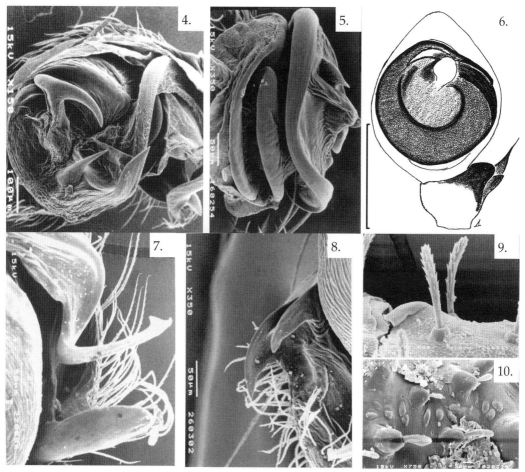

Fig. 4-10. Diagnostic characters of Coriarachnini sensu Ono. **Fig. 4.** *Xysticus cristatus* (Switzerland, Ospizio) male palp with tegular apophysis. **Fig. 5.** *Ozyptila praticola* (Finland, Korppoo; male palp with tegular apophysis). **Fig. 6.** *Demogenes* aff. n. sp. (Malaysia, Cameron Highlands) male palp with "median" apophysis. **Fig. 7.** *Xysticus cristatus* (Finland, Naantali) male palpal tutaculum. **Fig. 8.** *Psammitis sabulosus* (Finland, Utsjoki) male tutaculum. **Fig. 9.** *Narcaeus* n. sp. (Malaysia, Kalimantan Timur, Samarinda Ulu) tibial clavate setae. **Fig. 10.** "*Phrynarachne*" *clavigera* (Madagascar) femoral clavate setae.

Heriaeus Simon, 1875, but also a junior homonym. Type species of *Pellysticus* by original designation *Xysticus pellax* O. Pickard-Cambridge, 1894 from Central America and of *Lassysticus* by original designation *Xysticus lassanus* Chamberlin, 1925 from Texas.

Differential diagnosis of *Xysticus* s.str.: Males are very easily differentiated from other genera of *Xysticus* s.lat. by the presence of tegular apophyses (usually two, sometimes bifurcate).

Close relatives of the type species have paired epigynal fovea, while the epigyne of other species groups is variable. There is usually a central ridge, if the fovea is unpaired. The subgenera *Pellysticus* and *Lassysticus* were characterized by Schick (1965) by minor details in the structure of the male genital organs, including also the type of tutaculum. Their status is not further discussed here. However, their subgeneric status is not disputed, as they might represent valid taxa. There are no confirmed cases of clavate setae on the body or legs, as in many species of *Bassaniodes*, but most of the poorly known species have not been checked for this character.

Males are differentiated from the apophysate species of *Ozyptila* by wider than long MOT, often also by a smaller size and lack of clavate setae. The tegular apophyses in *Ozyptila* are often single and, if double, of entirely different structure (Figs. 4-5). Females lack the epigynal hood, typical of practically all species of *Ozyptila*, although the structure homologous to the hood may be widely modified in the *O. rauda*-group and some solitary species outside this group.

Anapophysate 'Xysticus'

Coriarachne is traditionally regarded as an independent genus by all authors using *Xysticus* s. lat. instead of *Xysticus* s.str. However, the strongly flattened body is just an adaptation to life under bark. The copulatory organs are close to those of *Psammitis* and *Spiracme*. Some large Asiatic species with a very simple colour pattern have sometimes been listed in *Ozyptila*, sometimes in *Xysticus*. Actually the pattern of their copulatory organs does not fit to *Ozyptila* at all and it will be necessary to create a new supraspecific taxon for this well-defined Asiatic group with at least five species.

This complex could be taxonomically treated in several different ways, depending on the emphasis laid on somatic and genital characters. When genital characters are regarded as most significant, even a classification with a single, somatically very variable genus, *Coriarachne*, could be chosen. This would necessarily provide the use of at least three subgenera, *Psammitis*, *Coriarachne* and a new subgenus for the *X. lugubris* group. Special evolution of the genital organs would probably necessitate the revalidation of the subgenus *Spiracme*, as well as a named supraspecific group for the *X. labradoriensis* group, at least. Even in this kind of classification the group *Bassaniana* would fall in synonymy with *Coriarachne* or with *Psammitis*. *Spiracme* could be regarded as a subgenus of either *Coriarachne* or *Psammitis*.

Proxysticus auct. (= *Bassaniodes*) is finally excluded here from *Xysticus* s.lat., as its anapophysate condition is shared by a majority of thomisids and no other essential characters than a fairly large size are similar to *Xysticus*.

Psammitis Menge, 1875
    Thomisus Hahn, 1831: 28, in part
    Xysticus C.L. Koch, 1838: 26, in part et auct. seq.
    Psammitis Menge, 1875: 449; Wunderlich, 1995: 751 (note), in part (*abramovi, ovadan, pseudoluctuosus, luctuosus, turkmenicus, tyshchenkoi, embriki, xysticiformis, zonshteini, lindbergi*) 762, in part only (*kempeleni, bicolor*), non Wunderlich, 1987.
    Xysticus (*Spiracme*) Gertsch, 1953: 450
    Xysticus (*Proxysticus*) Schick, 1965: 162, misidentification

Type species by original designation *Thomisus sabulosus* Hahn, 1832 from Central Europe. Numerous species from Europe, North Africa, Asia, and North America. *Psammitis* represents the most speciose group of the anapophysate *Xysticus* s.lat. No attempt has been made to transfer poorly known species to this genus. The concept of the anapophysate *Xysticus* by Wunderlich (1987, 1992, 1995) has changed without any explanation (cf. Platnick, 2001 under *Xysticus*). There are no true *Psammitis* in the Canary Islands, although Wunderlich (1987) listed four species. They all belong to *Bassaniodes*.

Differential diagnosis. Males of *Psammitis* can be separated from *Coriarachne* by normal body form and basally thinner embolus, and from *Spiracme* by non-screwed embolus. There are also differences in the average type of tutaculum and detailed pattern of the tibial apophyses. Females of *Psammitis* are characterized by an unpaired central epigynal cavity. *Coriarachne* spp. have paired epigynal pits, while *Spiracme* spp. have a distal pit on long central scape. The variation of dorsal pattern overlaps between most taxa of *Xysticus* s.lat., especially between *Xysticus* s.str. and *Psammitis*.

Spiracme Menge, 1875

> *Xysticus* L. Koch, 1870: 31, in part (*striatipes* only)
> *Spiracme* Menge, 1875: 447
> *Psammitis (Spiracme)* Ono, 1978: 285, non *Xysticus (Spiracme)* Gertsch, 1953

Type species of *Spiracme* by original designation *S. striata* Menge, 1875 = *Xysticus striatipes* L. Koch, 1870 from Central Europe (male and female from Burjatia, Siberia & female from Mongolia, coll. Y. Marusik examined). Some Central Asian samples may belong to another, related species.

Diagnosis. Embolus shorter and thicker than in *Psammitis*. The distally screwed embolus is easily differentiated also from that of *Coriarachne*. Origination of the ejaculatory duct is subcentral and subdistal and the duct itself is thicker than in the three related taxa. The tutacular apophysis is very distinct, triangle-shaped. VTA of male palp is simple, distally slightly hooked, while RTA has a distinct distal hook. The male abdomen is elongated, but the colour pattern is that of *Xysticus* s.str. The median septum of the epigyne ends in a pit. The pattern of the female abdomen is very simple.

Spiracme could be treated as a subgenus of *Coriarachne* or *Psammitis*. The latter alternative has been favoured by Ono (1978).

The concept of *Coriarachne*

When originally described by Thorell (1869), the flattened body was emphasized as the most important diagnostic character. This adaptive character was even used to characterize a thomisid tribe by Simon (1895: 1013) and Roewer (1954), although included genera with similar parallel adaptation had widely different genital structures. Gertsch (1953: 456) stated that 'the character used by Simon to isolate these groups has little validity' and emphasized the obvious relationship of *Coriarachne* to *Xysticus*. Actually, *Coriarachne* seems to be a sister group of *Psammitis* and more distantly related to *Xysticus* s.str. With some standards *Psammitis* and the other anophysate groups of *Xysticus* s.lat. except *Proxysticus* could be simply treated as subgenera of *Coriarachne*.

Coriarachne Thorell, 1869

The complex synonymic history of *Coriarachne* and *Bassaniana* was reviewed by Bowling & Sauer (1975) and Platnick (2001).

Type species of *Coriarachne* by original monotypy *Xysticus depressus* C.L. Koch, 1837 from Central Europe. Type species of *Bassaniana* by original designation and monotypy *B. aemula* O. Pickard-Cambridge, 1898 from Mexico. This species has later been regarded as junior synonym of *Coriarachne versicolor* Keyserling, 1880 from North America (Bowling & Sauer 1975). No material labelled as *B. aemula* has been studied by myself, but the detailed drawings by Gertsch (1953: figs. 60-61, 64, 67-68) depict two clearly different species. Type species of *Platyxysticus* Gertsch, 1932 by original designation *C. versicolor* Keyserling, 1880 from E. United States.

Coriarachne is traditionally regarded as an independent genus by all authors using *Xysticus* s.lat. instead of *Xysticus* s.str. However, the strongly flattened body is just an adaptation to life under bark on tree trunks. The genital organs are close to those of *Spiracme* and *Psammitis*. Ono (1985) stressed the body shape as a generic character and even treated *Bassaniana* with weaker flattening as a genus of its own. This act was accepted by Platnick (2001), although *Bassaniana* had been treated as a synonym of *Coriarachne* by Gertsch (1953) and Bowling & Sauer (1975). Here the taxon *Bassaniana* is regarded lower than subgenus.

The *Ozyptila*-problem

Parallel to *Xysticus* s.lat., the traditional genus *Ozyptila* also includes apophysate and non-apophysate groups. The North American species have been well treated by Gertsch (1953) and Dondale & Redner (1975), while a modern treatment of the European *Ozyptila* is lacking. The non-apophysate *floridana*-group in southern North America has been named *Modysticus* by Gertsch (1953). Gertsch used the lack of tegular apophysis and the wide MOT as the most important subgeneric characters. The European non-apophysate species have a much different type of copulatory organs.

The most deviating European non-apophysate species is *Ozyptila blackwalli* Simon, 1875, which should probably have status as a separate genus. Females of the Holarctic *rauda*-group also have lost their epigynal hood, while there is a large central cavity and an anterior fingerlike process, most probably homologous with the hood of other taxa of *Ozyptila* s.lat. (cf. Hippa et al. 1986). The group most probably deserves a new named taxon.

Dondale & Redner (1975) included all North American species outside *Modysticus* in a single *brevipes*-group, but according to the European species, it seems that a splitting to *trux*-, *atomaria*- and *praticola*-groups would, at least, reflect more naturally the relationships. Inclusion of the Mediterranean and Asiatic species will yield more groups and finally the creation of named taxa after careful phylogenetic analysis might be the most informative solution.

Ozyptila Simon, 1864

  Thomisus Walckenaer, 1826: 79; Walckenaer, 1837: 510

  Ozyptila Simon, 1864: 439; Bryant, 1930: 376; Strand, 1934: 273; Schick, 1965: and most later American authors, dominating since Platnick, 1993

  Oxyptila Thorell, 1869: 36 as an emendation of *Ozyptila* Simon; Gertsch, 1953: 463; most later European authors until recently

  Xysticus Thorell, 1872: 256, in part (*scabricula & pusio = simplex*); Thorell, 1875: 93 (*pullatus*)

  Coriarachne, Menge, 1875: 423

Type species by original designation *Thomisus claveatus* Walckenaer, 1837 from the Pyrenees and Egypt. This species was long regarded as a synonym of *Heriaeus hirtus* Latreille, 1819, referring to Walckenaer's large Egyptian species, previously misidentified by Savigny & Audouin (1825) as *Thomisus hirtus*. However, Walckenaer's original description fits to some small species with clavate hairs. Dondale & Redner (1975) checked all known Pyrenaean species with clavate hairs. They concluded that the description only fits to *Xysticus nigritus* Thorell, 1875 and designated a neotype for *Thomisus*

claveatus from Pyrenaean material. Blackwall (1861) misidentified *Oxyptila blackwalli* Simon, 1875 as *Thomisus claveatus* and Menge (1875) used the name *Coriarachne claveata* for *Thomisus [Oxyptila] scabriculus* Westring, 1851. The original author of *Ozyptila* later (Simon 1875) proposed a new type designation for *Ozyptila*, *Thomisus brevipes* Hahn, 1826 from Europe. This invalid taxonomic act was later accepted by many specialists (Gertsch, 1953; Schick, 1965, and still by Levy, 1985). *O. brevipes* was also listed as the type in catalogues by Roewer (1954) and Bonnet (1958). The large, mainly Holarctic genus *Ozyptila* s.lat. must, most probably, be split. The Holarctic *O. trux* group is distinctly related to the *O. brevipes*-group, while the *O. nigrita*-group is mainly Mediterranean (cf. Levy, 1985). On the other hand, species of *O. rauda*-group and *O. praticola*-group may have clavate hairs due to parallel adaptation. These groups are not discussed here in more detail.

Bassaniodes Pocock, 1903

  Thomisus Savigny & Audouin, 1825: 165, in part

  Xysticus C.L. Koch, 1838: 61, in part; Levy, 1985: 105, in part, et auct. seq.

  Oxyptila Simon, 1875: 218, in part (*blitea, albimana & bufo*); Berland, 1927: 13 (*albimana*)

  Bassaniodes Pocock, 1903: 198, type species *B. socotrensis*, Pocock, 1903

  Proxysticus Dalmas, 1922: 91, type species *Thomisus lalandii* Savigny & Audouin, 1825 from North Africa, syn.n., non *Xysticus (Proxysticus)* Schick, 1965 or Wunderlich, 1987

  Xysticus (Spiracme) Gertsch, 1953: 450, in part (no American spp.)

  Xysticus [Spiracme] (Proxysticus) Ono, 1978: 268

  Ozyptila Song & Hubert, 1983: 10, in part (*pseudoblitea*)

  Proxysticus Wunderlich, 1992: 495 (*canariensis, fuerteventurensis, grohi, lanzarotensis, ? madeirensis, pinocorticalis, squalidus & Xysticus asper* Lucas, 1838, nomen dubium)

  Coriarachne Jocque, 1993: 119 (*fienae*), non

Fig. 11-12. *Bassaniodes cribratus* (Turkey, Anatolia). **Fig. 11.** Female (left), male (right). **Fig. 12.** male palp.

Thorell, 1870

Psammitis Wunderlich, 1987: 255 (*clavulus, cribratus, tristrami*) , Mikhailov & Fet, 1994: 515, in part (*turanicus*); Wunderlich, 1995: 758, in part (*bliteus, bufo, tristrami, tenebrosus, graecus, sardiniensis, pseudorectilineus & fienae,*), non Menge, 1875, non *Xysticus (Psammitis)* Wunderlich, 1987

Type species of *Bassaniodes* by original monotypy *B. socotrensis* Pocock, 1903 from Socotra Island, type preservation unknown; this seems to be a junior synonym of some of the widespread species of the Mediterranean region. *Xysticus ferus* O. Pickard-Cambridge, 1876 is the most probable alternative, but the drawing of the epigyne by Pocock (cf. Pocock´s fig. 2 in Pl. 26 to fig. 118 of Levy, 1985) is too schematic for a confirmed synonymy. Cf. also the note by Wunderlich (1995 p. 761). Type species of *Proxysticus* by original designation *Thomisus lalandii* Savigny & Audouin, 1825 from Egypt. Berland & Fage in Simon (1932) erroneously listed [*Xysticus*] *albimanus* Simon, 1870 as the type species. The concept of *Xysticus (Proxysticus)* Schick, 1965 is entirely different from the real *Proxysticus*. These species are anaphophysate, but all species of both the *X. montanensis* group and the *X. luctuosus* group belong to *Psammitis* Menge, 1875. Actually, no confirmed records of *Bas-*

saniodes are known from the New World. Wunderlich (1995) simply placed all anapophysate species of *Xysticus* s.lat. to *Psammitis*.

Species of this mainly Mediterranean-West Asian group have, in the past, been listed under *Xysticus* as well as *Oxyptila* and *Coriarachne*, as the eye pattern does not fit to the definition of *Xysticus* s.lat. by many authors, e.g. Locket & Millidge (1951). A majority of *Xysticus* s.lat. from Central Asia and from dry Mediterranean habitats belong to *Bassaniodes.* The checking of all of them is out of the scope of this paper. All South African species described as *Xysticus* may belong to this genus or they might represent a supraspecific taxon of their own.

Diagnosis. General habitus of both sexes of *Xysticus*-type (Fig. 11). MOT often longer than wide, as in *Ozyptila*. Body setae clavate or blunt as in *Ozyptila*, but exceptionally thin and pointed, e.g., in *B. lalandei*. Male palpal tegulum without tegular apophyses (Fig. 12), but origin of the circumtegular ridge strongly sclerotized and a variously shaped tegular ridge is present close to that area (Fig. 12). Three tibial apophyses, VTA very large, often encircling a cavity.

Female epigyne with a posteriorly concave, well sclerotized anterior hood, most probably homologous with the anterior pit or hood of *Ozyptila* s.str. Some species habitually resemble *Ozyptila* spp.

Most species of *Bassaniodes* are as large or even larger than common species of *Xysticus* and therefore they have originally mostly been described as *Xysticus*. However, many genital (epigynal hood, complex tibial apophyses) and some somatic characters (eye pattern, structure of setae) are more similar to *Ozyptila*. Therefore a relationship with *Ozyptila* is suggested here.

Acknowledgements

For making it possible to study types and other thomisid material during recent visits to the following museums, I wish to acknowledge the generous hospitality of the curators: Museum National d'Histoire Naturelle, Paris (J. Heurtault & C. Rollard), Museo Civico di Storia Naturale, Genoa (G. Doria), Natural History Museum, Stockholm (T. Kronestedt), and International Rice Research Institute, Los Baños (K. Schoenly & A. Barrion). Some essential thomisid types were checked during earlier visits to the Natural History Museum, London (P. Hillyard), Zoologisches Institut and Zoologisches Museum, Hamburg (G. Rack), and American Museum of Natural History (N.I. Platnick). Thomisid material from the Zoologisches Museum Berlin was kindly made available by S. Nawai and J. Dunlop. The finishing of this work after my retirement would not have been possible without the generous support of the Finnish forest technology company, UPM-Kymmene Group. The role of the director of research, K. Ebeling, and the Vice President of International Affairs, O. Henriksen has been of tremendous importance in this respect, as well as the help of M. Halinen and the staff of the local April Company during my latest field work in Sumatra 1999.

The drawings were made by M. Mustonen, Y. Marusik, D. Logunov and A. Tanasevitch. Discussions with Marusik, Logunov and Tanasevitch about the taxonomic problems of Thomisidae have been fruitful for this work. The work of Ms Mustonen was made possible by the University of Turku (The Dean, H. Lönnberg), the Foundation for Turku University, and as a part of the whole taxonomic programme, by the Academy of Finland.

Y. Marusik as well as V. Rinne and S. Koponen, Zoological Museum, University of Turku, have helped with various technical and practical problems. Y. Marusik, V. Rinne and P. Soljala helped with computer problems. All these companies, organisations and people are gratefully acknowledged.

References

Audouin, V. 1827. Explication sommaire des planches d'arachnides de l'Egypte et de la Syrie publiées.... In: "Description de l'Egypte...". *Zoologie* 22, 291-430.

Barrion, A.T. & Litsinger, J.A. 1995. *Riceland spiders of South and Southeast Asia.* CAB International.

Blackwall, J. 1861. *A history of the spiders of Great Britain and Ireland.* Ray Society, London.

Berland, L. 1922. Araignées. In *Voyage de M. le Baron de Rothschild en Ethiopie et en Afrique orientale anglaise (1904-1905). Résultats scientifiques. Animeux articulés* Vol. 1, pp. 43-90. Paris.

Berland, L. 1927. Contributions à l'étude de la biologie des Arachnides (2e Mémoire). *Archives de zoologie expérimentale et générale. Notes et Revue* 66 (2), 7-29.

Bonnet, P. 1958. *Bibliographia Araneorum* Vol. II, 4eme partie. Douladoure, Toulouse.

Bowling, T.A. & Sauer, R.J. 1975. A taxonomic revision of the crab spider genus *Coriarachne* (Araneida, Thomisidae) for North America north of Mexico. *Journal of Arachnology* 2, 183-193.

Bristowe, W.S. 1931. A preliminary note on the spiders of Krakatau. *Proceedings of the Zoological Society of London* 1931 (4), 1387-1400.

Bryant, E.B. 1930. A revision of the American species of the genus *Ozyptila*. *Psyche* 37, 375-391

Caporiacco, L. di 1935. Aracnidi dell'Himalaia e del Karakoram, raccolti dalla Missione italiana al Karakoram (1929-VII). *Memorie della Societa entomologica italiana. Genova* 13, 161-163.

Caporiacco, L. di 1941. *Arachnida.* Missione Biol. Sagan-Omo 12 (Zool. 6), 1-159

Clerck, C. 1757. *Aranei Suecici, descriptionibus et figuris oeneis illustrati, ad genera subalterna redacti speciebus ultra LX determinati*. Stockholm.

Crome, W. 1962. Studien an Krabbenspinnen (Araneae: Thomisidae) 4. Bemerkungen zur praktischen Unterscheidung der beiden Genera *Xysticus* C.L.Koch, 1835 u. *Ozyptila* Simon, 1864. *Mitteilungen der Deutsche entomologische Gesellschaft*. 21 (3), 37-39.

Dahl, F. 1907. *Synaema marlothi*, eine neue Laterigraden-Art und ihre Stellung in System. *Mitteilungen aus dem zoologischen Museum im Berlin* 3, 369-395.

Dalmas, R. de 1922. Catalogue des Araignées récoltées par le Marquis G. Doria dans l'ile Giglio (Archipel toscan). *Annali del Museo civico di storia naturale di Genova* 50, 79-96.

Dondale, C.D. & Redner, J.H. 1975. The genus *Ozyptila* in North America (Araneida, Thomisidae). *Journal of Arachnology* 2, 129-181.

Emerton, J.H. 1893. New England spiders of the family Thomisidae. *Transactions of the Connecticut Academy of Arts and Sciences* 8, 359-381.

Gertsch, W.J. 1953. The spider genera *Xysticus, Coriarachne*, and *Oxyptila* (Thomisidae, Misumeninae) in North America. *Bulletin of the American Museum of Natural History* 102 (4), 417-482.

Hahn, C.W. 1831. *Die Arachniden*. Erster Band, Nürnberg.

Herman, O. 1879. *Magyarország Pók-faunája* Vol. 3. Budapest.

Hippa, H., Koponen, S. & Oksala, I. 1986. Revision and classification of the holarctic species of the *Ozyptila rauda* group (Araneae, Thomisidae). *Annales Zoologici Fennici* 23 (3), 321-328.

Jantscher, E. 2002. The significance of male pedipalpal characters for the higher systematics of the crab spider genus *Xysticus* C.L. Koch, 1835 (Araneae: Thomisidae). In *European Arachnology 2000* (S. Toft & N. Scharff eds.), pp. 329-336. Aarhus University Press, Aarhus.

Jézéquel, J.-F. 1964. Araignées de la savane de Singrobo (Côte d'Ivoire) 3. Thomisidae. *Bull. Inst. fond. Afr. noire* 26 (A), 1103-1143.

Jézéquel, J.-F. 1966. Araignées de la savane de Singrobo (Côte d'Ivoire) 5. Note complémentaire sur les Thomisidae. *Bulletin du Museum d'histoire naturelle, Paris* 37, 613-630.

Jocqué, R. 1993. A new species of *Coriarachne* from Spain. *Bulletin de l'institut Royal de Sciences naturelles de Belgique, Entomologie* 63, 199-222.

Karsch, F. 1878. Exotisch-araneologisches. *Zeitschrift für die gesammten Naturwissenschaften* 51, 771-828.

Kaston, B.J. 1948. Spiders of Connecticut. *Bulletin. State of Connecticut. Geological and Natural History Survey* 70, 1-874.

Keyserling, E. 1880. *Die Spinnen Amerikas, vol. 1. Laterigradae*. Nürnberg.

Keyserling, E. 1884. Neue Spinnen aus America. V. *Verhandlungen der zoologische-botanischen Gesellschaft in Wien* 33, 649-684.

Koch, C.L. 1835. Arachniden. In *Faunae Insectorum Germaniae initia* (G.W.F. Panzer ed.), Heft 128-131 [129; fol. 12-24]. Regensburg.

Koch, C.L. 1838. *Die Arachniden*, Band 4. Nürnberg

Koch, L. 1867. Beschreibungen neuer Arachniden und Myriapoden. *Verhandlungen der zoologische-botanischen Gesellschaft in Wien* 17, 173-250.

Koch, L. 1870. Beiträge zur Kenntniss der Arachnidenfauna Galiziens. *Jahrbuch der k.k. gelehrten Gesellschaft in Krakau* 41, 1-56.

Koch, L. 1874. *Die Arachniden Australiens, nach der Natur beschrieben und abgebildet*. Vol. 1, pp. 473-576. Nürnberg.

Kroneberg, A.I. 1875. Araneae. In *Puteshestvie v Tourkestan. Reise in Turkestan*. Zoologischer Teil, 2 (A.P. Fedtschenko ed.), pp. 1-58. Moscow.

Lawrence, R.F. 1928. Contributions to a knowledge of the fauna of South-West Africa V. Arachnida. *Annals of the South African Museum* 25 (1), 1-75 .

Lawrence, R.F. 1936. Scientific results of the Vernay-Lang Kalahari Expedition, March to September 1930. Spiders (Ctenizidae excepted). *Annals of the Transvaal Museum* 17 (2), 145-158.

Lawrence, R.F. 1952. New spiders from the eastern half of South Africa. *Annals of the Natal Museum* 12, 183-226.

Levy, G. 1985. Araneae: Thomisidae. In *Fauna Palaestina. Arachnida* II. Israel Academy of Sciences and Humanities, Jerusalem.

Linné, C., 1758: *Systema Naturae per regna tria naturae, secundum classes, ordines, genera, species cum characteribus differentiis, synonymis, locis.* Editio decima, reformata. Vol. 1, Holmiae (Stockholm).

Locket, G.H. & Millidge, A.F. 1951. *British Spiders* I. Ray Society, London.

Logunov, D. & Marusik, Y., 1998: A new species of the genus *Xysticus* from the mountains of South Siberia (Araneae, Thomisidae). *Bulletin of the British Arachnological Society* 11(3), 103-106.

Marusik, Y. & Logunov, D. 1991. Poorly known spider species of the families Salticidae and Thomisidae (Aranei) from the Far East of USSR. *Entomological investigations in the North-East part of USSR, Vladivostok; FEB AS USSR* 2, 131-140.

Marusik, Y. & Logunov, D. (1994) 1995. The crab spiders of Middle Asia (Aranei, Thomisidae). *Beiträge zur Araneologie* 4, 133-175.

Menge, A., 1875. Preussische Spinnen. VIII. Fortsetzung. *Schriften der naturforschenden Gesellschaft in Danzig* (N. F.) 3, 423-454.

Mikhailov, K.G. & Fet, V. 1994: Fauna and zoogeography of spiders (Aranei) of Turkmenistan. In: *Biogeography and Ecology of Turkmenistan* (V. Fet & K.I. Atamuradov eds.), pp. 499-524. Kluwer Academic Publ., Netherlands.

Ono, H. 1978. Thomisidae aus dem Nepal-Himalaya. I. Das Genus *Xysticus* C.L. Koch, 1835 (Arachnida: Araneae). *Senckenbergiana Biologica* 59(3-4), 267-288.

Ono, H. 1979. Thomisidae aus dem Nepal-Himalaya. II. Das Genus *Lysiteles* Simon, 1895. *Senckenbergiana Biologica* 60, 91-108.

Ono, H. 1985. Revision einiger Arten der Familie Thomisidae (Arachnida, Araneae) aus Japan. *Bulletin of the National Science Museum Tokyo (Zool.)* 11 (1), 19-39.

Ono, H. 1988. A revisional study of the spider family Thomisidae (Arachnida, Araneae) of Japan. *Tokyo, National Science Museum,* 1-252.

Platnick, N.I., 1993. *Advances in spider taxonomy 1988-1991. With synonymies and transfers 1940-1980.* New York Entomological Society.

Platnick, N.I., 2001. The World Spider Catalog (electronic version) Thomisidae [213 pp.] AMNH [http://research.amnh.org/entomology/spiders/catalog81-87/INTRO2.ht]

Pocock, R.I. 1903. Arachnida. In: *The Natural History of Sokotra and Abd-el-Kuri* (H.O. Forbes ed.). Liverpool, Special Bulletin of the Liverpool Museums under the City Council., pp. 175-208

Roewer, C.F. 1938. Araneae. In: Résultats scientifiques du Voyage aux indes orientales néerlandaises de LL. AA. RR. de Belgique. *Mémoires du Musée royal d'histoire naturelle de Belgique* 3 (19), 1-94.

Roewer, C.F. 1954. *Katalog der Araneae von 1758 bis 1940, bzw. 1954.* Vol. 2b. Bruxelles.

Savigny, J.C. & Audouin, V. 1825. Explication sommaire des Planches d'Arachnides de l'Egypte et de la Syrie, publiées par Jules-César Savigny, membre de l'Institut; offrant un exposé des caractères naturels des genres avec la distinction des espèces. In *Description de l'Egypte... Histoire Naturelle* Vol. 1 (4). Paris.

Schick, R.X. 1965. The crab spiders of California (Araneida: Thomisidae). *Bulletin of the American Museum of Natural History* 129 (1), 1-180.

Schrank, F. von Paula 1803. *Fauna Boica. Durchgedachte Geschichte der in Baiern einheimischen und Zahmen Tiere* Vol. 3 (1). Landshut.

Simon, E. 1864. *Histoire naturelle des Araignées (Aranéides).* Paris.

Simon, E. 1875. *Les Arachnides de France.* Part 2. Paris.

Simon, E. 1895. *Histoire naturelle des Araignées.* Part 1 (4). Paris.

Simon, E. 1909. Étude sur les Arachnides du Tonkin (1re partie). *Bulletin scientifique de la France et de la Belgique* 42, 69-147.

Simon, E. 1932. *Les Arachnides de France. Synopsis générale et Catalogue des espèces françaises de l'ordre des Araneae.* Part 6 (4) (Final posthumous edition by L. Berland & L. Fage). Paris.

Sørensen, W. 1898. Arachnida Groenlandica (Acaris exceptis). *Videnskabelige Meddelelser fra den Naturhistoriske Forening i Kjöbenhavn* 1898, 176-235.

Song, D.-X. & Hubert, M. 1983. A redescription of the spiders of Beijing described by E. Simon in 1880. *J. Huizhou Teacher's College* 2, 1-23.

Strand, E. 1934. Miscellanea nomenclatorica zoologica et palaeontologica VI. *Folia zoologica et hydrobiologica.* 6 (2), 271-277

Taczanowski, L. 1872. Les Aranéides de la Guyane française. *Horae Societatis entomologicae Rossicae. Saint-Pétersbourg* 8, 32-132 .

Thorell, T. 1869. On European spiders. Part I. Review of the European genera of spiders, preceded by some observations on Zoological Nomenclature. *Acta regiae Societatis scientiarum Upsaliensis. Stockholm* 3 (7), 1-108.

Thorell, T. 1870. On European spiders. *Acta regiae Societatis scientiarum Upsaliensis. Stockholm* 3 (7), 109-242.

Thorell, T. 1872. *Remarks on synonyms of European spiders.* Part III, pp. 229-374. Uppsala.

Thorell, T. 1875. Verzeichnis südrussischer Spinnen. *Horae Societatis entomologicae Rossicae. Saint-Pétersbourg* 11, 39-122.

Turnbull, A.L., Dondale C.D. & Redner, J.H. 1965. The spider genus *Xysticus* C.L. Koch (Araneae: Thomisidae) in Canada. *Canadian Entomologist* 97 (12), 1233-1280.

Walckenaer, C.A., 1826: Aranéides. In: *Faune française ou Histoire naturelle générale et particulière des animaux qui se trouvent en France.* Vol. 11-12. Paris.

Walckenaer, C.A. 1837. *Histoire naturelles des Insectes. Aptères.* Vol 1. Paris.

Wunderlich, J. 1987. *Die Spinnen der Kanarischen Inseln und Madeiras: Adaptive Radiation, Biogeographie, Revisionen und Neubeschreibungen.* Triops, Langen.

Wunderlich, J. 1992. Die Spinnen-Fauna der Makaronesischen Inseln. *Beiträge zur Araneologie* 1, 1-620.

Wunderlich, J. 1995. Zur Kenntnis West-Paläarktischer Arten der Gattungen *Psammitis* Menge, 1875, *Xysticus* C.L. Koch 1835 und *Ozyptila* Simon 1864 (Arachnida: Araneae: Thomisidae). *Beiträge zur Araneologie* 4, 749-774.

European Arachnology 2000 (S. Toft & N. Scharff eds.), pp. 329-336.
© Aarhus University Press, Aarhus, 2002. ISBN 87 7934 001 6
(Proceedings of the 19th European Colloquium of Arachnology, Århus 17-22 July 2000)

The significance of male pedipalpal characters for the higher systematics of the crab spider genus *Xysticus* C.L. Koch, 1835 (Araneae: Thomisidae)

ELKE JANTSCHER
Institute of Zoology, Karl-Franzens-University, Universitätsplatz 2, A-8010 Graz, Austria
(Jantsche@kfunigraz.ac.at)

Abstract

Male pedipalp characters support the hypothesis that there are potentially at least three clades within European *Xysticus* s.l. C.L. Koch, 1835 (Araneae: Thomisidae). These clades correspond with the older, but currently synonymised, genera *Proxysticus* Dalmas, 1922 and *Psammitis* Menge, 1875. Synapomorphies in the male pedipalp shared with other thomisid genera (e.g. *Coriarachne* Thorell, 1870, *Ozyptila* Simon, 1864) suggest that *Xysticus* s.l. represents a paraphyletic group.

Key words: Thomisidae, *Xysticus*, *Psammitis*, *Proxysticus*, systematics, phylogeny

INTRODUCTION

Xysticus C.L. Koch, 1835 is the most species-rich genus within the family Thomisidae (crab spiders) and contains more than 340 species (Ono 1988). These are widely distributed, but occur mainly in the northern hemisphere. It is not the number of species but the enormous heterogeneity in genital and morphological characters within the *Xysticus* species that makes it difficult to resolve relationships between them. Despite the efforts of previous authors to subdivide it (see below), the taxonomic status of this crab spider genus is uncertain and it is currently regarded as a large *Xysticus* sensu lato group (Brignoli 1983, Platnick 1997). Close investigation of middle European species suggests that structures in the male palp include apomorphic characters which may be used to define distinct groups, at least within the European *Xysticus* s.l. Furthermore, some of these apomorphic characters appear to be present in other thomisid genera and thus draw into question the monophyly of *Xysticus* s.l.

HISTORY

During the last 150 years there have already been other arachnologists who realised that *Xysticus* is heterogeneous and tried to solve the problem by splitting the genus or establishing subtaxa. Some of the main points in the history of *Xysticus*, since it was first described by C.L. Koch in 1835, are the separating by Thorell of *Coriarachne* Thorell, 1870, the separation of two new genera from *Xysticus* by Menge (*Psammitis* Menge, 1875 and *Spiracme* Menge, 1875) and the creation by Dalmas of the new genus *Proxysticus* Dalmas, 1922.

Thorell (1870) considered *Coriarachne* as a distinct genus not only because of the flattened body, but also because of the different arrangement of the eyes. Obviously impressed by the screw-like embolus and the flat tegulum without any protruding parts, Menge (1875) estab-

Table 1. Previous suggestions of subgroups and subgenera in *Xysticus* s.l.

Simon 1932: *Xysticus*, 4 subgroups
cristatus
longipes
sabulosus
robustus
Gertsch 1939: *Xysticus*, 5 groups
cristatus
cunctator
concursus
sabulosus
robustus
Gertsch 1953: *Xysticus*, 2 subgenera
Xysticus
Spiracme
Schick 1965: *Xysticus*, 5 subgenera
Pellysticus
Lassysticus
Xysticus
Psammitis
Proxysticus

lished the new genus *Spiracme*. He also stated in his diagnosis of the genus that the body is elongate and the eyes are similar to *Xysticus*, but that the lateral eyes are somewhat larger. For *Psammitis*, the second genus he established, he referred to the disk-like tegulum and the embolus surrounding it. Again he also used the different size of the lateral eyes to distinguish *Psammitis* from *Xysticus*. Dalmas (1922) described the males of his new genus *Proxysticus* as lacking any tegular apophyses, but with a protruding area in the middle of the bulb and an evenly twisted embolus.

As Ono (1988) stated, the tendency after that time was rather to establish subgroups or subgenera, (e.g. Simon 1932; Gertsch 1939, 1953; Schick 1965; see Table 1). In particular Schick (1965) intensively studied male palp characters not only of *Xysticus*, but of other nearctic thomisids too. Based on the recognition of the different types of tutacula, emboli and tegular apophyses he suggested the subgenera *Pellysticus* Schick, 1965, *Xysticus* C.L. Koch and *Lassysticus* Schick, 1965, belonging to the apophysate (i.e. apophyses present) division, and the subgenera *Proxysticus* Dalmas and

Psammitis Menge, belonging to the anapophysate division. Later Wunderlich (1987, 1994) resurrected *Psammitis* as a distinct genus. He characterised the males of *Psammitis* by a tegulum with 1 or 2 humps/'pockets' and/or one apophysis (but never with two apophyses as in *Xysticus*), with a long or short embolus and 2-4 tibial apophyses. On the contrary he defined *Xysticus* by the presence of two tegular apophyses, never with a sickle-shaped pocket (as in *Psammitis*), two tibial apophyses and a long and thin embolus (Wunderlich, 1994, p. 751). This was not followed by other arachnologists and despite all earlier efforts in the history of *Xysticus*, the spider catalogues (Brignoli 1983, Platnick 1997) still list *Xysticus* as a single, large species-complex. For example Platnick (1997) rejected both *Proxysticus* and *Psammitis*, noting the absence of comprehensive studies and regarded *Xysticus* as a large *sensu lato* group.

MATERIALS AND METHODS

This study is based on alcohol preserved material belonging to different museum collections (e.g. Basel, Berne, Berlin, Frankfurt, Stockholm, Paris, Vienna) and to private collections. More than 30 different *Xysticus* species, all of which are found in central Europe, have been examined and the structure of the male pedipalp has been studied intensively. The species examined were: *X. acerbus* Thorell, 1872, *X. albomaculatus* Kulczyński, 1891, *X. apricus* L. Koch, 1875, *X. audax* (Schrank, 1803), *X. bifasciatus* C.L. Koch, 1837, *X. bonneti* Denis, 1938, *X. cor* Canestrini, 1873, *X. cristatus* (Clerck, 1757), *X. desidiosus* Simon, 1875, *X. embriki* Kolosvary, 1935, *X. erraticus* (Blackwall, 1834), *X. ferrugineus* Menge, 1875, *X. gallicus* Simon, 1875, *X. ibex* Simon, 1875, *X. kempeleni* Thorell, 1872, *X. kochi* Thorell, 1872, *X. lanio* C. L. Koch, 1835, *X. lineatus* (Westring, 1851), *X. luctator* L. Koch, 1870, *X. luctuosus* (Blackwall, 1836), *X. macedonicus* Šilhavý, 1944, *X. marmoratus* Thorell, 1875, *X. ninnii* Thorell, 1872, *X. obscurus* Collett, 1877, *X. paniscus* L. Koch, 1875, *X. robustus* (Hahn, 1832), *X. sabulosus* (Hahn, 1832), *X. secedens* L. Koch,

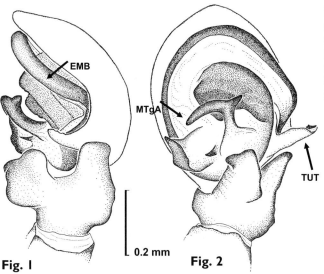

Fig. 1-2. Left pedipalp, *Xysticus cristatus* (Clerck, 1757), priv. Coll. Peter Horak (Tirol, nr. Innsbruck, 590 m, Meyer leg.) 1: retrolateral, 2: ventral. Abbreviations: EMB: embolus, MTgA: median tegular apophysis, TUT: tutaculum.

1875, *X. striatipes* L. Koch, 1870, *X. tortuosus* Simon, 1932, *X. ulmi* (Hahn, 1831), *X. viduus* Kulczyński, 1898.

The similar-looking genera *Coriarachne* and *Ozyptila* Simon, 1864 were also examined, since provisional observations suggested that male pedipalpal characters seen in *Xysticus* are also present in these taxa. For outgroup comparison the pedipalp morphology of various thomisid genera (including the species *Misumena vatia* (Clerck, 1757), *Misumenops tricuspidatus* (Fabricius, 1775), *Synaema globosum* (Fabricius, 1775), *Coriarachne depressa* (C.L. Koch, 1837), *Coriarachne fulvipes* (Karsch, 1879), *Ozyptila trux* (Blackwall, 18469 and *Thomisus onustus* Walckenaer, 1806), was studied from the literature (e.g. Ono 1988; Roberts 1995) or from specimens deposited at the Museum für Naturkunde, Berlin, and my own private collection.

RESULTS

My studies on central European representatives of *Xysticus* s.l. suggest that there exist at least three different groups within Europe, which can potentially be diagnosed on autapomorphies derived from the male pedipalp (Table 2).

In *Xysticus* the pedipalp is usually charac-terised by the presence of tibial apophyses (mostly ventral and retrolateral tibial apophyses, sometimes intermediate tibial apophysis – respectively VTiA, RTiA, ITiA) and tegular apophyses with various forms and shapes from hook-like structures to simple ridges. A tutaculum is present at the retrolateral edge of the bulbus. Its size and shape are variable.

Group 1 - 'Xysticus sensu stricto' (Figs. 1, 2)

Xysticus cristatus (Clerck, 1757) is a common representative of this group, which can be defined by a very complex tegular structure with at least two distinct tegular apophyses. The embolus is elongate, whip-like and its tip lies within a simply constructed tutaculum (Fig. 2). The tibia bears two apophyses (VTiA, RTiA). *Xysticus audax*, the type species of the genus *Xysticus*, also belongs here, being part of the *cristatus* group (e.g. Jantscher 2001).

Group 2 - 'Proxysticus' (Figs. 3, 4)

The drawings show the pedipalp of *Xysticus robustus* (Hahn, 1831). This group also has a long embolus originating at the basal part of the bulb. Again the tutaculum is simply built, but the tegulum is different. It does not carry distinct apophyses but is merely slightly structured in the centre. Three apophyses (VTiA,

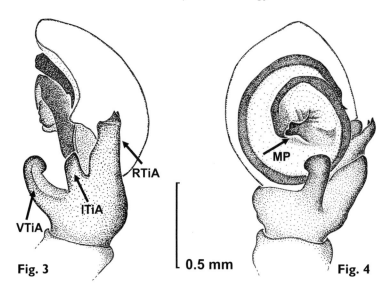

Fig. 3   0.5 mm   Fig. 4

Fig. 3-4. Left pedipalp, *Xysticus robustus* (Hahn, 1831), Natural History Museum Basel 438c, 3: retrolateral, 4: ventral. Abbreviations: ITiA: intermediate tibial apophysis, MP: median projection, RTiA: retrolateral tibial apophysis, VTiA: ventral tibial apophysis.

ITiA and RTiA) are present on the tibia, each emerging from its own base.

Group 3 - 'Psammitis' (Figs. 5, 6)

This group, represented here by *Xysticus sabulosus* (Hahn, 1831), is characterised by the com-

pletely flat tegulum without any special structures but at best simple ridges. The embolus originating at the apical part of the bulbus is, in comparison to those of the other groups, very short and often characteristically shaped (e.g. screw-like). In cases where the embolus is

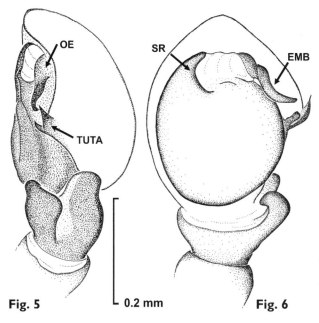

Fig. 5   0.2 mm   Fig. 6

Fig. 5-6. Left pedipalp, *Xysticus sabulosus* (Hahn, 1831), Natural History Museum Basel 1894d, 5: retrolateral, 6: ventral. Abbreviations: EMB: embolus, OE: origin of embolus, SR: sclerotised ridge, TUTA: tutacular apophysis.

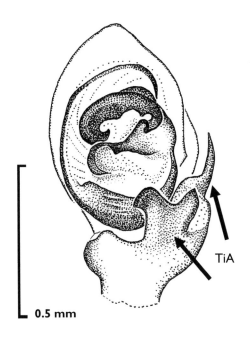

0.5 mm

Fig. 7. Left pedipalp, *Ozyptila trux* (Blackwall, 1846), after Roberts (1995, p. 167), ventral. Abbreviations: TiA = tibial apophyses

TiA

ingly, this structure is more complex than in the other two groups and often even two pointed tips (tutaculum and tutacular apophysis) are present instead of one simple structure. The tibia bears the usual ventral and retrolateral apophyses.

Ozyptila (Fig. 7)

This genus, here represented by *Ozyptila trux* (Blackwall, 1846), has a structured tegulum with more or less distinct apophyses. The tibial apophyses are highly complex and do not fit clearly into the VTiA-RTiA scheme but possibly the VTiA and RTiA share a common base.

Coriarachne (Figs. 8-9)

The drawings show *Coriarachne fulvipes* (Karsch, 1879). *Coriarachne* is characterised by a very flat tegulum without any apophyses but simply a sickle-shaped ridge in the upper part of the bulb. Two tibial apophyses (VTiA, RTiA) are present. The tutaculum is small and far away from the embolus tip. The embolus is short and originates in the apical area of the bulb. Note that the overall shape is very similar to 'Psammitis*.

twisted, it can reach a considerable length, but always originates in the apical area of the bulbus. In most cases the terminal part of the embolus does not reach the tutaculum. Surpris-

To date there is no published cladogram of the Thomisidae at the genus level, thus the sister

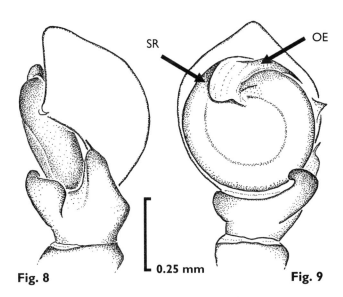

SR

OE

Fig. 8

0.25 mm

Fig. 9

Fig. 8-9. Left pedipalp, *Coriarachne fulvipes* (Karsch, 1879), after Ono (1988, figs. 60-61), 8: retrolateral, 9: ventral. Abbreviations: OE: origin of embolus, SR: sclerotised ridge.

Table 2. Autapomorphies for the groups identified here and the related genera

'Xysticus s.s.':	two or more distinct bulbar apophyses
'Psammitis':	modified tutaculum
'Proxysticus':	three tibial apophyses
Ozyptila:	small size, modified body spination
Coriarachne:	flat body

group of *Xysticus* s.l. remains to be determined. However, a flattened tegulum occurs widely in other thomisid genera (see e.g. figs. in Roberts 1995) such as *Thomisus* Walckenaer, 1805, *Synaema* Simon, 1864, *Misumenops* F.O.P.-Cambridge, 1900, *Pistius* Simon, 1875, *Heriaeus* Simon, 1875, *Misumena* Latreille, 1804 and *Runcinia* Simon, 1875 and is therefore potentially plesiomorphic for *Xysticus*.

The plesiomorphic number of tibial apophyses is more difficult to polarise but two simple apophyses occur in a number of thomisid genera, e.g. *Thomisus*, *Synaema*, *Runcinia*. This character state may therefore be plesiomorphic for Thomisidae, but in other genera there are clearly three distinct apophyses (see 'Proxysticus', Fig. 3) or complex combinations in which the apophyses seem to fuse with one another (see *Ozyptila*, Fig. 7). These modifications are potential apomorphies compared to the simple morphology noted above.

A short summary of the autapomorphies for these probably monophyletic *Xysticus* groups, as well as for the related genera *Ozyptila* and *Coriarachne*, is given in Table 2.

DISCUSSION

A significant result of this study is that it was not possible to find an obvious autapomorphy in the pedipalp to support the monophyly of the currently accepted *Xysticus* s.l. group. By contrast, possible clades within *Xysticus* s.l. have been identified, based on distinct, apomorphic male genital characters: 'Psammitis', 'Xysticus sensu stricto' and 'Proxysticus' (Table 2). Whether these palpal characters justify resurrecting *Psammitis* and *Proxysticus* is unclear at present, but the names remain available. Currently I consider Menge's genus *Spiracme* as

a possible synonym of 'Psammitis' although the type of the female epigyne does not match with those of other representatives of 'Psammitis'. Furthermore, these clades potentially share synapomorphies in the pedipalp with other thomisid genera, i.e. *Coriarachne* and *Ozyptila* in this study. This supports the hypothesis that *Xysticus* s.l. is a paraphyletic assemblage of common crab spiders.

One provisional clade is (*Coriarachne* + 'Psammitis') the synapomorphy for which could be the modified embolus originating in the apical part of the bulbus (Figs. 5-6, 8-9, OE), although this character must be tested against other taxa. There are further thomisid genera where the tegulum and embolus also show remarkable similarities to the (*Coriarachne* + 'Psammitis') synapomorphy suggested here, e.g. *Misumena*, *Pistius*, and these could belong to this group too. The second provisional clade is ('Xysticus s.s.' ('Proxysticus' + *Ozyptila*)). This entire clade is defined by the structured tegulum (Figs. 2, 4, 7), as opposed to the plesiomorphic flat tegulum (Figs. 6, 9). The synapomorphy of ('Proxysticus' + *Ozyptila*) is the presence of complex tibial apophyses. 'Proxysticus' is characterised by three tibial apophyses and *Ozyptila* also shows a highly complex tibia (Fig. 8), but the apophyses do not each emerge from a distinct and separate base. So far it is not clear which parts are homologous to the different apophyses. Both these clades, if correct, imply that *Xysticus* s.l. is not a monophyletic group.

Female genital characters, which are very complex and need to be studied in more detail, also support the hypothesis of three groups within *Xysticus* s.l. Corresponding with the three potential groups, three types of epigynes can be found, but these characters are not comprehensively discussed here. However, included here is a brief survey of the three major types of epigynes. In the 'Xysticus s.s.' group the epigyne is usually characterised by the presence of one or two deep depressions, the genital atria. In the latter case the atria are separated from each other by a distinct median septum, e.g. in *Xysticus cristatus* (e.g. Roberts 1985, text fig. 39a). 'Psammitis' on the contrary merely shows one slight depression mostly in combination with a groove (e.g. Roberts 1985, text fig. 40d). The genital opening is not so obvious as it is in *Xysticus* s.s. Females of 'Proxysticus' have a highly characteristic epigyne: heavily sclerotised large bulges and folds surrounding a pocket-like structure in the middle of the epigynal plate (e.g. Roberts 1985, text fig. 41b).

Other characters in *Xysticus*, like body shape or spination, should also be closely investigated and could contribute further characters. These provisional results must also be tested against non-European material, in which other potential subgroups might be expected.

ACKNOWLEDGEMENTS

I thank the following museum curators for access to material in their collections: P. Hillyard (London), T. Szuts (Budapest), D. Logunov (formerly Novosibirsk), C. Rollard (Paris), J. Gruber (Vienna), A. Hänggi (Basel), C. Kropf (Berne), M. Grasshoff (Frankfurt), T. Kronestedt (Stockholm), L. Tiefenbacher (Munich), J. Dunlop (Berlin) and A. Kůrka (Prague). For additional material used in this study I also thank P. Horak (Graz), J. Wunderlich (Straubenhardt), C. Deltshev (Sofia), C. Muster (Dresden), K. Thaler (Innsbruck) and C. Komposch (Graz). I am very grateful to J. Dunlop (Berlin) and two anonymous referees for important comments on the manuscript. This study was partly supported by grants from the Faculty of Natural Sciences, University of Graz, the Steiermärkische Landesregierung and the Österreichische Forschungsgemeinschaft.

REFERENCES

Brignoli, P.M. 1983. *A catalogue of the Araneae described between 1940 and 1981.* Manchester University Press, New York.

Dalmas, R. de. 1922. Catalogue des Araignées récoltées par le Marquis G. Doria dans l'île Giglio (Archipel toscan). *Annali del Museo Civico di Storia Naturale* (Genova) 10, ser. 3a (50), 79-96.

Gertsch, W.J. 1939. A revision of the typical crab spiders (Misumeninae) of America north of Mexico. *Bulletin of the American Museum of Natural History* 76, 277-442.

Gertsch, W. J. 1953. The spider genera *Xysticus, Coriarachne,* and *Oxyptila* (Thomisidae, Misumeninae) in North America. *Bulletin of the American Museum of Natural History* 102, 413-482.

Koch, C.L. 1835. Arachniden. In: *Faunae Insectorum Germaniae initia* (G.W.F. Panzer & G.W. Herrich-Schäfer eds.) 129. Regensburg.

Jantscher, E. 2001. Diagnostic characters of *Xysticus cristatus, X. audax* and *X. macedonicus* (Araneae: Thomisidae). *Bulletin of the British Arachnological Society* 12, 17-25.

Menge, A. 1875. Preussische Spinnen. VIII. Fortsetzung. *Schriften der Naturforschenden Gesellschaft in Danzig* (N.F.) 3, 423-454.

Ono, H. 1988. *A revisional Study of the Spider Family Thomisidae (Arachnida, Araneae) of Japan.* National Science Museum, Tokyo.

Platnick, N.I. 1997. *Advances in spider taxonomy 1992-1995. With redescriptions 1940-1980.* New York Entomological Society.

Roberts, M.J. 1985. *The spiders of Great Britain and Ireland.* Vol 1. Harley Books, Colchester.

Roberts, M.J. 1995. *Spiders of Britain and Northern Europe.* HarperCollins, London.

Schick, R.X. 1965. The crab spiders of California (Araneida, Thomisidae). *Bulletin of the American Museum of Natural History* 129, 1-180.

Simon, E. 1864. *Histoire naturelle des Araignées (Aranéides).* Roret, Paris.

Simon, E. 1932. *Les Arachnides de France.* 6(4), 773-978.

Thorell, T. 1870. On European spiders. Part. 1.

Nova Acta Regiae Societatis Scientiarum Upsaliensis 3 (7), 109-242.

Wunderlich, J. 1987. *Die Spinnen der Kanarischen Inseln und Madeiras. Adaptive Radiation, Biogeographie, Revisionen und Neube-schreibungen.* Triops Verlag, Langen, Germany.

Wunderlich, J. 1994. Zur Kenntnis westpaläarktischer Arten der Gattungen *Psammitis* Menge, 1875, *Xysticus* C.L. Koch 1835 und *Ozyptila* Simon 1864 (Arachnida: Araneae: Thomisidae). *Beiträge zur Araneologie* 4, 749-774.

European Arachnology 2000 (S. Toft & N. Scharff eds.), pp. 337-344.
© Aarhus University Press, Aarhus, 2002. ISBN 87 7934 001 6
(Proceedings of the 19th European Colloquium of Arachnology, Århus 17-22 July 2000)

A superspecies in the genus *Amaurobius* on Crete, and additional records from Greece (Araneae: Amaurobiidae)

KONRAD THALER & BARBARA KNOFLACH

Institute of Zoology and Limnology, University of Innsbruck, Technikerstrasse 25, A-6020 Innsbruck, Austria
(konrad.thaler@uibk.ac.at)

Abstract

Amaurobius candia n.sp. and *A. geminus* n.sp. are described from eastern Crete, together with the female of *A. cretaensis* Wunderlich. The three species, which seem closely related and exhibit a parapatric distribution pattern, should be grouped together as a superspecies. Additional records are presented for four further species. *A. atticus* Thaler & Knoflach is proposed as a junior synonym of *A. pelops* Thaler & Knoflach. The range of *A. pelops* extends therefore from north-eastern Peloponnese to Evvoia and Mt. Iti.

Key words: *Amaurobius*, taxonomy, Crete, endemism, superspecies

INTRODUCTION

During the last decade the inventory of Greek *Amaurobius* species has increased stepwise through the discovery of six new and probably endemic species from Peloponnese and the mainland. Three further species were diagnosed from Corfu, Crete, and Naxos (Thaler & Knoflach 1991, 1993, 1995, 1998, Wunderlich 1995). On the mainland the new species show a vicariance pattern covering the main mountain regions, see our distribution map (1998). More recent excursions in 1998-2000 produced only a slight modification, expanding the range of *A. pelops* Thaler & Knoflach, 1991 (= *A. atticus* Thaler & Knoflach, 1995, n.syn.) northwards to Evvoia and Mt. Iti. The fauna of the islands is less known. The search for the hitherto unknown female of *A. cretaensis* Wunderlich, 1995 led surprisingly to the discovery of two further new species in western Crete, closely allied to *A. cretaensis*. These three species show allopatry at a narrow geographic scale, probably forming a superspecies (Mayr 1953).

Specimens have been deposited as follows: CB Bosmans collection, CD Deeleman collection, CTh Thaler collection; MHNG Muséum d'histoire naturelle, Genève; NMB Naturhistorisches Museum Basel; NHMC Natural History Museum of Crete, Iraklio; NMW Naturhistorisches Museum Wien; NRS Naturhistoriska Riksmuseet Stockholm. Specimens were collected by the authors, if not indicated otherwise. Several juveniles were kept alive until they reached maturity, as indicated additionally. All measurements are in mm.

DESCRIPTIVE PART

Amaurobius cretaensis Wunderlich, 1995 (Figs. 3-4, 11-13, 17, 20, 24)

Wunderlich (1995): n.sp., ♂. Greece, Crete, type locality Mesa Potami, road Tzermiado/Neapoli, 2♂ Feb. 1982, leg. Malicky (Holotype and paratype in collection J. Wunderlich, not examined).

Material examined: Eastern Crete, Dikti Mts.: Lassithiou region, Mt. Selena 1550 m a.s.l., in scree 2♂ 2♀ (CTh) 7 April 1998. Tzermiado, track to Mt.

Figs 1-6. *Amaurobius geminus* n.sp. (1-2, Mt. Afendis Stavromenos), *A. cretaensis* Wunderlich (3-4, Kera), *A. candia* n.sp. (5-6, Skinaras). Epigynum, ventral (1, 3, 5), aboral (2, 4, 6). In Fig. 2 measurement of width of median plate is indicated. Scale lines 0.20 mm. - FD fertilization duct; MP median plate; PP primary pore; R seminal receptacle; SL side lobe.

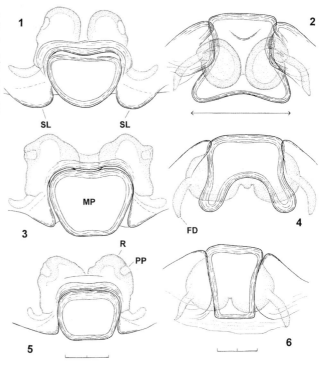

Selena, around Ag. Timios 1100-1200 m a.s.l., 3♀ (CTh) 10 April 1998, 1♀ (NMW) 21 Sept. 1998, 1♂ (NMW), adult Oct. 1998, 1♀ (NMW), adult 4 Oct. 1998. Tzermiado, litter in dry river bed, 7♂ 14♀ (CB) 17 Oct. 1998, leg. Bosmans. Kera 700 m, 1♂ 4♀ (CTh) 8 April 1998. Krasi, litter in dry river bed, 1♂ 3♀ (CB) 16 Oct. 1998. Katharo plain above Kritsa 1000 m, 3♂ 10♀ (CTh, NMB) 1 Oct. 1998. Males, Sarakinas gorge 200 m a.s.l., 1♂ 2♀ (MHNG) 28 Sept. 1998, 4♀, adult Dec. 1998 - Feb. 1999.

Diagnosis: This species can be readily distinguished by the dorsal apophysis of the male palpal tibia and by the lack of an intermediate apophysis (Figs 11-13). The epigynum is characterized in aboral view by strong and even protruding dorsal angles of the median plate (Fig. 4).

Description (♀, n = 5): Total length 5.4-6.6, prosoma length 2.5-3.05, width 1.7-2.05, femur I 2.1-2.4. Prosoma and legs brownish yellow, legs not annulated. Abdomen greyish, with indistinct pattern of light spots. Epigynum and vulva: Figs 3-4. Median plate and side lobes strongly developed, median plate in ventral view cordiform, in aboral view rectangular, but concave, with strongly projecting dorsal angles. Copulatory ducts very short, receptacles globular.

♂ palp: Figs 11-13, 17, 20, 24. Tibia without intermediate apophysis, prolateral apophysis strong, distally tapering to a finger-like process, its anterior surface smooth and concave. Bulbus: tegulum globular, tegular process triangular, tegular apophysis strong, conductor fleshy, embolus short and bent.

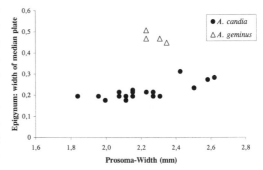

Fig. 7. Relation between prosoma width and width of median plate of the epigynum (aboral view, mm, see Fig. 2) in *Amaurobius candia* n.sp. and *A. geminus* n.sp.

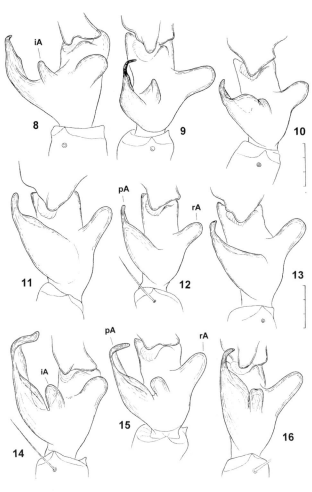

Figs 8-16. *Amaurobius geminus* n.sp. (8-9, Mt. Afendis Stavromenos, same specimen), *A. cretaensis* Wunderlich (11-13, Kera [11], Katharo plain [12], Sarakinas [13]), *A. candia* n.sp. (10 Kato Chorio, 14-16, Youchtas [14], Afrati [15], Skinaras [16]). Male palpal tibia, retrolateral (8), dorsal (9-16). Scale lines: 0.30 mm. - iA, pA, rA intermediate, prolateral and retrolateral tibial apophysis.

Affinities: *A. cretaensis* is very similar and probably closely related to other two taxa from eastern and central Crete, which are here described, *A. geminus* n.sp. and *A. candia* n.sp. These three species differ clearly in their prolateral tibial apophysis and tegular process from *A. ferox* (Walckenaer, 1830), which was cited as a possible relative of *A. cretaensis* by Wunderlich (1995). However, there is apparently an overall similarity in the epigyna. Remarkably, in *A. candia* n.sp. and *A. geminus* n.sp. an intermediate apophysis is present in the male palpal tibia, as in most other members of *Amaurobius*, the *A. ferox* group excepted. The intermediate apophysis therefore has presumably been lost secondarily in *A. cretaensis*. The species of the *ferox* group are apparently concentrated in peninsular Italy and in the southern Alps (Pesarini 1991), Italy being probably the origin of *A. ferox*, which is now widely distributed owing to its synanthropic occurrence.

Distribution: Fig. 27, eastern Crete. The localities are grouped in the northern and eastern parts of Dikti region, with Tzermiado standing in the centre, at the border of Lassithiou plain. Krasi, Kera and Mesa Potami are situated on the northern and eastern slopes of Mt Selena respectively, while Katharo plain and Sarakinas gorge are on the eastern slope of Dikti Mts. Specimens were confined to sheltered microhabitats in an arid landscape, such as a gorge, rock crevices and the scree system, which provide a humid atmosphere and moderate temperature.

Amaurobius candia n.sp. (Figs 5-6, 10, 14-16, 18, 21-22, 25-26)

Type material: 1♂ Holotype (NMW): Greece, central Crete, Ida Mts., road Anogia to Ideo Andro, bifurcation to Skinaras *c.* 1400 m, 23 Sept. 1998; Paratypes: 3♂ 14♀ , same data as holotype (2♀ NMW; 1♂ 2♀ MHNG; 2♀ NHMC; 2♀ NMB; 2♀ NRS; 2♂ 4♀ CTh).

Other material examined: Central Crete: Iraklio, wall outside town, ruderal vegetation, 1♀ (NRS) 15 April 1975, leg. Waldén. Mt. Youchtas, 1♂ (CTh) Jan. 1996, leg. Chatzaki. Ida Mts., road Anogia to Ideo Andro, 1300 m, 3♀ 23 Sept. 1998. Eastern Crete: Kastelli, litter bordering irrigated garden, 1♂ (CB) 19 Oct. 1998, leg. Bosmans. Road Kastelli/Viannos, Afrati, in olive grove, 1♂ (CTh) 5 April 1998, adult 25 Dec. 1998. Dikti Mts., road Kaminaki/Embaros, stones in grassland, 1♂ (CB) 19 Oct. 1998, leg. Bosmans. Pefkos, pine forest, 2♀ (CD) 6 May 1986, leg. C.L. & P.R. Deeleman. Base of Thripti Mts., near Kato Chorio 100 m, overgrown stone wall on agricultural land, 1♂ (CTh) 26 Sept. 1998, adult Dec. 1998.

Etymology: *Candia*, noun in apposition, hence invariable. Obsolete name for Iraklio (and Crete).

Diagnosis: *A. candia* n.sp. can be clearly distinguished from *A. cretaensis* by the absence of projecting angles on the median plate of the epigynum (Fig. 6) and by the presence of an intermediate apophysis of the male palpal tibia (Figs 14-16). It is separated from *A. geminus* n.sp. by the width of the median plate (aboral view, Fig. 7) and by the prolateral tibial apophysis of the male, with inner margin broadened (Figs 9 vs. 14-16).

Description (♂/♀, n = 4/5, specimens from type locality): Total length 5.6-6.6/5.6-8.2, prosoma length 3.0-3.4/2.8-3.8, width 2.1-2.3/1.8-2.6, femur I 3.3-3.8/2.5-3.2. Colour and pattern as in *A. cretaensis*. Palp: Figs. 16, 18, 22, 26. Tibia with intermediate apophysis, prolateral apophysis broad, distally oblique, its inner edge prolonged into a   finger-like process, anterior surface smooth and concave. Bulbus: tegulum, tegular process, tegular apophysis, conductor, and embolus similar to *A. cretaensis*. Epigynum and vulva: Figs. 5-6. Close to *A. cretaensis*, but dorsal

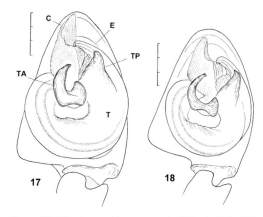

Figs 17-18. *Amaurobius cretaensis* Wunderlich (17, Kera), *A. candia* n.sp. (18, Skinaras). Male palpal organ, ventral (17-18). Scale lines: 0.30 mm. - C conductor; E embolus; T tegulum; TA tegular apophysis; TP tegular process.

angles of median plate not projecting, median plate narrower than in *A. geminus* n.sp. (Fig. 7).

Variation: The distal finger-like process of the prolateral apophysis of the male palp is longer in specimens from Mt Youchtas (Fig. 14) and from the western slopes of Dikti Mts. (Fig. 15) than from Ida Mts. (fig. 16). The intermediate apophysis is more closely attached to the prolateral apophysis in the peripheric ♂ from Kato Chorio (Fig. 10) than in other ♂. All these ♂ are accepted here as conspecific, as in the corresponding females no further separation was indicated (fig. 7).

Affinities: Closely related to *A. cretaensis* and *A. geminus* n.sp.

Distribution: Fig. 27. Localities are mainly grouped along the lowland between Dikti Mts. and Ida: around Iraklio, at the base of the western (Kastelli, Afrati, Embaros) and southern (Pefkos) slopes of Dikti Mts., and in Ida Mts. *c.* 1400 m a.s. l. One male from a peripheric site at base of Thripti Mts., *c.* 30 km East of Pefkos, was also assigned to this species. *A. candia* n.sp. is probably widely distributed along the southern coast of Crete. Like *A. cretaensis*, specimens were found at microhabitats providing humid and moderate conditions in an arid environment, as at Ida Mts. among scree in a semi-cavern and at the base of an overhanging rock, from where a film of water was still pouring out in late Sept. Other speci-

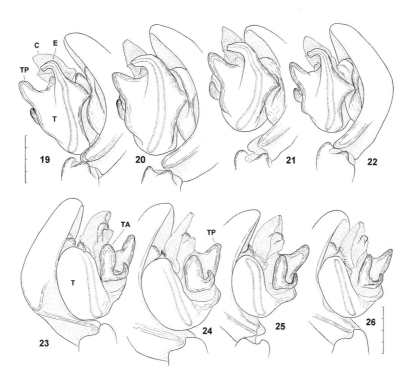

Figs 19-26. *Amaurobius geminus* n.sp. (19, 23, Mt. Afendis Stavromenos), *A. cretaensis* Wunderlich (20, 24, Kera), *A. candia* n.sp. (21-22, 25-26, Youchtas [21, 25], Skinaras [22, 26]). Male palpal organ, prolateral (19-22), retrolateral (23-26). Scale lines 0.40 mm. - C conductor; E embolus; T tegulum; TA tegular apophysis; TP tegular process.

mens were taken at low elevation and on cultivated land.

Amaurobius geminus n.sp. (Figs 1-2, 8-9, 19, 23)
Type material: 1♂ Holotype (NMW): Greece, eastern Crete, Thripti Mts., Mt. Afendis Stavromenos 1470 m a.s.l., 25 Sept. 1998. Paratypes: 2♂ 4♀, same data as holotype (2♀ NMW; 1♂ 2♀ CTh).

Etymology: The specific name is a latin adjective, meaning twin.

Diagnosis: *A. geminus* n.sp. can be clearly distinguished from its sibling species by the median plate of the epigynum, which is broad, without projecting angles (Figs 2, 7). Males can be separated from *A. candia* n.sp. by the prolateral tibial apophysis of the palp, which is gradually tapering (Fig. 9).

Description (♂/♀, n = 2/4): Total length 5.8, 6.2/7.2-8.2, prosoma length 2.8, 3.3/3.4-3.6, width 2.1, 2.4/2.2-2.4, femur I 3.2, 3.6/2.7-2.9. Colour and pattern as in *A. cretaensis*. Palp: Figs 8-9, 19, 23. Tibia with intermediate apophysis, prolateral apophysis more slender than in its allies, almost cylindrical, gradually tapering. Bulbus: close to *A. cretaensis*. Epigynum and vulva: Figs 1-2. Close to *A. cretaensis*, but dorsal angles of median plate not projecting, lateral lobes broad, median plate wider than in *A. candia* n.sp. (Fig. 7).

Affinities: Closely allied to *A. cretaensis* and *A. candia* n.sp., probably a restricted endemic species in Thripti Mts.

Distribution: Fig. 27. Known only from the type locality, the summit region of Mt. Afendis. Specimens were found at its crest, in rock crevices and holes facing north, where moisture is provided by the clouds coming from the Cretan sea.

Amaurobius deelemanae Thaler & Knoflach, 1995
Thaler & Knoflach (1995): n.sp., ♂♀. Greece, type locality Naxos Island, 2♂ 1♀ 23 April 1984, leg. C.

L. & P.R. Deeleman (male holotype in MHNG). Additional specimens from Crete, Perama, and Rhodes, all leg. Deeleman.

Material examined: Western Crete, Georgiopouli 10-250 m a.s.l., in phrygana, pasture, 2♂ 2♀ (CTh) 27 March 1999; in olive grove, 1♂ 4♀ (NMW) 28 March 1999; in rift, in dense shrubs 4♂ 11♀ (CTh, MHNG, NMB) 29 March 1999. Lefka Ori 1650 m a.s.l., pitfalls, 4♂ (NHMC) 1992, leg. Lymberakis & Chatzaki. Hania, Vamvakades, 1♀ (CB) 10 May 1994, leg. J. Van Keer. Central Crete, road Anogia/Gonies, grassland along dry rivulet, 3♀ (CB) 17 Oct. 1998, leg. Bosmans.

Differentiation: *A. deelemanae* resembles *A. cretaensis* and its relatives in the tegular process. Males differ clearly in the embolus, which is long and strongly bent, and in the configuration and shape of the tibial apophyses. Females differ in the epigynum, the median plate being transverse, not cordiform, the side lobes less developed.

Remarks: Hitherto only known from islands in the Aegean Sea: Naxos (type locality, Kiklades Islands), Rhodes (further records in Thaler & Knoflach 1998) and Crete, apparently absent in the eastern part of this island, where it is replaced by *A. cretaensis* and *A. candia* n.sp. In western Crete it occurs across a wide range of altitude (10-1650 m), selecting moist microhabitats (Fig. 27).

Amaurobius erberi (Keyserling, 1863)
Material examined: Evvoia: Eretria, in stand of pines 100 m a.s.l., 1♂ (CTh) 19 Sept. 1997. Sporades Is., Skiathos: Moni, Troulos, pine forest 200 m a.s.l., 2♀ (CD) 28-30 April 1986, leg. P.R. Deeleman.

Remarks: Holomediterranean according to the literature, confined to lowland habitats and probably absent on Peloponnese and on the Aegean islands, where it is replaced by *A. deelemanae*.

Amaurobius pelops Thaler & Knoflach, 1991
A. atticus Thaler & Knoflach, 1995: Revue suisse Zool. 102: 48 (♂), **nov. syn.** (male holotype in MHNG)
A. cf. *pelops* - Thaler & Knoflach (1995): Revue suisse Zool. 102: 46 (♂♀).

Thaler & Knoflach (1991): n.sp., ♂♀. Greece, type locality Peloponnese, Feneos plain (male holotype in NMW).
Thaler & Knoflach (1998): ♂, further captures along Feneos plain.

Material examined: Evvoia: Dirfis Mts, Macrovouni, above Ag. Vlasios 1000 m, in stand of Greek fir (*Abies cephalonica* Loud.) 1000 m a.s.l., 6♂ 2♀ (CTh) 20 Sept. 1997; road Steni/Stropones, eastern slope, in stand of Greek fir 1000 m a.s.l., 46♂ 40♀ (CTh, MHNG, NMB, NMW) 21 Sept. 1997. Central Greece, Mt. Iti, above Ipati, in sparse pine forest 1800 m a.s.l., 2♂ 10♀ (CTh) 18 Sept. 1997.

Remarks: The ♂ holotype of *A. pelops* presents an aberrant intermediate apophysis. This was recognized from additional captures around the type locality, Feneos basin in northern Peloponnese (Thaler & Knoflach 1995, 1998). We now feel sure that *A. atticus*, which we described from 1♂ taken in a cave in Attiki and separated from *A. pelops* on details of its tibial apophyses, merely represents another variant and should be placed as a synonym of *A. pelops*. The new captures extend its distribution to the north. The species apparently occupies a narrow range from northeastern Peloponnese to Evvoia and Mt. Iti. Its neighbours are *A. longipes* Thaler & Knoflach, 1995 in the south, *A. ossa* Thaler & Knoflach, 1993 in the north and *A. timidus* in Pindos Mts. At Mt. Iti *A. pelops* and *A. timidus* occur in sympatry. There we found *A. timidus* at lower sites around 1400 m a.s.l. (Thaler & Knoflach 1998).

Amaurobius timidus Thaler & Knoflach, 1995
Thaler & Knoflach (1995): n.sp., ♂♀. Greece, type locality Peloponnese, Erimanthos, Kalendzi 1♀ 27 Sept. 1991, 3♂ 4♀ 20 Sept. 1992, leg. authors (male holotype in MHNG).

Material examined: Kefallonia: Mt. Enos, in stands of Greek fir 1200 m a.s.l., 1♂ 1♀ (CTh) 20 Sept. 1999, 1600 m a.s.l. 15♂ 21♀ (CTh, MHNG, NMB, NMW) 20-23 Sept. 1999.

Remarks: Widely distributed in mountain regions of northern Greece, in Pindos Mts. and Chalkidiki, also on Ionian islands. Southernmost station (and type locality) at Erimanthos Mts., north-western Peloponnese, northernmost station

Fig. 27. Distribution of *Amaurobius* species in Crete: I *A. deelemanae* Thaler & Knoflach, 2 *A. candia* n.sp., 3 *A. cretaensis* Wunderlich, 4 *A. geminus* n.sp.

in Bulgaria. The northern limits of its range are not yet known.

CONCLUDING REMARKS

The thirteen species of the genus *Amaurobius* now established for Greece fall into various groups according to their ranges. Only two of them are common in Europe, *A. erberi* in the mediterranean region, and *A. fenestralis* (Ström, 1768) in mid- and northern Europe respectively. Another two show comparatively wide ranges, whose limits are not yet fully established: *A. timidus* in northern Greece and in Pindos Mts., *A. deelemanae* in the Aegean islands. Only known from their type localities are *A. phaeacus* Thaler & Knoflach, 1998 (Corfu) and *A. ausobskyi* Thaler & Knoflach, 1998 (Athos peninsula). Restricted endemic species occur in the mountains of Peloponnese and mid Greece: *A. pelops*, *A. paon* Thaler & Knoflach, 1993 (Taygetos Mts.), *A. longipes* (Parnon Mts.), *A. ossa* (Mt. Ossa, Mt. Olympos). All these species are well defined and do not form obvious groupings.

Surprisingly in eastern Crete three taxa occur, which appear closely allied in genitalic characters and show a parapatric distribution pattern (Fig. 27). Two are confined to mountain areas: *A. cretaensis*, which is most distinct, to the northern and eastern parts of Dikti Mts., *A. geminus* n.sp. to the summit region of Thripti Mts. The third species, *A. candia* n.sp., occurs along the southern and western slopes of Dikti Mts. and extends to Iraklio and also to Ida Mts. As *Amaurobius* species are confined to sheltered microhabitats providing humidity and moderate temperatures, their

ranges should have oscillated in the Pleistocene according to the changes of macroclimate. In the more remote past at the end of the Tertiary, Crete was even separated into islands by marine transgression (Rögl & Steininger 1983). Therefore *A. cretaensis* and its allies might have speciated in the Pleistocene, each in a refugial area. They can be grouped together into a superspecies, according to their close affinities and parapatric distribution pattern.

ACKNOWLEDGEMENTS

For the loan or donation of specimens we thank Ms. Maria Chatzaki (Irakleio, Crete), Dr. Christa Deeleman-Reinhold (Ossendrecht), Dr. R. Bosmans (Gent) and Dr. T. Kronestedt (Stockholm); for linguistic revision Dr. P. Merrett (Swanage).

REFERENCES

Mayr, E. 1953. *Methods and principles of systematic zoology.* McGraw-Hill Book Company, New York, Toronto, London.

Pesarini, C. 1991. The Amaurobiidae of Northern Italy (Araneae). *Atti della Società Italiana di Scienze naturali e del Museo civico di Storia naturale di Milano* 131(1990), 261-276.

Rögl, F. & Steininger, F.F. 1983. Vom Zerfall der Tethys zu Mediterran und Paratethys. Die neogene Paläogeographie und Palinspastik des zirkum-mediterranen Raumes. *Annalen des Naturhistorischen Museums Wien* 85 A, 135-163.

Thaler, K. & Knoflach, B. 1991. Eine neue *Amaurobius*-Art aus Griechenland (Arachnida:

Araneae, Amaurobiidae). *Mitteilungen der schweizerischen entomologischen Gesellschaft* 64, 265-268.

Thaler, K. & Knoflach, B. 1993. Two new *Amaurobius* species (Araneae: Amaurobiidae) from Greece. *Bulletin of the British arachnological Society* 9, 132-136.

Thaler, K. & Knoflach, B. 1995. Über Vorkommen und Verbreitung von *Amaurobius*-Arten in Peloponnes und Ägäis (Araneida: Amaurobiidae). *Revue suisse de Zoologie* 102, 41-60.

Thaler, K. & Knoflach, B. 1998. Two new species and new records of the genus *Amaurobius* (Araneae, Amaurobiidae) from Greece. In: *Proceedings of the 17th European Colloquium of Arachnology, Edinburgh 1997* (P. Selden ed.), pp. 107-114. British Arachnological Society, Burnham Beeches, Bucks.

Wunderlich, J. 1995. Beschreibung einer bisher unbekannten Art der Gattung *Amaurobius* C. L. Koch 1837 von Kreta (Arachnida: Araneae: Amaurobiidae). *Beiträge zur Araneologie* 4, 729-730.

European Arachnology 2000 (S. Toft & N. Scharff eds.), pp. 345-354.
© Aarhus University Press, Aarhus, 2002. ISBN 87 7934 001 6
(Proceedings of the 19th European Colloquium of Arachnology, Århus 17-22 July 2000)

Character states and evolution of the chelicerate claws

JASON A. DUNLOP

Institut für Systematische Zooloigie, Museum für Naturkunde der Humboldt-Universität zu Berlin, Invalidenstraße 43, D-10115 Berlin, Germany (jason.dunlop@rz.hu-berlin.de)

Abstract

Outgroups of Chelicerata have an apotele in which two smaller claws insert on a larger median claw. A three-clawed plesiomorphic state is retained in basal Pycnogonida and the Palaeozoic xiphosuran *Weinbergina*. Modifications or reductions from this pattern are interpreted here as apomorphic character states. A single apotele element occurs in the crown group Xiphosurida and in the extinct taxa Eurypterida and Chasmataspida. The digitigrade, 'eurypteroid' apotele of *Allopalaeophonus*-like fossil Scorpiones may not be the plesiomorphic condition for the group since the most basal clade, *Palaeoscorpius*, has an apotele more like *Weinbergina* and the outgroups. Among the other arachnids Palpigradi retain the most plesiomorphic apotele morphology with three claws on all postcheliceral appendages. Unequivocal homologies between the claws in different arachnid orders are difficult to resolve, especially in relation to the complex apoteles seen among the Acari. However, further apomorphic apotele states in arachnids include the development of the empodial region between the claws into a pulvillus in adults of basal Amblypygi, in Solifugae and Pseudoscorpiones and among the mites in the (Opilioacariformes + Parasitiformes) clade, but *not* in basal Acariformes.

Key words: Apotele, claw, ungue, empodium, pulvillus, Chelicerata, phylogeny

INTRODUCTION

The terminal element of the postcheliceral limbs in Chelicerata is called the apotele. This apotele has been modified in arachnids to form the claws, while in some taxa (e.g. solifuges, some amblypygids) the membranous region between the claws - the empodium - has been further modified to form a complex, typically eversible structure which is usually called the pulvillus. Chelicerate limb morphology, including apotele character states, has been reviewed by authors such as Barrows (1925), van der Hammen (1989) and Shultz (1989). However, these authors restricted their surveys to Recent euchelicerates and did not consider the basal pycnogonid group (sea spiders), extinct taxa (e.g. eurypterids) or fossil representatives of scorpions and xiphosurans. Palaeozoic fossils of the latter two taxa preserve character states different from extant forms, which probably better reflect the ground pattern of these clades. The aim of the present paper is to give an overview of apotele morphology which integrates the fossil data and to try and identify potential synapomorphies for clades within the chelicerates.

MATERIALS AND METHODS

Wherever possible specimens of Recent taxa were drawn from life from material in the collections of the Muesum für Naturkunde, Berlin, supplemented by descriptions in the literature as detailed below. Well preserved eurypterids and trigonotarbids from the Natural History

Museum, London were also examined. Terminology generally follows Barrows (1925) and/or Shultz (1989).

RESULTS

Outgroups (Trilobita)

The sister group of Chelicerata (including pycnogonids) has not been satisfactorily established, but Trilobita, and various arachnomorph or trilobite-like taxa, have emerged as potential outgroups for polarising chelicerate characters. Appendage morphology can only be determined with certainty in a few fossils showing exceptional preservation. Studies of *Triarthrus* by Cisne (1975) and Whittington & Almond (1987), of *Agnostus* by Müller & Walossek (1987) and of *Phacops* by Bruton & Haas (1999) consistently show an apotele morphology in which two lateral claws appear to insert into a slightly larger central claw (Fig. 1a). These three apotele structures are potentially homologous with the three claws seen in many chelicerates (see below) and provide a plesiomorphic condition against which the chelicerate apotele can be compared.

Pycnogonida

Most phylogenetic studies have concluded that Pycnogonida are basal chelicerates (Weygoldt & Paulus 1979; Wheeler & Hayashi 1998), representing the sister group of all other chelicerates: the Euchelicerata. The palps (limb II) of Pycnogonida are variable within the group and are reduced or absent in adults of certain taxa. Nevertheless, a distinct apotele in the palp appears to be lacking, at least in adults. By contrast, the oviger (limb III) - also absent in some taxa - ends in a single claw in most groups where it is present (Arnaud & Bamber 1987). The legs of Pycnogonida (limbs IV-VII) typically end in three claws (Fig. 1b): a main claw plus a pair of auxiliary claws (Arnaud & Bamber 1987). The relative proportions of these claw elements can vary, even intraspecifically (e.g. Helfer & Schlottke 1935, fig. 44). The main claw is usually longer than the auxiliary claws, but see e.g. *Ammothea biunguiculata* where the

main claw is tiny, and in contrast to the trilobite condition (see above) the auxiliary claws do not insert into the main claw, but arise from the membrane above it (Fig. 1b). The auxiliary claws are lost in more derived clades, e.g. Pycnogonidae.

Xiphosura

Following Anderson & Selden (1997), the class Xiphosura (horseshoe crabs) can be divided into a series of stem group plesion taxa, the synziphosurines, plus a monophyletic crown group, Xiphosurida. Appendages are rarely preserved in Palaeozoic Xiphosura, but significantly the Lower Devonian synziphosurine *Weinbergina opitzi* (Fig. 1c) lacks the chelate postcheliceral appendages with a single apotele characteristic for living species. Stürmer & Bergström (1981) redescibed *W. opitzi* and although there are inconsistencies in their interpretative drawings, their plates (especially their fig. 7a) indicate a trifurcate apotele with all three elements approximately the same size. These three elements appear to emerge adjacent to each other (Fig. 1c) and the lateral elements do not insert on the central one as in the trilobites. By contrast, extant Xiphosura have only a single apotele element in all their postcheliceral limbs which forms the movable finger of a distal chela (Fig. 1d). This apparently apomorphic condition arose in the Xiphosurida by the Carboniferous, having been recently described in the fossil genus *Euproops* by Schultka (2000, pl. 1, fig. 2). In Recent taxa this chela is larger in appendages II-V, but rather small in appendage VI and is essentially subchelate in appendages III and IV of the extant species *Tachypleus tridentatus*.

Eurypterida

In most reconstructions the extinct Eurypterida (sea scorpions) are shown with a trifurcate end to the legs. Jeram (1998) referred to the lateral elements as 'tarsal spurs' and suggested that homologous structures occur in fossil scorpions and that they are in turn homologous with the lateral claws or ungues of other arachnids (see

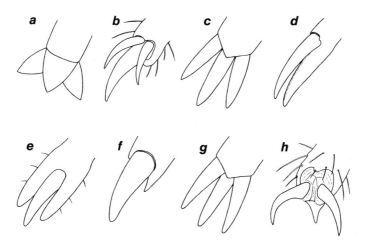

Fig. 1. (a) *Triarthrus* sp. (Trilobita). Ordovician, after Whittington & Almond (1987). **(b)** *Achelia echinata* (Pycnogonida). Recent, from life. **(c)** *Weinbergina optzi* (Xiphosura: 'synziphosurine'). Devonian, after Stürmer & Bergström (1981). **(d)** *Limulus polyphemus* (Xiphosura: Xiphosurida). Recent, from life. **(e)** *Baltoeurypterus tetragonophthalmus* (Eurypterida). Silurian, after Selden (1981). **(f)** *Diploaspis casteri* (Chasmataspida). Devonian, from original material. **(g)** *Palaeoscorpius devonicus* (Scorpiones: stem group). Devonian, after Kjellesvig-Waering (1986). **(h)** *Buthus occitanus* (Scorpiones: Buthidae). Recent, from life. All drawings not to scale.

below). However, when examined in detail - see e.g. Selden's (1981) study of *Baltoeurypterus tetragonophthalmus* - the apotele of eurypterids is represented by a single claw. The spinous elements either side of it, when present, are derived from the *preceding* podomere (Fig. 1e) either as fixed spines (see e.g. Selden 1981, figs. 26-29 for *Baltoeurypterus*) or as socketed spines (see e.g. Clarke & Ruedemann 1912, pl. 28 for *Carcinosoma*) and thus their homology with the ungues of arachnids is questionable. In stylonurid eurypterids, which lack paddles and which probably represent the more basal taxa (S. Braddy pers. comm.), the spines on this podomere preceding the apotele are either absent or only weakly developed; see e.g. Tollerton (1989, fig. 9).

Chasmataspida

Chasmataspida are a rare group of extinct chelicerates. Although initially interpreted as unusual xiphosurans, their phylogenetic position is uncertain (see Anderson & Selden 1997) since they share apomorphies with both Xiphosura

and Eurypterida (Dunlop in press). Their appendages are poorly known, but Caster & Brooks (1956) described an isolated limb with chelate, distal podomeres, resembling those of Xiphosurida (see above). A better preserved limb described by Størmer (1972) and Dunlop et al. (2002) ends in a short, slightly curving element with a weakly developed spine on the preceding podomere. As in Eurypterida there is no evidence for lateral claws (Fig. 1f) and Chasmataspida should also provisionally be scored as retaining only a single element in their apotele.

Scorpiones

Among the best known Palaeozoic scorpions are the Silurian *Palaeophonus* / *Allopalaeophonus* species which have been figured as having crab-like legs ending in a single, large apotele - Pocock's (1901, p. 295) 'clawless terminal segment' - similar to that seen in many eurypterids (see above). However, in the phylogeny of Jeram (1998), the Lower Devonian species *Palaeoscorpius devonicus*, discovered in a marine palaeoenvironment, emerged as the most basal scorpion taxon. Interestingly, *P. devonicus* has an apotele morphology (Fig. 1g) similar to both the outgroups and *Weinbergina* (see above) with three similar-sized claw elements. Note that the original description by Lehmann (1944) failed to note the middle claw. The lateral elements of the apotele in fossil scorpions have been called tarsal spurs (e.g. Jeram 1998), but in contrast to those of eurypterids (see above) they appear to be post-tarsal, apotele elements and Jeram homologised them with the lateral claws or ungues of Recent arachnids. Extant Scorpiones typically have a median claw shaped like a small, ventrally-pointing spike while the ungues are relatively large and strongly curved (Fig. 1h). A pulvillus-like structure has also been reported between the ungues in at least one buthid scorpion (Millot & Vachon 1949: fig. 191). The apotele of the scorpion pedipalp (limb II) has been lost in the development of the large chela which is formed from the tibia and the tarsus (Shultz 1990).

Other arachnids

Details of apotele morphology in the remaining arachnids can be found in comparative studies, in particular Barrows (1925). Fossils of the remaining orders do not present different character states compared to extant taxa. Barrows (1925) defined the arachnid claw (= the apotele) as the entire terminal segment operated by two antagonsitic tendons. The insertion points of the muscles of these tendons are phylogenetically informative and were detailed by Shultz (1989, 1990). The ventral sclerotised region of the arachnid apotele attached to the flexor tendon was called the tendon plate, the small claw which occurs here in some taxa was called the pseudonychium (other authors have simply called it the onychium or the median claw) and the larger paired claws seen in many arachnids were called the ungues (Barrows 1925); although the less appropriate term 'ungules' has sometimes been used.

Palpigradi / Araneae / Fossil orders

Palpigradi retain three claws on both the pedipalps (limb II) (Fig. 2d) and the legs (limbs III-VI) (Fig. 2e) (see also Hansen & Sørensen 1897). In Araneae (spiders) and the extinct Trigonotarbida (see e.g. Shear et al. 1987) three claws are retained on the legs (Fig 2b), but the pedipalp apotele is reduced to a single claw (Fig. 2a). Barrows (1925) described the development of the male palpal organ (a spider autapomorphy) from a so-called 'claw fundament' on the pedipalp and further noted that some Lycosidae curiously retain additional claw-like structures on the palp of mature males. Apotele morphology is unknown in the extinct orders Phalangiotarbida and Haptopoda.

Amblypygi / Thelyphonida / Schizomida

Shultz (1999, characters 14 and 17) scored Amblypygi (whip spiders), Thelyphonida (whip scorpions) and Schizomida (schizomids) as having synapomorphically fused the tarsus to the apotele in both the pedipalps (limb II) and the first pair of legs (limb III). In fact the pedipalpal apotele is retained as a distinct element both in Schizomida, see e.g. Werner (1935, fig. 26) for *Trithyreus*, and in some Amblypygi such as the basal genus *Charon*, see e.g. Weygoldt (2000, figs. 7-8). In Thelyphonida the pedipalpal apotele is barely discernible as an immovable distal structure (Werner 1935, fig. 27; personal observations). Furthermore, in contrast to Shultz's coding, an apotele *is* retained on leg 1, albeit as a pair of highly reduced claws in some Amblypygi, see Igelmund (1987) and Weygoldt (2000, fig. 81) for details. The remaining limbs (IV-VI) of thelyphonids and schizomids have a spider-like morphology with a small median claw and larger lateral claws. Amblypygids are similar, but in some taxa in place of the empodial claw there is a distinct, fleshy pad: the pulvillus (Fig. 2c). This soft, slightly folded structure presents a flattened, oval, distal face with a slight dorsal indentation in some taxa, and some sclerotised supporting elements in a more proximal position. Weygoldt (1996) interpreted the pulvillus as homologous with the

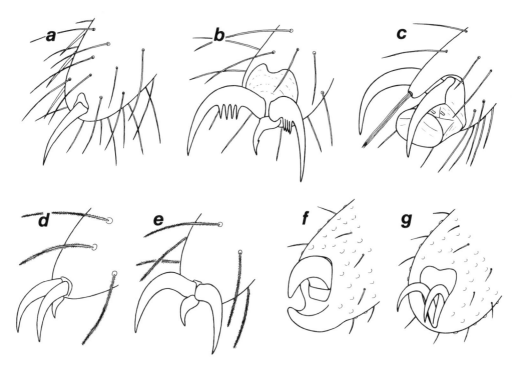

Fig. 2. (a) *Liphistius desultor* (Araneae: Liphistiidae), ♀ pedipalp. Recent, from life. **(b)** Leg of the same. **(c)** *Charon grayi* (Amblypygi: Charontidae). Recent from life. **(d)** *Eukoenenia mirabilis* (Palpigradi), pedipalp. Recent, after Hansen & Sørensen (1897). **(e)** Leg of the same. **(f)** *Ricinoides westermanni* (Ricinulei), pedipalp. Recent, from life. **(g)** Leg of the same. All drawings not to scale.

empodial claw and, since those Amblypygi without a pulvillus also lack an empodial claw, he regarded absence of a pulvillus as apomorphic within Amblypygi and a character defining an Apulvillata clade. Furthermore, Quintero (1981) noted that juveniles of the apullvilate genus *Phrynus* retain a pulvillus-like structure which may help the animal cling to its mother.

Ricinulei / Opiliones

Ricinulei retain a single apotele claw in the pedipalps (limb II) as the free finger of the minute claw (Fig. 2f). In the legs (limbs III-VI) Ricinulei appear to have lost the median claw having only two claws - ?the ungues - which can be withdrawn through a heart-shaped distal opening into a cavity at the tip of the tarsus (Fig. 2g). Basal Opiliones, i.e. Cythophthalmi, have only a single tarsal claw on both the pedipalps (limb II) (Fig. 3a) - where it is highly re-

duced in some taxa Shultz (1998, character 8) - and the legs (limbs III-VI) (Fig. 3b). However, the situation within Opiliones is more complex since the suborder Laniatores is characterised by having a single claw on the anterior two pairs of legs but *two* claws on the posterior two pairs of legs, further modified in other laniatore taxa; see e.g. Giribet et al. (1999, characters 11-12) for alternative character states. Note that some laniatore nymphs have also been described with a pulvillus-like structure in the apotele (Munzo-Cuevas 1971).

Psudoscorpiones / Solifugae

Pseudoscorpiones have lost the apotele on their large, chelate pedipalp (limb II). Pseudoscorpions retain lateral claws on the legs (limbs III-VI), but in the place of the empodial claw there is a fleshy structure (Fig. 3c) which is usually called the pedal arolium; see Harvey (1992) for arolium character states within Pseudoscor-

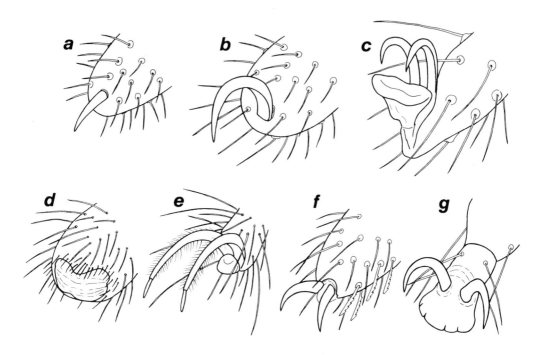

Fig. 3. (a) *Siro* sp. (Opiliones: Cythophthalmidae), pedipalp. Recent, from life. **(b)** Leg of the same. **(c)** *Neobisium* sp. (Pseudoscorpiones: Neobisiidae), leg. Recent, partly after Eisenbeis & Wichard (1985). **(d)** *Galeodes* sp. (Solifugae: Galeodidae), pedipalp. Recent, from life. **(e)** Leg of the same. **(f)** *Opilioacarus texanus* (Acari: Opilioacariformes), pedipalp. Recent, after van der Hammen (1989). **(g)** Leg of the same. All drawings not to scale.

piones. This structure was called the empodium by Shultz (1989, 1990). The pseudoscorpion arolium is basically similar in all taxa (Roewer 1936). It originates from a narrow base below the ungules and widens distally to form a pale, membranous, somewhat funnel-shaped structure with a slightly concave distal face. The arolium is extensible and Roewer (1936) suggested that extension occurs through blood pressure. The arolium functions as an attachment organ and allows pseudoscorpions to walk on vertical surfaces.

The pedipalps (limb II) of Solifugae end in a unique (autapomorphic) structure, the adhesive palpal organ (Fig. 3d). In gross morphology it forms a membranous series of 'lips' at the ventro-distal tip of the tarsus and is probably homologous with the apotele, since Roewer (1934) described articulating condyles between it and the tarsus. Lateral claws are absent, but vestigial claw bases on the palpal organ were described and figured by both Barrows (1925, fig. 28) and Roewer (1934, fig. 75). Leg 1 (limb III) of Solifugae has all but lost its locomotory function and the claws here are either weakly developed or absent (Roewer 1934). The remaining legs (limbs IV-VI) bear large, curved lateral claws (Fig. 3e) which are uniquely subdivided (an autapomorphy) in all solifuge taxa (Roewer 1934), with the tip of each claw articulating against the rest of the ungue. Highly setous ungues (Fig. 3e) are apomorphic for Galaeodidae only (Roewer 1934) and not Solifugae in general. Like pseudoscorpions, solifuges also have a fleshy structure in place of the empodial claw (Fig. 3e). Roewer (1934) homologised these structures in these two orders and called it the arolium, a homology accepted by Shultz (1989, 1990) who called it the empodium (see above). In Solifugae this structure is a rounded or distally widening, fleshy organ, divided into two distinct lobes in some taxa (e.g. Roewer 1934). Again extension occurs through blood pressure and an adhesive function has been suggested.

Acari

Acari are widely interpreted as monophyletic (see Lindquist 1984) and are currently divided into three major taxa: Opilioacariformes ('primitive' mites), Parasitiformes ('ticks') and Acariformes ('mites'). Acari show the largest number of empodial character states within Chelicerata and more detailed descriptions can be found in Barrows (1925), Lindquist (1984), van der Hammen (1989) and Evans (1992). It is thus difficult to score a single apotele state as 'typical' for all Acari and to homologise all the various mite structures, some of which may be setal derivatives (Evans 1992), either within Acari or in relation to other chelicerates. Lindquist (1984) tentatively regarded a three-clawed leg as plesiomorphic and this certainly seems to be true for Acariformes, see e.g. van der Hammen's (1989, fig. 49) account of the basal taxon *Alycus*. Three claws occur in many oribatid mites while other Acariformes show quite unique developments such as a single claw on a fleshy caruncle in some Astigmata (Evans 1992, fig. 2.17). By contrast in the Opilioacariformes and the Parasitiformes there is a cushion-like pulvillus between the paired lateral claws (Fig. 3g), similar to the condition seen in amblypygids, pseudoscorpions and solifuges. In Opilioacariformes the pedipalps have two claws (Fig. 3f) and this morphology is seen in many other mites too (Evans 1992).

DISCUSSION

Outgroup taxa (e.g. Trilobita), basal Pycnogonida and fossil examples of basal Xiphosura and Scorpiones all imply that the plesiomorphic apotele condition in Chelicerata was three rather spine-like elements (Figs. 1a-c, g) and not a single terminal segment as assumed by

Barrows (1925) in his otherwise excellent paper written before the discovery of well-preserved appendages in key fossil taxa. Reductions or modifications of this 3-element pattern should be treated as potential apomorphies. Recent Xiphosurida, e.g. *Limulus*, are sometimes used to polarise character states in arachnids, but the fossil *Weinbergina* suggests that the plesiomorphic apotele condition is retained in the ground pattern of Xiphosura (Fig. 1c) and that the reduction to a single apotele element in the Xiphosurida (sensu Anderson & Selden 1987) (Fig. 1d) is apomorphic only for crown group horseshoe crabs. This apomorphic character state can also be scored for the extinct Eurypterida and Chasmataspida (Figs. 1e-f) and represents a potential synapomorphy of these taxa.

In this respect Eurypterida differ from Scorpiones (Figs. 1g-h) which retain all three apotele elements as a central 'claw' plus a pair of tarsal spurs (e.g. Jeram 1998). These spurs are minute in the Silurian *Allopalaeophonus*-like scorpions and some authors have given the impression that these pointed, crab-like, digitigrade legs with a large, central claw are a ground pattern character state for Scorpiones; see e.g. Kjellesvig-Waering's (1986) suggestion that 'eurypteroid' legs in the fossils supported a close relationship between scorpions and eurypterids. However, in the most basal scorpion genus, *Palaeoscorpius*, the tarsal spurs are similar in size to the median element of the apotele (Fig. 1g). Resolving which of these conditions is plesiomorphic for Scorpiones is difficult, but there is an apparent trend in the Siluro-Devonian fossils towards enlargement of the tarsal spurs to form the typical ungues of arachnids and a corresponding reduction of the median, empodial element to form together an apotele essentially similar to that in Recent scorpions (see e.g. Jeram 1998, character 19, fig. 2: node C). This modification of the apotele *within* the scorpion clade is paralleled in the claws of other arachnids.

Although difficult to prove unequivocally, the three claws in the arachnid apotele - the onychium or empodial claw and its associated tendon plate (see Barrows 1925) plus the paired ungues - are potentially homologous with the three apotele elements seen in the outgroups, e.g. the main claw plus the auxiliary claws in Pycnogonida (Fig. 1b). In this model Palpigradi retain the most plesiomorphic apotele state of any arachnid expressing all three apotele elements on *all* postcheliceral appendages (Figs. 2d-e). Apomorphic apotele states within the arachnids would include the apparent (?homoplastic) loss of the ungues on the pedipalp in most arachnid orders, although as Barrows (1925) and Roewer (1934) noted there are potentially homologous elements associated with the palpal organ (Fig. 3d) of Solifugae, while among the mites Opilioacariformes (Fig. 3f) retain two ungue-like structures on the palp (e.g. van der Hammen 1989, fig. 134).

One of the most interesting modifications of the apotele in the legs of arachnids is the development of the empodial region between the ungues to form a pad-like pulvillus in adults of basal Amblypygi (Fig. 2c), in Solifugae (Fig. 3e) and Pseudoscorpiones (Fig. 3c) and among the mites in the (Opilioacariformes (Fig. 3g) + Parasitiformes) clade, but *not* in basal Acariformes. This character was scored as an 'eversible empodium' by Shultz (1990, character 38) as a synapomorphy for (Solifugae + Pseudoscorpiones). By contrast Weygoldt (1996) used the term 'pulvillus' for all taxa with a modified empodial region and this apomorphic state should probably be scored for all arachnid higher taxa where it is present. When superimposed on the main alternative phylogenetic hypotheses (Weygoldt & Paulus 1979; van der Hammen 1989; Shultz 1990) it emerges as a homoplastic character, probably developing as an adaptation for gripping the substrate (Roewer 1936; Weygoldt 1996).

ACKNOWLEDGEMENTS

I thank Lyall Anderson, Claudia Arango, Simon Braddy, Greg Edgecombe, Mark Judson, Jörg Wunderlich and three anonymous reviewers for valuable comments. Andrew Ross kindly provided access to specimens from the Natural History Museum, London.

REFERENCES

Anderson, L.I. & Selden, P.A. 1997. Opisthosomal fusion and phylogeny of Palaeozoic Xiphosura. *Lethaia* 30, 19-31.

Arnaud, F. & Bamber, R.N. 1987. The biology of Pycnogonida. *Advances in Marine Biology* 24, 1-96.

Barrows, W.M. 1925. Modification and development of the arachnid palpal claw with especial reference to spiders. *Annals of the Entomological Society of America* 18, 483-516.

Bruton, D.L. & Haas, W. 1999. The anatomy and functional morphology of *Phacops* (Trilobita) from the Hunsrück Slate (Devonian). *Palaeontographica A* 253, 29-75.

Caster, K.E. & Brooks, H.K. 1956. New fossils from the Canadian-Chazyan (Ordovician) hiatus in Tennessee. *Bulletins of American Paleontology* 36, 157-199.

Clarke, J.M. & Ruedemann, R. 1912. The Eurypterida of New York. *Memoirs of the New York State Museum* 14, 1-439.

Cisne, J.L. 1975. Anatomy of *Triarthrus* and the relationships of the Trilobita. *Fossils and Strata* 4, 45-63.

Dunlop, J.A. In press. Arthropods from the Lower Devonian Severnya Zemlya Formation of October Revolution Island, Russia. *Geodiversitas*.

Dunlop, J.A., Poshmann, M. & Anderson, L.I. 2002. On the Emsian (Early Devonian) arthropods of the Rhenish Slate Mountains: 3. The chasmataspidid *Diploaspis*. *Paläontologische Zeitschrift* 75, 253-269.

Eisenbeis, G. & Wichard, W. 1985. *Atlas zur Biologie der Bodenarthropoden*. Gustav Fischer Verlag, New York.

Evans, G.O. 1992. *Principles of Acarology*. C.A.B. International, Wallingford.

Giribet, G. Rambla, M., Carranza, S., Bagunà J., Riutort, M. & Ribera, C. 1999. Phylogeny of the arachnid order Opiliones (Arthropoda) inferred from a combined approach of complete 18S and partial 28S ribosomal DNA sequences and morphology. *Molecular Phylogenetics and Evolution* 11, 296-307.

Hammen, L. van der 1989. *An introduction to comparative Arachnology.* SPB Academic Publishing bv, The Hague.

Hansen, H.J. & Sørensen, W. 1897. The order Palpigradi Thor. (*Koenenia mirabilis* Grassi) and its relationship to the other Arachnida. *Entomologisk Tidskrift* 1897, 223-240.

Harvey, M.S. 1992. The phylogeny and classification of the Pseudoscorpionida (Chelicerata: Arachnida). *Invertebrate Taxonomy* 6, 1373-1435.

Helfer, H. & Schlottke, E. 1935. Pantopoda. In: *Dr. H.G. Bronns Klassen und Ordnung des Tierreichs. Band 5, Abteilung IV, Buch 2.* Akademische Verlagsgesellschaft m. b. H., Leipzig.

Igelmund, P. 1987. Morphology, sense organs and regeneration of the forelegs (whips) of the large whip spiders *Heterophrynus elephas* (Arachnida: Amblypygi). *Journal of Morphology* 193, 75-89.

Jeram, A.J. 1998. Phylogeny, classification and evolution of Silurian and Devonian scorpions. In: *Proceedings of the 17th European Colloquium of Arachnology, Edinburgh 1997* (P.A. Selden ed.), pp. 17-31. British Arachnological Society, Burnham Beeches, Bucks.

Kjellesvig-Waering, E.N. 1986. A restudy of the fossil Scorpionida of the world. *Paleontographica Americana* 55, 1-287.

Lehmann, W.M. 1944. *Palaeoscorpius devonicus* n.g., n.sp., ein Skorpion aus dem rheinischen Unterdevon. *Neues Jahrbuch für Paläontologie, Monatshefte* 1944, 177-185.

Millot, J. & Vachon, M. 1949. Ordre des Scorpions. In: *Traite de Zoologie VI* (P.P. Grassé ed.), pp. 386-436. Masson et Cie, Paris.

Müller, K.J. & Walossek, D. 1987. Morphology, ontogeny and life habit of *Agnostus pisiformis* from the Upper Cambrian of Sweden. *Fossils and Strata* 19, 1-124.

Munzo-Cuevas, A. 1971. Étude du tarse, de l'apotele et de la formation des griffes cours du développment postembryonnaire chez *Pachylus quinamavidensis* (Arachnida, Opil-

iones, Gonyleptidae). *Bulletin du Muséum National d'Histoire Naturelle*, 42, 1027-1036.

Pocock, R.I. 1901. The Scottish Silurian scorpion. *The Quarterly Journal of Microscopical Science* 44, 291-311.

Quintero, D. Jr., 1981. The amblypygid genus *Phrynus* in the Americas (Amblypygi, Phrynidae). *Journal of Arachnology* 9, 117-166.

Roewer, C.F. 1936. Chelonethi oder Pseudoskorpione. In: *Dr. H.G. Bronns Klassen und Ordnung des Tierreichs. Band 5, Abteilung IV, Buch 6.* Akademische Verlagsgesellschaft m.b.H., Leipzig.

Schultka, S. 2000. Zur Paläkologie der Euproopiden im Nordwestdeutschen Oberkarbon. *Mitteilungen aus dem Museum für Naturkunde Berlin, Geowissenschaftliche Reihe* 3, 87-98.

Selden, P.A. 1981. Functional morphology of the prosoma of *Baltoeurypterus tetragonophthalmus* (Fischer) (Chelicerata: Eurypterida). *Transactions of the Royal Society of Edinburgh: Earth Sciences* 72, 9-48.

Shear, W.A., Selden, P.A., Rolfe, W.D.I., Bonamo, P.M. and Grierson. J.D. 1987. New terrestrial arachnids from the Devonian of Gilboa, New York (Arachnida, Trigonotarbida). *American Museum Novitates* 2901, 1-74.

Shultz, J.W. 1989. Morphology of locomotor appendages in Arachnida: evolutionary trends and phylogenetic implications. *Zoological Journal of the Linnean Society* 97, 1-56.

Shultz, J.W. 1990. Evolutionary morphology and phylogeny of Arachnida. *Cladistics* 6, 1-38.

Shultz, J.W. 1998. Phylogeny of Opiliones (Arachnida): an assessment of the 'Cyphopalpatores' concept. *The Journal of Arachnology* 26, 257-272.

Shultz, J.W. 1999. Muscular anatomy of a whipspider, *Phrynus longipes* (Pocock) (Arachnida, Amblypygi), and its evolutionary significance. *Zoological Journal of the Linnenan Society* 126, 81-116.

Størmer, L. 1972. Arthropods from the Lower

Devonian (Lower Emsian) of Alken an der Mosel, Germany. Part 2: Xiphosura. *Senckenbergiana Lethaea* 53, 1-29.

Stürmer, W. & Bergström, J. 1981. *Weinbergina*, a xiphosuran arthropod from the Devonian Hunsrück Slate. *Paläontologische Zeitschrift* 55, 237-255.

Tollerton, V.P. 1989. Morphology, taxonomy, and classification of the order Eurypterida Burmeister, 1843. *Journal of Paleontology* 63, 642-657.

Werner, F. 1935. Scorpiones, Pedipalpi. pp. 317-490 In: *Dr. H.G. Bronns Klassen und Ordnung des Tierreichs. Band 5, Abteilung IV, Buch 8*. Akademische Verlagsgesellschaft m.b.H., Leipzig.

Weygoldt, P. 1996. Evolutionary morphology of whip spiders: towards a phylogenetic system (Chelicerata: Arachnida: Amblypygi). *Journal of Zoological and Systematic Evolution and Research* 34, 185-202.

Weygoldt, P. 2000. *Whip Spiders (Chelicerata: Amblypygi)*. Apollo Books, Stenstrup.

Weygoldt, P. & Paulus H.F. 1979. Untersuchungen zur Morphologie, Taxonomie und Phylogenie der Chelicerata. *Zeitschrift für zoologisches Systematik und Evolutionsforschung* 17, 85-116, 177-200.

Wheeler, W.C. & Hayashi, C.Y. 1998. The phylogeny of the extant chelicerate orders. *Cladistics* 14, 173-192.

Whittington, H.B. & Almond, J.E. 1987. Appendages and habits of the Upper Ordovician trilobite *Triarthus eatoni*. *Philosophical Transactions of the Royal Society of London B* 317, 1-46.

European Arachnology 2000 (S. Toft & N. Scharff eds.), pp. 355-358.
© Aarhus University Press, Aarhus, 2002. ISBN 87 7934 001 6
(Proceedings of the 19th European Colloquium of Arachnology, Århus 17-22 July 2000)

Ant mimicry by spiders and spider-mite interactions preserved in Baltic amber (Arachnida: Acari, Araneae)

J. WUNDERLICH

Hindenburgstr. 94, D-75334 Straubenhardt, Germany (joergwunderlich@t-online.de)

Abstract

Three examples of interactions between arachnids in fossil Baltic amber from the Early Tertiary are presented and discussed: a myrmecomorph spider (Araneae: Corinnidae), a parasitic mite larva (Acari: Trombidoidea) on a spider (Araneae: Agelenidae) and a predatory mite (Acari: Prostigmata: ?Labidostemmidae) on a spider (Araneae: Theridiidae).

Key words: Arachnida, Araneae, Acari (parasitic Trombidoidea, predatory Prostigmata), fossils, Baltic amber, myrmecomorphy

INTRODUCTION

During the last decades several new observa-tions on fossils in Baltic amber inclusions were made. The well preserved Baltic amber fossils allow us to present hypotheses on the 'frozen behaviour' of Early Tertiary Arachnida. Here I give some initial short notes on selected fossils whose age is about 50 million years: the oldest discoveries of (1) an ant-mimicing fossil spider (Fig. 1), (2) a spider parasitized by a mite (Fig. 2), and (3) a spider attacked by a mite (Fig. 3). The material is kept in the private collection of the author. More detailed descriptions of the amber pieces are in preparation. The mites were identified by M. Judson (Paris) and A. Wohltmann (Berlin).

FOSSIL MYRMECOMORPH SPIDER

Interpretative drawing (Fig. 1): The body length of this beautifully preserved male spider is 6 mm. It is an undescribed member of the family Corinnidae. The slender body and legs, a saddle-shaped constriction of the opistho-soma and white hairs in this area give the illu-sion of a three-segmented body (ant-like).

The first fossil myrmecomorph spiders were described 15 years ago from Dominican amber (Wunderlich 1986, 1988). Ant-mimicry is characteristic for most extant members of the spider family Corinnidae. One may speculate about possible frozen ant-like behaviour in the fossil spider (Fig. 1): the forelegs are raised in antenna-like fashion, and the abdomen too. The resemblance of the spider to its ant model(s) may be placed between grades 2 and 3 in the sense of Wunderlich (1995).

Myrmecomorph spiders could not evolve before the origin of ants. The earliest unequi-vocal occurence of ants is in the Cretaceous (Grimaldi & Agosta 2000). Rust & Andersen (1999) wrote '...early Tertiary records of true ants like *Pachycondyla* support the hypothesis of a pre-Tertiary origin of the ant lineage (and of caste differentiation and social organisation of populations) and a relatively rapid diversi-fication of extant sub-families during the early Tertiary'.

I have found myrmecomorph spiders in the Early Tertiary Baltic amber, representing Bate-sian mimicry (Foelix 1992; Wunderlich 1995).

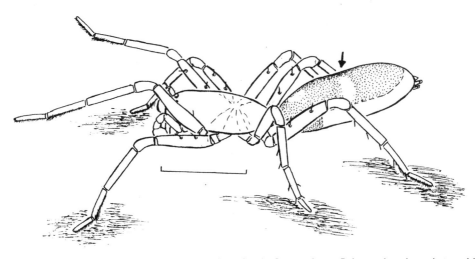

Fig. 1. A fossil male myrmecomorph member of the family Corinnidae in Baltic amber, lateral view. Note the raised forelegs and abdomen. The arrow indicates the saddle-shaped constriction of the abdomen. Scale 2.0 mm.

Fossil examples are found in the family Zodariidae and even more distinct, frequent and diverse examples has been documented in the family Corinnidae. No myrmecomorph spiders representing the families Salticidae or Gnaphosidae have yet been found in the Early Tertiary Baltic amber. However, myrmeco-morph Salticidae are known from Young Tertiary Dominican amber (about 20 million years old) (Wunderlich 1988). From the data at hand, we can conclude that myrmecomorphy in Corinnidae is at least 50 million years old.

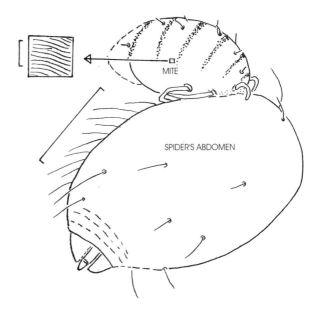

Fig. 2. A fossil mite (Acari: Trombidoidea) in Baltic amber sucking body fluids from the abdomen of a juvenile spider (Araneae: Agelenidae), lateral view. Scale 0.2 mm. Enlarged: The cuticle structure of the mite's abdomen.

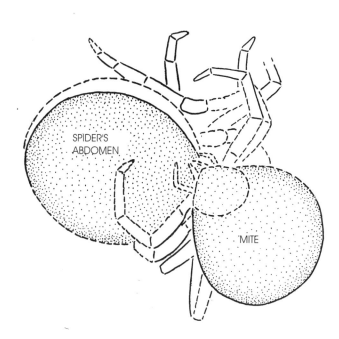

Fig. 3. A fossil predatory/parasitic mite (Acari: Prostigmata: ?Labidostemmidae) in Baltic amber, attacking a female spider (Araneae: Theridiidae), on the left side, dorsal view. Body length about 1 mm.

PARASITIC MITE LARVA SUCKING BODY FLUIDS FROM A SPIDER

Interpretative drawing (Fig. 2): A parasitic thin-legged mite, a larva of the Trombidoidea, body length 0.8 mm, was found attached to a spider, a juvenile member of the spider family Agelenidae, body length 2.4 mm. The abdomen of both arachnids is soft. The mite is situated antero-dorsally on the spider's abdomen. The enlarged abdomen of the larva suggest that it had already been feeding for a while.

Parasitic mites can occasionally be observed on extant spiders and the dorsally-frontally position of the mite (Fig. 2) on the spider's abdomen is characteristic for most extant parasitic mites as well as parasitic Hyme-noptera larvae. In this position the spider cannot use their fangs for defense and cannot use their legs to wipe off the mite from the abdomen. Thus by Early Tertiary mites already used the same feeding position as extant relatives. Fossils including parasitic mites are rare. To date, I have only found three parasitic mites among the 100,000 fossils that I have investigated.

PREDATORY OR PARASITIC MITE ATTACKING A SPIDER

Interpretative drawing (Fig. 3): The small adult predatory/parasitic mite of the Prostigmata: ? Labidostemmidae shown on the right side of Fig. 3 has a body length of about 1 mm and is attacking a female of the spider family Theridiidae. Both arachnids are heavily armoured and both have large dorsal punctuated scuta (Fig. 3).

Both animals are heavily armoured. Extant members of both taxa are known as predators/parasites and they often attack animals of their own size. Blaszak et al. (1990) suggested that the armour serves as protection against predators.

REFERENCES

Blaszak, C., Ehrnsberger, R. & Schuster, R. 1990. Beiträge zur Kenntnis der Lebensweise der Litoralmilbe *Macrocheles superbus* Hull, 1918 (Acarina: Gamasina). *Osnabrücker naturwissenschaftlicher Mitteilungen* 16, 51-62.

Foelix, R. F. 1992. *Biologie der Spinnen*. Georg Thieme Verlag, Stuttgart & New York.

Grimaldi, D. & Agosti, D. 2000. The oldest ants are Cretaceous, not Eocene: comment. *The Canadian Entomologist* 132, 691-693.

Rust, J & Andersen, N. M. 1999. Giant ants from the Paleogene of Denmark with a discussion of the fossil history and early evolution of ants (Hymenoptera: Formicidae). *Zoological Journal of the Linnean Society* 125, 331-348.

Wunderlich, J. 1986. *Spinnenfauna gestern und heute. Fossile Spinnen in Bernstein und ihre heute lebenden Verwandten.* Publishing House J. Wunderlich, D-75334 Straubenhardt, Germany.

Wunderlich, J. 1988. Die fossilen Spinnen im Dominikanischen Bernstein. *Beiträge zur Araneologie* 2, 1-378. Publishing House J. Wunderlich, D-75334 Straubenhardt.

Wunderlich, J. 1995. Über "Ameisenspinnen" in Mitteleuropa (Arachnida: Araneae). *Beiträge zur Araneologie* 4 (1994), 447-470. Publishing House J. Wunderlich, D-75334 Straubenhardt.